116 292
53 ms

2 277
6 ms

112 283
177 s

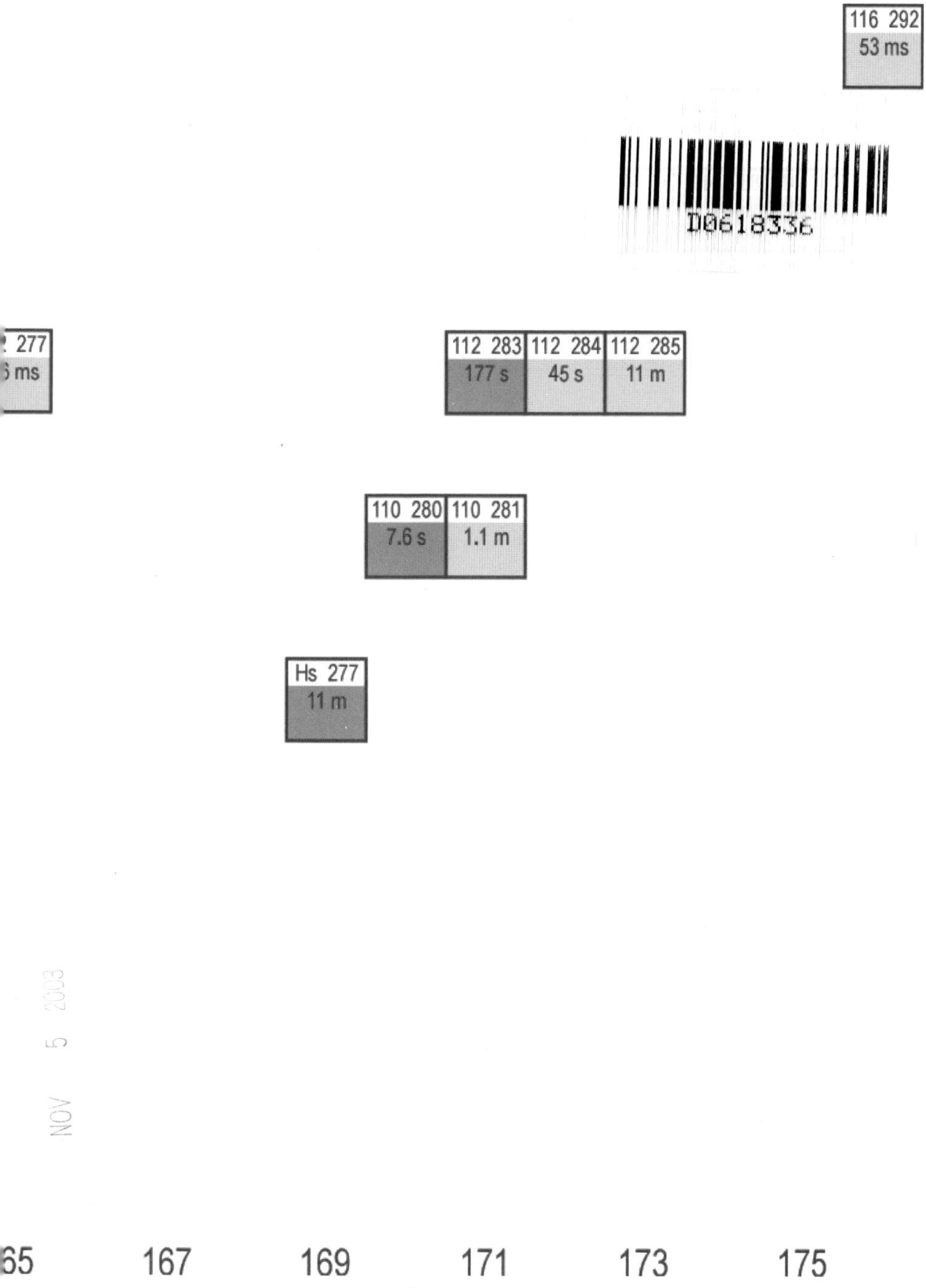

112 284
45 s

112 285
11 m

110 280
7.6 s

110 281
1.1 m

Hs 277
11 m

65 167 169 171 173 175

The Chemistry of Superheavy Elements

The Chemistry of Superheavy Elements

Edited by

MATTHIAS SCHÄDEL

Gesellschaft für Schwerionenforschung mbH (GSI),
Darmstadt, Germany

KLUWER ACADEMIC PUBLISHERS
DORDRECHT / BOSTON / LONDON

A C.I.P. Catalogue record for this book is available from the Library of Congress.

ISBN 1-4020-1250-0

Published by Kluwer Academic Publishers,
P.O. Box 17, 3300 AA Dordrecht, The Netherlands.

Sold and distributed in North, Central and South America
by Kluwer Academic Publishers,
101 Philip Drive, Norwell, MA 02061, U.S.A.

In all other countries, sold and distributed
by Kluwer Academic Publishers,
P.O. Box 322, 3300 AH Dordrecht, The Netherlands.

Printed on acid-free paper

Printed in the Netherlands.

Contents

Chapter 1

Properties and Syntheses of Superheavy Elements

S. Hofmann

Gesellschaft für Schwerionenforschung mbH, Planckstasse 1, D-64221 Darmstadt, Germany

1. Introduction

Searching for new chemical elements is an attempt to answer questions of partly fundamental character: How many elements may exist? How long is their lifetime? Which properties determine their stability? How can they be synthesized? What are their chemical properties? How are the electrons arranged in the strong electric field of the nucleus?

Searching for new elements beyond uranium by the process of neutron capture and succeeding β^- decay, O. Hahn and F. Straßmann [1] discovered the possibility that a heavy nucleus might "divide itself into two nuclei." This was the correct interpretation given by L. Meitner and O.R. Frisch [2], and the term "fission" was coined for this process. By applying the existing charged liquid-drop model of the nucleus [3,4], nuclear fission was explained quite naturally, and it was shown that fission will most likely limit the number of chemical elements. At that time, the maximum number of elements was expected to be about 100. This number results from the balance of two fundamental nuclear parameters, the strength of the attractive nuclear force which binds neutrons and protons together and creates a surface tension and the repulsive electric force.

M. Schädel (ed.), The Chemistry of Superheavy Elements, 1-29.

The properties of nuclei are not smooth uniform functions of the proton and neutron numbers, but show non-uniformities as evidenced by variations in the measured atomic masses. Like the electrons in an atom, also the nucleons in a nucleus - described by quantum mechanical laws - form closed shells called "magic" numbers. At the magic proton or neutron numbers 2, 8, 20, 28, 50, and 82, the nuclei have an increased binding energy relative to the average trend. For neutrons, N = 126 is also identified as a magic number. However, the highest stability is observed in the case of the "doubly magic" nuclei with a closed shell for both protons and neutrons. Amongst other special properties, the doubly magic nuclei are spherical and resist deformation.

The magic numbers were successfully explained by the nuclear shell model [5,6], and an extrapolation into unknown regions was reasonable. The numbers 126 for the protons and 184 for the neutrons were predicted to be the next shell closures. Instead of 126 for the protons also 114 or 120 were calculated as closed shells. The term superheavy elements, SHEs, was coined for these elements, see also Chapter 8.

The prediction of magic numbers, although not unambiguous, was less problematic than the calculation of the stability of those doubly closed shell nuclei against fission. As a consequence, predicted half-lives based on various calculations differed by many orders of magnitude [7-12]. Some of the half-lives approached the age of the universe, and attempts have been made to discover naturally occurring SHEs, see Chapter 8. Although discoveries were announced from time to time, none could be substantiated after more detailed inspection.

There was also great uncertainty of the production yields for SHEs. Closely related to the fission probability of SHEs in the ground-state, the survival of the compound nuclei formed after complete fusion was difficult to predict. Even the best choice of the reaction mechanism, fusion or transfer of nucleons, was critically debated. However, as soon as experiments could be performed without technical limitations, it turned out that the most successful methods for the laboratory synthesis of heavy elements are fusion-evaporation reactions using heavy-element targets, recoil-separation techniques, and the identification of the nuclei by generic ties to known daughter decays after implantation into position-sensitive detectors [13-15].

In the following sections a detailed description is given of the set-ups of the physical experiments used for the investigation of SHEs. (The instrumentation based on chemical methods for the study of heavy elements is presented in subsequent chapters of this book.) Experiments are presented, in which cold and hot fusion reactions were used for the synthesis of SHEs.

Preface

This book is the first to treat the chemistry of superheavy elements, including important related nuclear aspects, as a self contained topic. It is written for those – students and novices -- who begin to work and those who are working in this fascinating and challenging field of the heaviest and superheavy elements, for their lecturers, their advisers and for the practicing scientists in the field – chemists and physicists - as the most complete source of reference about our today's knowledge of the chemistry of transactinides and superheavy elements. However, besides a number of very detailed discussions for the experts this book shall also provide interesting and easy to read material for teachers who are interested in this subject, for those chemists and physicists who are not experts in the field and for our interested fellow scientists in adjacent fields. Special emphasis is laid on an extensive coverage of the original literature in the reference part of each of the eight chapters to facilitate further and deeper studies of specific aspects. The index for each chapter should provide help to easily find a desired topic and to use this book as a convenient source to get fast access to a desired topic.

Superheavy elements – chemical elements which are much heavier than those which we know of from our daily life – are a persistent dream in human minds and the kernel of science fiction literature for about a century. This book describes in Chapter 1 how today this dream becomes true at a few accelerator laboratories, what the tools are to synthesize these elusive, man-made elements in heavy-ion nuclear reactions and how to detect the specific nuclear decays which terminate their existence shortly after they are created. The current status of experimental and theoretical insights into this very unique region of nuclear stability is briefly reviewed. The last chapter outlines historical developments, from first scientifically sound ideas about the existence of superheavy elements, which surfaced during the mid-50s, all the way to the beginning of the current research programs described in Chapter 1. It also discusses experimental attempts and prospects of the search for superheavy elements in Nature.

Today, one century after Ernest Rutherford and Frederick Soddy postulated that in the radioactive decay one chemical element transmutes into a new one, we know of 112 chemical elements. The discoveries of elements 114 and 116 are currently waiting to be confirmed and experimentalists are embarking to discover new and heavier elements. Now where are superheavy elements located on a physicist's chart of nuclides and on the Periodic Table of the Elements -- the most basic chart in chemistry?

The term "superheavy elements" was first coined for elements on a remote "island of stability" around atomic number 114 (Chapter 8). At that time this island of stability was believed to be surrounded by a "sea of instability". By now, as shown in Chapter 1, this sea has drained off and sandbanks and rocky footpaths, paved with cobblestones of shell-stabilized *deformed* nuclei, are connecting the region of shell-stabilized *spherical* nuclei around element 114 to our known world.

Perfectly acceptable, some authors are still using the term "superheavy element" in its traditional form; others have widened this region and have included lighter elements. It is generally agreed that the term "superheavy element" is a synonym for elements which exist solely due to their nuclear shell effects. From this point of view there are good arguments to begin the series of superheavy elements with element 104, rutherfordium. Because of the extra stability from nuclear shell-effects the known isotopes of rutherfordium exhibit half-lives of up to one minute. This is 16 orders of magnitude longer than the expected nuclear lifetime of 10^{-14} s these isotopes would survive without any extra shell stabilization. Taking 10^{-14} s as a realistic limit for a minimum lifetime of a system which can be called a chemical element, and assuming the absence of any shell effects, the world of chemical elements would be terminated at the end of the actinides. The appealing aspect of having the superheavy elements begin at element 104 is that this is identical with the beginning of the series of transactinide elements. The terms "superheavy elements" and "transactinide elements", in short "transactinides", are used with an equal meaning in this book.

One of the most important and most fascinating questions for a chemist is the one about the position of the superheavy elements in the Periodic Table of the Elements; how well accommodates the Periodic Table these elements as transition metals in the seventh period. Do the rules of the Periodic Table still hold for the heaviest elements? What is a valid architecture of the Periodic Table at its upper end? The main body of information to answer this question from our today's knowledge of the chemistry of superheavy or transactinide elements is embraced between the two mainly "nuclear" oriented chapters at the beginning and at the end.

One century after the beginning of most dramatic changes in physics and chemistry, after the advent of quantum theory and in the year of the 100th anniversary of Paul A.M. Dirac, modern relativistic atomic and molecular calculations clearly show the very strong influence of direct and indirect relativistic effects not only on electronic configurations but also on chemical properties of the heaviest elements. The actual state of the theoretical chemistry of the heaviest elements is comprehensively covered in Chapter 2. It does not only discuss most recent theoretical developments and results, where especially up to date molecular calculations dramatically increased our insights over the last decade, but it also relates these results to experimental observations.

The chemistry of superheavy elements always faces a one-atom-at-a-time situation – performing separations and characterizations of an element with single, short-lived atoms establishes one of the most extreme limits in chemistry. While large numbers of atoms and molecules are deeply inherent in the statistical approach to understanding chemical reactions as dynamic, reversible processes Chapter 3 discusses specific aspects how the behavior of single atoms mirrors properties of macro amounts.

A large variety of tools, from manual separation procedures to very sophisticated, automated computer-controlled apparatuses have been developed and are now at hand to study the chemical properties of these short-lived elements one-atom-at-a-time in the liquid phase and in the gas phase. It is demonstrated in Chapter 4 how this can be achieved, what kinds of set-ups are presently available and what the prospects are for future developments to further expand our knowledge.

The known chemical properties of superheavy elements are presented and discussed in Chapter 5 and 7 based upon experimental results obtained from the liquid phase and from the gas phase, respectively. It is quite natural that there is a large body of information on group-4 element 104, rutherfordium, and group-5 element 105, dubnium, which are now under investigation for three decades. However, recent detailed studies demonstrate that these elements still hold many surprises. They sometimes exhibit rather unexpected properties. The chemistry of element 106, seaborgium, was first tackled in 1995 followed by a series of experiments in the aqueous and the gas phase. While most of them revealed a "surprisingly normal" behavior, at least one experiment indicated a deviation from an extrapolation in group 6. Even more challenging, because of the only very few numbers of atoms produced per day, were recent investigations on elements 107, bohrium, and 108, hassium, performed in one gas phase experiment each. This is presented in Chapter 7 together with an attempt to get a first glimpse of the chemical property of element 112. Will it chemically react like mercury or will it be much more inert; presumably due to strong influence of relativistic effects?

Empirical models are frequently applied in chemistry to relate experimental observations to physicochemical or thermodynamical quantities. This has extensively been used over several decades for the interpretation of experimental results obtained from gas phase adsorption processes and is still used to interpret the gas chromatographic results discussed in Chapter 7. These empirical procedures and correlations are outlined in Chapter 6 for a deeper understanding of one of the possible ways to interpret experimental findings from gas phase chemistry.

All the authors of the individual chapters are describing the up-to-date ongoing research in their field where they are leading experts and give a thorough and comprehensive review of our today's knowledge. The individual chapters were finished between mid of the year and November of the year 2002. Pictures of the people involved in many of the described experiments, photos of the instruments and more details on experiments and results can be found on the web-page http://www.gsi.de/kernchemie.

I wish to acknowledge the contributions of Jan Willem Wijnen and Emma Roberts from the Kluwer Academic Publishers who started (JWW) and finalized (ER) this project with me. Many thanks go to the authors of the individual chapters who enthusiastically agreed to contribute to this book and who spent so much time and effort to collect, judge and write up extensive amounts of material. Only thanks to them was it possible to provide such a far-reaching coverage of the chemistry of superheavy elements. Last, but definitely not least, its a great pleasure to thank Brigitta Schausten very much for helping me and the authors with hundreds of smaller or larger details which came up during the preparation of this book, and especially for her work on some of the graphics and for preparing the final format.

Matthias Schädel
Darmstadt and Wiesbaden
December 2002

These experiments resulted in the identification of elements 107 to 112 at the Gesellschaft für Schwerionenforschung, GSI, in Darmstadt, and in the recent synthesis of elements 114 and 116 at the Joint Institute for Nuclear Research, JINR, in Dubna. We also report on a search for element 118, which started in 1999 at Lawrence Berkeley National Laboratory, LBNL, in Berkeley. In subsequent sections a theoretical description follows discussing properties of nuclei in the region of SHEs and phenomena, which influence the yield for the synthesis of SHEs. Empirical descriptions of hot and cold fusion nuclear reaction systematic are outlined. Finally, a summary and outlook is given.

2. Experimental Techniques

2.1 TARGETS AND ACCELERATORS

Transuranium elements are always man-made. Up to fermium, neutron capture in high-flux reactors and successive β^- decay made it possible to climb up the Periodic Table element by element. While from neptunium to californium, some isotopes can be produced in amounts of kilograms or at least grams, the two heaviest species, ^{254}Es and ^{257}Fm, are available only in quantities of micrograms and picograms, respectively. At fermium, however, the method ends due to the lack of β^- decay and too short α and fission half-lives of the heavier elements. Sufficiently thick enough targets cannot be manufactured from these elements.

The region beyond fermium is best accessible using heavy-ion fusion reactions, the bombardment of heavy-element targets with heavy ions from an accelerator. The cross section is less than in the case of neutron capture and values are considerably below the geometrical size of the nuclei. Moreover, only thin targets of the order of 1 mg/cm^2 can be used. This limitation arises from the energy loss of the ion beam in the target, which results (using thicker targets) in an energy distribution that is too wide for both the production of fusion products and their in-flight separation. On the other hand, the use of thin targets in combination with well defined beam energies from accelerators results in unique information about the reaction mechanism. The data are obtained by measuring excitation functions, the yield as a function of the beam energy.

Various combinations of projectiles and targets are in principle possible for the synthesis of heavy elements: actinide targets irradiated by light projectiles of elements in the range from neon to calcium, targets of lead and bismuth irradiated by projectiles from calcium to krypton, and symmetric combinations like tin plus tin up to samarium plus samarium. Also inverse reactions using e.g. lead or uranium as projectile are possible and may have technical advantages in specific cases.

Historically, the first accelerators used for the production of heavy elements were the cyclotrons in Berkeley, California, and later in Dubna, Russia. They were only able to accelerate light ions up to about neon with sufficient intensity and up to an energy high enough for fusion reactions. Larger and more powerful cyclotrons were built in Dubna for the investigation of reactions using projectiles near calcium. These were the U300 and U400, 300 and 400 centimeter diameter cyclotrons. In Berkeley a linear accelerator HILAC (Heavy Ion Linear ACcelerator), later upgraded to the SuperHILAC, was built. The shutdown of this accelerator in 1992 led to a revival of heavy element experiments at the 88-inch cyclotron. Aiming at the acceleration of ions as heavy as uranium, the UNILAC (UNIversal Linear ACcelerator) was constructed in Darmstadt, Germany, during the years 1969-74.

In order to compensate for the decreasing cross sections of the synthesis of heavy elements, increasing beam currents are needed from the accelerators. This demands a continuous development of ion sources in order to deliver high currents at high ionic charge states. Beam currents of several particle microamperes ($1\ \mu A_{part} = 6.24\times10^{12}$ particles/s) are presently reached. Such high currents, in turn, demand a higher resistance of the targets. An efficient target cooling and chemical compounds with higher melting points are presently tested. The developments in the laboratories in Berkeley, Dubna, and also in Finland, France, Italy and Japan are similar and are usually made in close collaboration and exchange of know-how.

2.2 RECOIL-SEPARATION TECHNIQUES AND DETECTORS

The identification of the first transuranium elements was by chemical means. In the early 1960s physical techniques were developed which allowed for detection of nuclei with lifetimes of less than one second at high sensitivity. A further improvement of the physical methods was obtained with the development of recoil separators and large area position sensitive detectors. As a prime example for such instruments, we will describe the velocity filter SHIP (Separator for Heavy-Ion reaction Products) and its detector system, which were developed at the UNILAC. The principle of separation and detection techniques used in the other laboratories is comparable.

In contrast to the recoil-stopping methods, as used in He-jet systems or mass separators, where ion sources are utilized, recoil-separation techniques use the ionic charge and momentum of the recoiling fusion product obtained in the reaction process. Spatial separation from the projectiles and other reaction products is achieved by combined electric and magnetic fields. The

separation times are determined by the recoil velocities and the lengths of the separators. They are typically in the range of 1-2 microseconds. Two types of recoil separator have been developed:
(1) The gas-filled separators use the different magnetic rigidities of the recoils and projectiles traveling through a low pressure (about 1 mbar) gas-filled volume in a magnetic dipole field [16]. In general, helium is used in order to obtain a maximum difference in the rigidities of slow reaction products and fast projectiles. A mean charge state of the ions is achieved by frequent collisions with the atoms of the gas.
(2) Wien-filter or energy separators use the specific kinematic properties of the fusion products. The latter are created with velocities and energies different from the projectiles and other reaction products. Their ionic charge state is determined when they escape from a thin solid-state target into vacuum. Ionic-charge achromaticity is essential for high transmission. It is achieved by additional magnetic fields or symmetric arrangements of electric fields. An example of such a separator used in experiments for the investigation of heavy elements is the velocity filter SHIP in Darmstadt [13] shown in Figure 1.

Recoil separators are designed to filter out those nuclei with a high transmission, which are produced in fusion reactions. Since higher overall yields result in increased background levels, the transmitted particles have to be identified by detector systems. The detector type to be selected depends on the particle rate, energy, decay mode, and half-life. Experimental as well as theoretical data on the stability of heavy nuclei show that they decay by α emission, electron capture or fission, with half-lives ranging from micro-seconds to days. Therefore silicon semiconductor detectors are well suited for the identification of nuclei and for the measurement of their decay properties. If the total rate of ions striking the focal plane of the separator is low, then the particles can be implanted directly into the silicon detectors. Using position-sensitive detectors, one can measure the local distribution of the implanted particles. In this case, the detectors act as diagnostic elements to optimize and control the ion optical properties of the separator.

Given that the implanted nuclei are radioactive, the positions measured for the implantation and all subsequent decay processes are the same. This is the case because the recoil effects are small compared with the range of implanted nuclei, emitted α particles or fission products, and detector resolution. Recording the data event by event allows for the analysis of delayed coincidences with variable position and time windows for the identification of the decay chains [14].

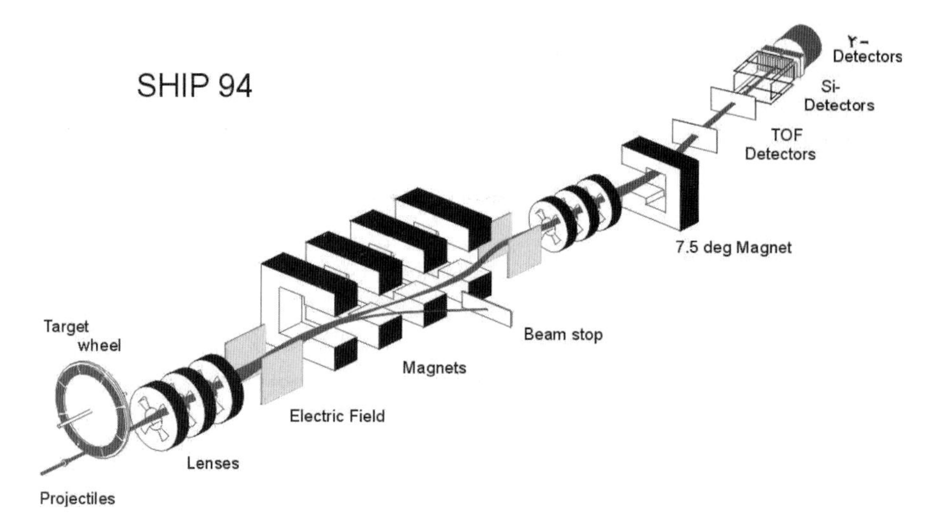

Fig. 1. The velocity filter SHIP (Separator for Heavy Ion reaction Products) and its detection system [13-15]. The drawing is approximately to scale; however, the target wheel and the detectors are enlarged by a factor of two. The length of SHIP from the target to the detector is 11 m. The target wheel has a radius up to the center of the targets of 155 mm. It rotates synchronously with beam macrostructure at 1125 rpm [17]. The target thickness is usually 450 μg/cm^2. The detector system consists of three large area secondary-electron time-of-flight detectors [18] (only two are shown in the graph) and a position-sensitive silicon-detector array (see text). The flight time of the reaction products through SHIP is 2 μs. The filter, consisting of two electric and four magnetic dipole fields plus two quadrupole triplets, was extended by a fifth deflection magnet, allowing for positioning of the detectors away from the straight beam line and further reduction of the background. Figure reproduced from Reference [21] with permission from IOP Publishing Ltd. (2002).

The presently used detector system is composed of three time-of-flight detectors, seven identical 16-strip silicon wafers, and germanium detectors [15]. A schematic view of the detector arrangement is shown in the focal plane of SHIP in Figure 1. Three secondary-electron foil detectors in front of the silicon detectors are used to measure the velocity of the particles [18]. They are mounted 150 mm apart from each other. The detector signals are also used to distinguish implantation from radioactive decays of previously implanted nuclei. Three detectors are used to increase the detection efficiency.

A time-of-flight signal and an energy signal from the silicon detector provide the information for switching off the beam after detection of an implanted residue [19]. After a response time of 20 μs a subsequent time window of preset duration opens for counting a preset number of α particles of the decay chain. If the desired conditions are fulfilled, the beam-off period is prolonged up to the expected measurable end of the decay chain by opening a third time window. This improvement considerably reduces the background during the measuring period of the decay chain and allows for the safe detection of signals from long lived decays. The sequence of three time windows is needed because time-of-flight and energy signals alone would trigger the switching off process for the beam too often due to background events in the corresponding windows.

3. Experimental Results

3.1 ELEMENTS PRODUCED IN COLD-FUSION REACTIONS

In this section, we present results dealing with the discovery of elements 107 to 112 using cold fusion reactions based on lead and bismuth targets. A detailed presentation and discussion of the decay properties of elements 107 to 109 and of elements 110 to 112 was given in previous reviews [15,20,21]. Presently known nuclei are shown in the partial chart of nuclides in Figure 2.

Bohrium, element 107, was the first new element synthesized at SHIP using the method of in-flight recoil separation and generic correlation of parent-daughter nuclei. The reaction used was $^{54}Cr + ^{209}Bi \rightarrow ^{263}107^*$. Five decay chains were observed [22].The next lighter isotope, ^{261}Bh, was synthesized at a higher beam energy [23]. Additional data were obtained from the α decay of ^{266}Mt [24], and the isotope ^{264}Bh was identified as granddaughter in the decay chain of $^{272}111$ [19,25]. The isotopes ^{266}Bh and ^{267}Bh were produced using the hot fusion reaction $^{22}Ne + ^{249}Bk \rightarrow ^{271}Bh^*$ [26]. These nuclei were identified after chemical separation, see Subsection 3.2 and Chapter 7.

Hassium, element 108, was first synthesized in 1984 using the reaction $^{58}Fe + ^{208}Pb$. The identification was based on the observation of three atoms [27]. Only one α-decay chain was measured in the irradiation of ^{207}Pb with ^{58}Fe. The measured event was assigned to the even-even isotope ^{264}Hs [28]. The results were confirmed in a later work [21,29], and for the decay of ^{264}Hs, a fission branching of 50 % was also measured. The isotope ^{269}Hs was discovered as a link in the decay chain of $^{277}112$ [19,30], and ^{270}Hs was identified in a recent chemistry experiment [31], see again Chapter 7.

Meitnerium, element 109, was first observed in the year 1982 in the irradiation of ^{209}Bi with ^{58}Fe by a single α-decay chain [32,33]. This result was confirmed later [34]. In the most recent experiment [24] twelve atoms of ^{266}Mt were measured, revealing a complicated decay pattern, as could be concluded from the wide range of α energies from 10.5 to 11.8 MeV. This property seems to be common to many odd and odd-odd nuclides in the region of the heavy elements. The more neutron-rich isotope ^{268}Mt was measured after α decay of $^{272}111$ [19,25].

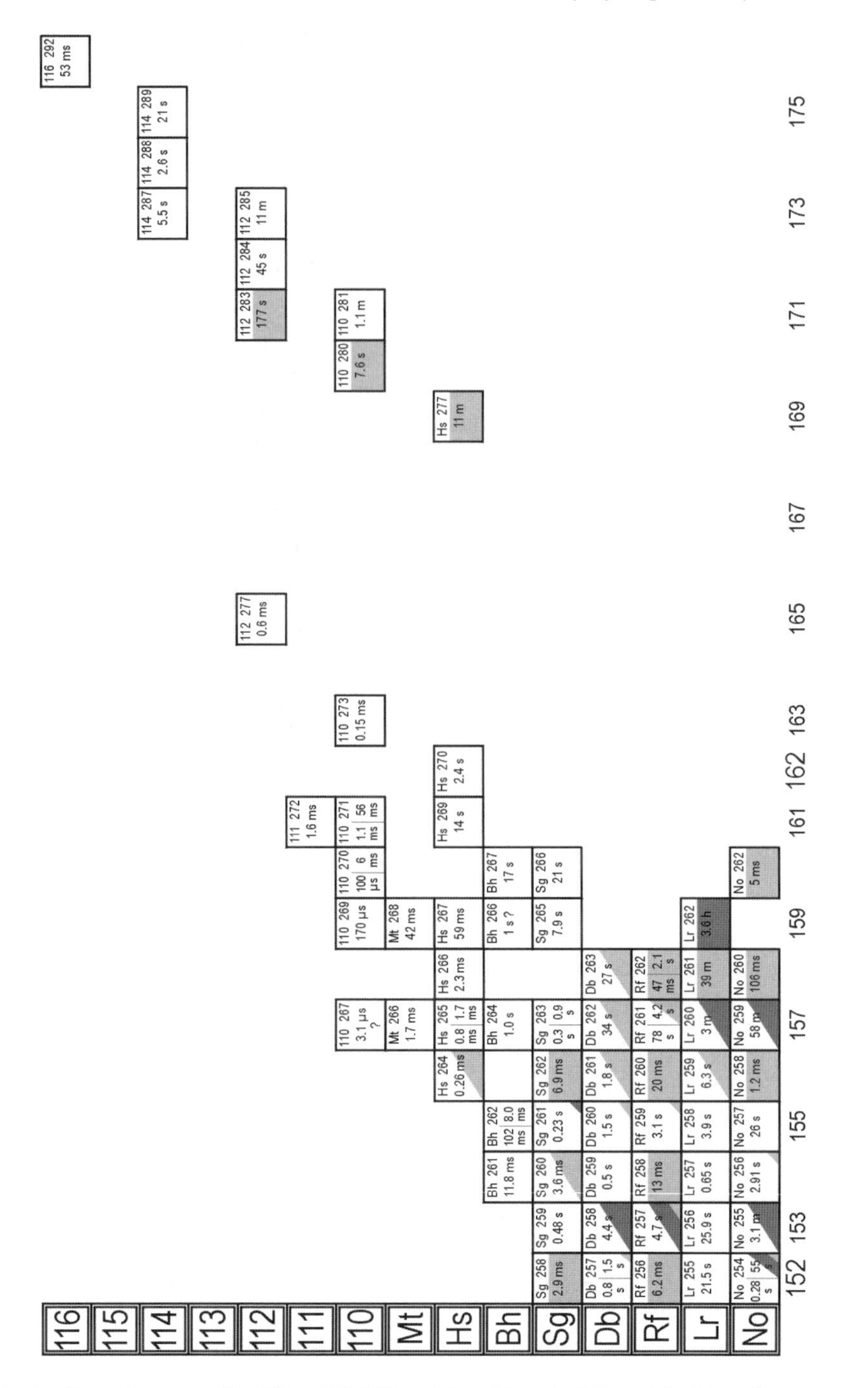

Fig. 2. Presently known N≥152 and Z≥102 nuclei and their half-lives (blank box for α decay, light grey for spontaneous fission decay and dark grey for β decay; see back book cover for a color version). Decay chains starting from element 110, 111 and 112 were measured at GSI in Darmstadt [15,20,21], that from 114 and 116 at JINR in Dubna [35].

Element 110 was discovered in 1994 using the reaction $^{62}Ni + ^{208}Pb \rightarrow ^{269}110 + 1n$ [29]. The main experiment was preceded by a thorough study of the excitation functions for the synthesis of ^{257}Rf and ^{265}Hs using beams of ^{50}Ti and ^{58}Fe in order to determine the optimum beam energy for the production of element 110. The data revealed that the maximum cross section for the synthesis of element 108 was shifted to lower excitation energy, different from the predictions of reaction theories. The heavier isotope $^{271}110$ was synthesized with a beam of the more neutron-rich isotope ^{64}Ni [21]. The important result for the further production of elements beyond meitnerium was that the cross section was enhanced from 2.6 pb to 15 pb by increasing the neutron number of the projectile by two, which gave hope that the cross sections could decrease less steeply with more neutron-rich projectiles. However, this expectation was not proven in the case of element 112.

Two more isotopes of element 110 have been reported in the literature. The first is $^{267}110$, produced at LBNL in the irradiation of ^{209}Bi with ^{59}Co [36]. The second isotope is $^{273}110$, reported to be observed at JINR in the irradiation of ^{244}Pu with ^{34}S after the evaporation of five neutrons [37]. Both observations need further experimental clarification.

The even-even nucleus $^{270}110$ was synthesized using the reaction $^{64}Ni + ^{207}Pb$ [38]. A total of eight α-decay chains were measured during an irradiation time of seven days. Decay data were obtained for the ground-state and a high spin K isomer, for which calculations predict spin and parity 8^+, 9^- or 10^-. The new nuclei ^{266}Hs and ^{262}Sg were identified as daughter products after α decay. Spontaneous fission of ^{262}Sg terminates the decay chain.

Element 111 was synthesized in 1994 using the reaction $^{64}Ni + ^{209}Bi \rightarrow ^{273}111^*$. A total of three α chains of the isotope $^{272}111$ were observed [25]. Another three decay chains were measured in a confirmation experiment in October 2000 [19].

Element 112 was investigated at SHIP using the reaction $^{70}Zn + ^{208}Pb \rightarrow ^{278}112^*$ [30]. The irradiation was performed in January-February 1996. Over a period of 24 days, a total of 3.4×10^{18} projectiles were collected. One α-decay chain, shown in the left side of Figure 3, was observed resulting in a cross section of 0.5 pb. The chain was assigned to the one neutron-emission channel. The experiment was repeated in May 2000 aiming to confirm the synthesis of $^{277}112$ [19]. During a similar measuring time, but using slightly higher beam energy, one more decay chain was observed, also shown in Figure 3. The measured decay pattern of the first four α decays is in agreement with the one observed in the first experiment.

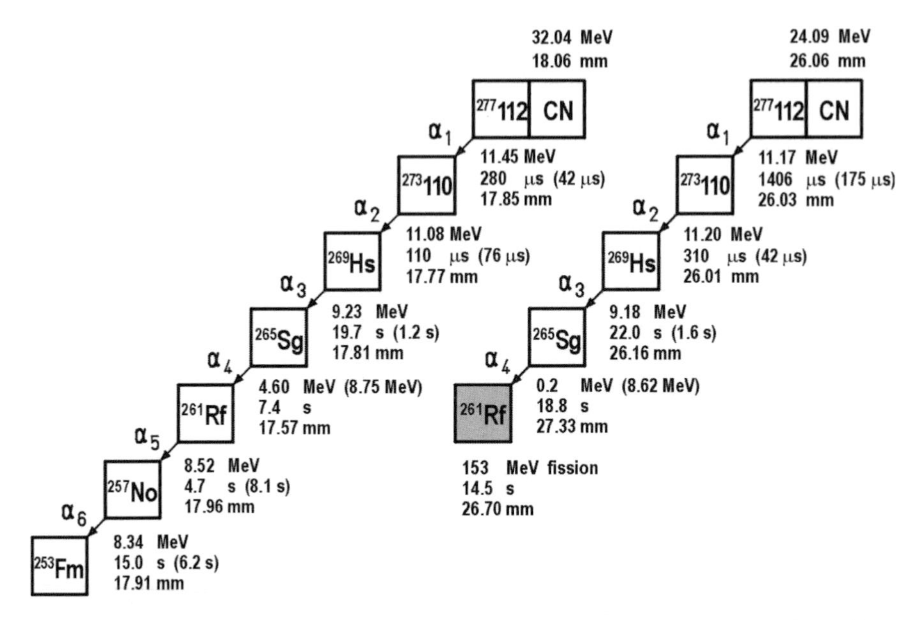

Fig. 3. Two decay chains measured in experiments at SHIP in the cold fusion reaction $^{70}Zn + ^{208}Pb \rightarrow ^{278}112^*$. The chains were assigned to the isotope $^{277}112$ produced by evaporation of one neutron from the compound nucleus. The lifetimes given in brackets were calculated using the measured α energies. In the case of escaped α particles the alpha energies were determined using the measured lifetimes.

A new result was the occurrence of fission which ended the second decay chain at ^{261}Rf. A spontaneous-fission branch of this nucleus was not yet known, however, it was expected from theoretical calculations. The new results on ^{261}Rf were proven in a recent chemistry experiment [31], in which this isotope was measured as granddaughter in the decay chain of ^{269}Hs, see the following Subsection and Chapter 7.

A reanalysis of all decay chains measured at SHIP since 1994, a total of 34 decay chains was analyzed, revealed that the previously published first decay chain of $^{277}112$ [30] (not shown in Figure 3) and the second of the originally published four chains of $^{269}110$ [29] were spuriously created. Details of the results of the reanalysis are given in [19].

Results from an experiment at the 88-inch cyclotron in Berkeley aiming to synthesize element 118 were published in 1999 [39]. Using the new BGS (Berkeley Gas-filled Separator) the reaction $^{86}Kr + ^{208}Pb \rightarrow ^{293}118^*$ was investigated. From three decay chains consisting of six subsequent α decays a surprisingly high cross section of 2 pb was deduced for the one neutron emission channel.
In order to confirm the data obtained in Berkeley, the same reaction was investigated at SHIP in the summer of 1999. The experiment is described in

detail in [15]. During a measuring time of 24 days a beam dose of 2.9×10^{18} projectiles was collected which was comparable to the Berkeley value of 2.3×10^{18}. No event chain was detected, and the cross section limit resulting from the SHIP experiment for the synthesis of element 118 in cold fusion reactions was 1.0 pb. The Berkeley data were retracted in the summer of 2001 [40] after negative results of a repetition experiment performed in the year 2001 in Berkeley itself and after a reanalysis of the data of the first experiment, which showed that the three reported chains were not in the 1999 data. A comparison of the measured cross section limit for this reaction with predictions from theoretical models is given in Section 5.

3.2 ELEMENTS PRODUCED IN HOT-FUSION REACTIONS

Hot fusion reactions are based on targets made from actinide elements. A number of differences exist compared with reactions using lead or bismuth targets. Probably the most significant is the excitation energy of the compound nucleus at the lowest beam energies necessary to initiate a fusion reaction. Values are at 10 - 20 MeV in reactions with lead targets and at 35 - 45 MeV in reactions with actinide targets, which led to the widely used terminology of cold and hot fusion reactions. Due to the lack of targets between bismuth and thorium, a gradual change from cold to hot fusion cannot be studied experimentally. A second important difference of actinide target based reactions is the synthesis of more n-rich isotopes compared with a cold fusion reaction leading to the same element, e.g. ^{269}Hs from a ^{248}Cm target (see below) and ^{264}Hs from a ^{208}Pb target.

Actinides served already as targets, when neutron capture and subsequent β^- decay were used for the first synthesis of transuranium elements. Later, up to the synthesis of seaborgium, actinides were irradiated with light-ion beams from accelerators. At that time it was already known that cold fusion reactions yield higher cross sections for heavy element production.

The argumentation changed again when elements 110 to 112 had been discovered in cold fusion reactions. The combination of actinide targets with beams as heavy as ^{48}Ca became promising to study more neutron rich isotopes, which are closer to the region of spherical SHEs and for which also longer half-lives were expected. In addition the lowest excitation energies of compound nuclei from fusion with actinide targets are obtained with beams of ^{48}Ca.

The experimental difficulty with using a ^{48}Ca beam is the low natural abundance of only 0.19 % of this isotope, which makes enrichment very expensive. Therefore, the development of an intense ^{48}Ca beam at low consumption of material in the ion source and high transmission through the accelerator was the aim of the work accomplished in Dubna during a period of about two years until 1998 [41].

The experiments at the U400 cyclotron were performed at two different recoil separators, which had been built during the 1980s. The separators had been upgraded in order to improve the background suppression and detector efficiency. The energy-dispersive electrostatic separator VASSILISSA was equipped with an additional deflection magnet [42,43]. The gas-filled separator GNS was tuned for the use of very asymmetric reactions with emphasis on the irradiation of highly radioactive targets [44]. Both separators were equipped with time-of-flight detectors and with an array of position-sensitive Si detectors in an arrangement similar to the one shown in Figure 1.

At the separator VASSILISSA attempts were undertaken to search for new isotopes of element 112 by irradiation of ^{238}U with ^{48}Ca ions [45]. The irradiation started in March, 1998. Two fission events were measured resulting in a cross section of 5.0 pb. The two events were tentatively assigned to the residue $^{283}112$ after 3n evaporation.

The experiments were continued in March 1999. The reaction $^{48}Ca + ^{242}Pu \rightarrow ^{290}114^*$ was investigated [46]. It was expected that, after evaporation of three neutrons, the nuclide $^{287}114$ would be produced and would decay by α emission into the previously investigated $^{283}112$. Over a period of 21 days, a total of four fission events were detected. Two of them could be assigned to fission isomers. The other two fission signals were preceded by signals from α particles (one was an escape α of 2.31 MeV) and implantations. A cross section of 2.5 pb was obtained for the two events. They were assigned to the nuclide $^{287}114$. The four events, two from 112 and two from 114 of the ^{238}U and ^{242}Pu irradiation with ^{48}Ca, are consistent. The fission lifetimes are within the limits given by statistical fluctuations. Fission was measured again after α decay, when the target was changed from ^{238}U to ^{242}Pu. The low background rate in the focal plane of VASSILISSA makes mimicking by chance coincidences unlikely. However, further investigation is needed for an unambiguous assignment.

At GNS a search for element 114 was started in November-December, 1998. The experiments were performed in collaboration between the Flerov Laboratory of Nuclear Reactions, FLNR, and the Lawrence Livermore National Laboratory, LLNL, Livermore, California. A ^{244}Pu target was irradiated for a period of 34 days with a ^{48}Ca beam. One decay chain was extracted from the data. The chain was claimed to be a candidate for the decay of $^{289}114$. The measured cross section was 1 pb [47].

The ^{48}Ca + ^{244}Pu experiment was repeated in June-October, 1999. During a period of 3.5 months, two more α-decay sequences, terminating in spontaneous fission, were observed [48]. The two chains were identical within the statistical fluctuations and detector-energy resolution, but differed from the first chain measured in 1998. The two new events were assigned to the decay of ^{288}114, the 4n evaporation channel. The cross section was 0.5 pb.

An investigation of element 116 was started in June 2000. Using a ^{248}Cm target, the previously detected isotopes ^{289}114 or ^{288}114 were expected to be observed as daughter products from the decay of the corresponding element 116 parent nuclei produced after evaporation of 3 or 4 neutrons. The first decay chain which was assigned to ^{292}116 was measured after 35 days on July 19, 2000 [49]. The irradiation was continued, and two more decay chains were measured on May 2 and 8, 2001 [35]. All three decay chains are plotted in Figure 4. The cross section is about 0.6 pb deduced from a total beam dose of 2.25×10^{19}.

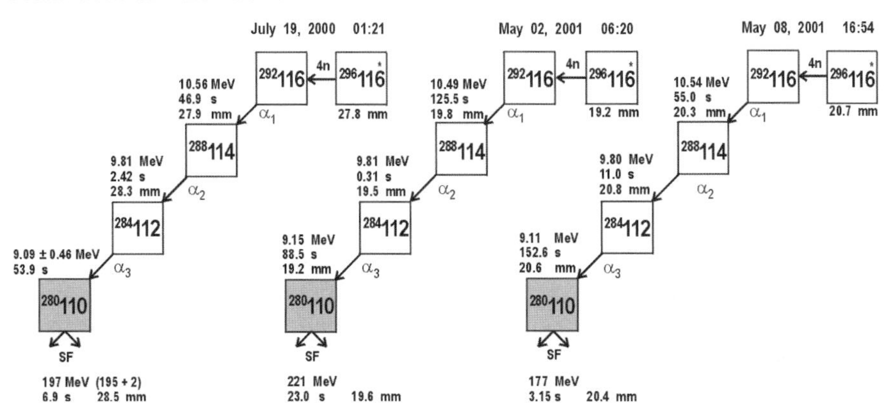

Fig. 4. Three decay chains measured in the reaction ^{48}Ca + ^{248}Cm → ^{296}116* [35]. After implantation of the evaporation residue und detection of the first α decay, the beam was switched off and the succeeding decays were measured under almost background free conditions. The decays assigned to the daughter nucleus ^{288}114 were in agreement with the data measured previously from two decay chains in the reaction ^{48}Ca + ^{244}Pu → ^{292}114*. Reproduced from Yu. Ts. Oganessian et al., Eur. Phys. J. A, Conf. Proc. ENAM 2001 (to be published), Chap. 1, copyright (2002), with permission from Springer-Verlag.

The newly measured chains are of high significance. The data reveal internal redundancy, and the lifetimes are relatively short, making an origin by chance events extremely unlikely. In particular, because all further decays in the chain, after the parent decay was observed, were measured during a beam free period. The beam was switched off using as a trigger the time-of-flight and energy signals from the implantation and the α decay from the parent. The assignment to the 4n channel is likely, but remains subject to further investigation until an unambiguous identification will become possible. As the chains end at $^{280}110$ by spontaneous fission, generic relations to known nuclei cannot be used. Other possible procedures which could help to establish a unique assignment could be measurements of excitation functions, further cross bombardments, direct mass measurements and chemical analysis of parent or daughter elements.

How well chemical properties can be used for the separation and identification of even single atoms was recently demonstrated in an experiment to study hassium [31], see Chapter 7. Using the hot fusion reaction $^{26}Mg + ^{248}Cm \rightarrow ^{274}Hs^*$, the isotope ^{269}Hs was produced after evaporation of five neutrons. The hassium atoms, a total of three decay chains were measured reacted with oxygen to form the volatile compound HsO_4. This way it was proven independently by chemical means that the produced atom belongs, like osmium which also forms a volatile tetroxide, to group 8 and thus to element 108 in the Periodic Table of the Elements. The measured decay properties of the separated atoms fully confirmed the data obtained from the decay $^{277}112$ [19].

Hot fusion reactions applied to synthesize long-lived nuclides of elements 104 through 108 for chemical studies are summarized in Table 1 [50]. Cross sections vary from about 10 nb to a few pb [26,31,51-54]. With typical beam intensities of 3×10^{12} ions per second on targets of about 0.8 mg/cm^2 thickness production yields range from a few atoms per minute for rutherfordium and dubnium isotopes to five atoms per hour for ^{265}Sg and even less for ^{267}Bh and heavier nuclides. Therefore, all chemical separations are performed with single atoms on an "atom-at-a-time" scale. Similar to the experiments with recoil separators characteristic α decays and time correlated α-α-decay chains are used to identify these nuclides in specific fractions or at characteristic positions after chemical separation.

Table 1. Nuclides from hot fusion reactions [50] used in chemical investigations.

Nuclide	$T_{1/2}$	Target	Beam	Evap.	Cross section	Production rate
^{261m}Rf	78 s	^{248}Cm	^{18}O	5n	≈ 10 nb	2 min^{-1}
		^{244}Pu	^{22}Ne	5n	4 nb	1 min^{-1}
^{262}Db	34 s	^{249}Bk	^{18}O	5n	6 nb	2 min^{-1}
		^{248}Cm	^{19}F	5n	1 nb	0.5 min^{-1}
^{263}Db	27 s	^{249}Bk	^{18}O	4n	10 nb	3 min^{-1}
^{265}Sg	7.4 s	^{248}Cm	^{22}Ne	5n	≈ 240 pb	5 h^{-1}
^{266}Sg	21 s	^{248}Cm	^{22}Ne	4n	≈ 25 pb	0.5 h^{-1}
^{267}Bh	17 s	^{249}Bk	^{22}Ne	5n	≈ 70 pb	1.5 h^{-1}
^{269}Hs	14 s	^{248}Cm	^{26}Mg	5n	≈ 6 pb	3 d^{-1}
^{270}Hs	2-7 s	^{248}Cm	^{26}Mg	4n	≈ 4 pb	2 d^{-1}

4. Nuclear Structure and Decay Properties

The calculation of the ground-state binding energy provides the basic step to determine the stability of SHEs. In macroscopic-microscopic models the binding energy is calculated as sum of a predominating macroscopic part (derived from the liquid-drop model of the atomic nucleus) and a microscopic part (derived from the nuclear shell model). This way more accurate values for the binding energy are obtained than in the cases of using only the liquid drop model or the shell model. The shell correction energies of the ground-state of nuclei near closed shells are negative which results in further decreased values of the negative binding energy from the liquid drop model - and thus increased stability. An experimental signature for the shell-correction energy is obtained by subtracting a calculated smooth macroscopic part from the measured total binding energy.

The shell-correction energy is plotted in Figure 5a using data from Reference [55]. Two equally deep minima are obtained, one at Z = 108 and N = 162 for deformed nuclei with deformation parameters $\beta_2 \approx 0.22$, $\beta_4 \approx -0.07$ and the other one at Z = 114 and N = 184 for spherical SHEs. Different results are obtained from self-consistent Hartree-Fock-Bogoliubov, HFB, calculations and relativistic mean-field models [56,57]. They predict for spherical nuclei shells at Z = 114, 120 or 126 (dashed lines in Figure 5a) and N = 184 or 172.

The knowledge of ground-state binding energies, however, is not sufficient for the calculation of partial spontaneous fission half-lives. Here it is necessary to determine the size of the fission barrier over a wide range of deformation. The most accurate data were obtained for even-even nuclei using a macroscopic-microscopic model [58]. Partial spontaneous fission half-lives are plotted in Figure 5b. The landscape of fission half-lives reflects the landscape of shell-correction energies, because in the region of SHEs the height of the fission barrier is mainly determined by the ground-state shell correction energy, while the contribution from the macroscopic liquid-drop part approaches zero for Z = 104 and beyond. Nevertheless we see a significant increase of spontaneous fission half-life from 10^3 s for deformed nuclei to 10^{12} s for spherical SHEs. This difference originates from an increasing width of the fission barrier which becomes wider in the case of spherical nuclei.

Partial α half-lives decrease almost monotonically from 10^{12} s down to 10^{-9} s near Z = 126, see Figure 5c. The valley of β-stable nuclei passes through Z = 114, N = 184. At a distance of about 20 neutrons away from the bottom of this valley, β half-lives of isotopes have dropped down to values of one second [59].

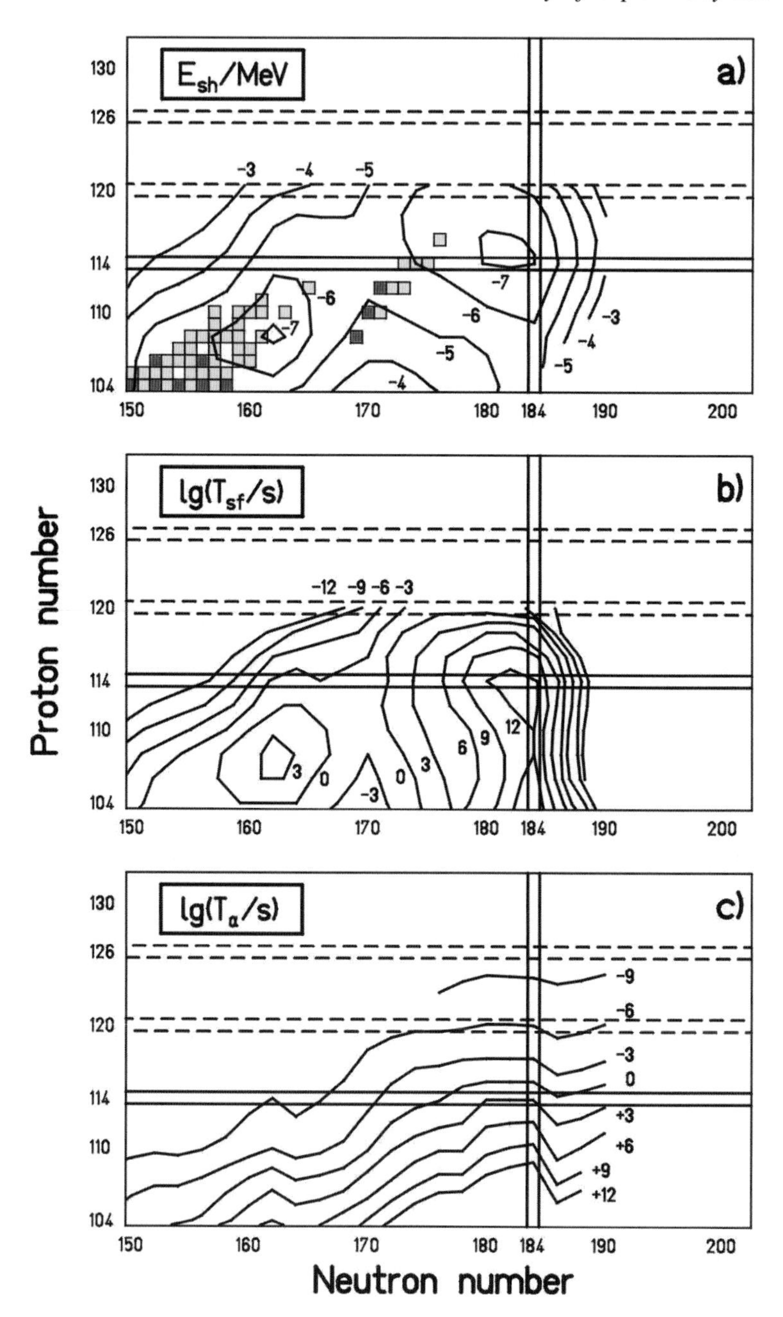

Fig. 5. Shell-correction energy (a) and partial half-lives for spontaneous fission (b) and α decay (c). See text for a detailed descriptions and for references.

Combining results from the individual decay modes one obtains the dominating partial half-life as shown in Figure 6 for even-even nuclei. The two regions of deformed heavy nuclei near N = 162 and spherical SHEs merge and form a region of α emitters surrounded by spontaneously fissioning nuclei. The longest half-lives are 1000 s for deformed heavy nuclei and 30 y for spherical SHEs. It is interesting to note that the longest half-lives are not reached for the doubly magic nucleus $^{298}_{184}114$, but for Z = 110 and N = 182. This is a result of continuously increasing Q_α values with increasing atomic number. Therefore, α decay becomes the dominant decay mode beyond element 110 with continuously decreasing half-lives. For nuclei at N = 184 and Z < 110 half-lives are determined by β^- decay.

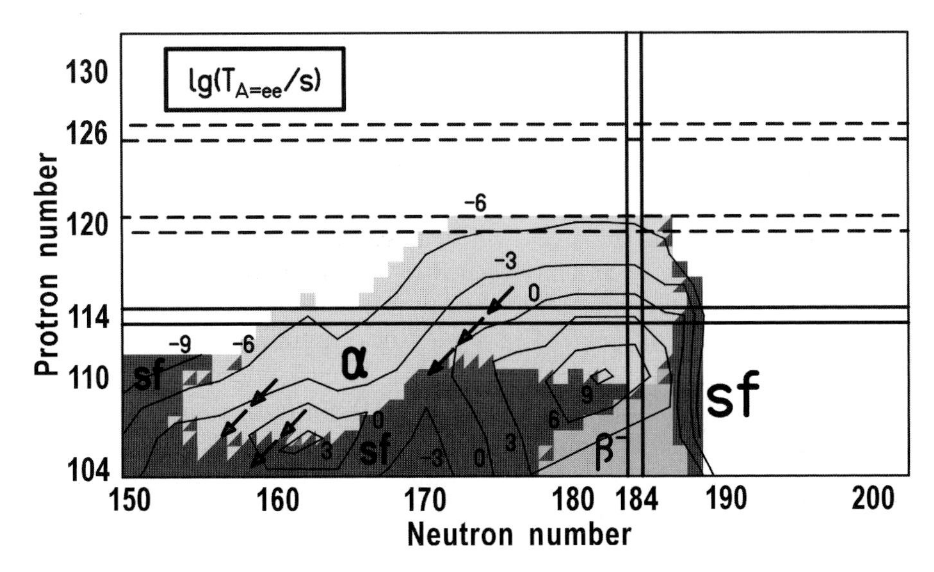

Fig. 6. Dominant half-lives for α, β^+/electron capture, β^- decay and spontaneous fission. The data are valid for even-even nuclei only. Arrows denote α-decay chains explained in the text.

The four member α-decay chain of $^{292}116$, the heaviest even-even nucleus, observed in the recent experiment in Dubna [35], is also drawn in Figure 6. The arrows follow approximately the 1-s contour line down to $^{280}110$. This is in agreement with the experimental observation. The nucleus $^{280}110$ is predicted to decay by spontaneous fission. The experimentally observed half-lives, see Section 3.2, are 53 ms - 2.6 s - 45 s - 7.6 s, respectively, which on average is by a factor of 10 longer than the calculated values. However, this deviation is well within the accuracy limits of the calculation. E.g., a change of the α energy of $^{288}114$ by 350 keV only changes the half-life by a factor of 10. The decay chains of two other recently synthesized even-even nuclei, $^{270}110$ [38] and ^{270}Hs [31], are also drawn in the figure. In these cases the decay chains end by spontaneous fission at ^{262}Sg and ^{262}Rf, respectively.

For odd nuclei, partial α and spontaneous fission half-lives calculated by R. Smolanczuk and A. Sobiczewski [55] have to be multiplied by a factor of 10 and 1000, respectively, thus making provisions for the odd particle hindrance factors. However, we have to keep in mind that fission hindrance factors show a wide distribution from 10^1 to 10^5, which is mainly a result of the specific levels occupied by the odd nucleon. For odd-odd nuclei, the fission hindrance factors from both the odd proton and the odd neutron are multiplied. For odd and odd-odd nuclei, the island character of α emitters disappears and for nuclei with neutron numbers 150 to 160 α-decay prevails down to rutherfordium and beyond. In the allegorical representation where the stability of SHEs is seen as an island in a sea of instability, see Chapter 8, even-even nuclei portray the situation at high-tide and odd nuclei at low-tide, when the island is connected to the mainland.

The interesting question arises, if and to which extent uncertainties related to the location of proton and neutron shell closures will change the half-lives of SHEs. Partial α and β half-lives are only insignificantly modified by shell effects, because their decay process occurs between neighboring nuclei. This is different for fission half-lives which are primarily determined by shell effects. However, the uncertainty related to the location of nuclei with the strongest shell-effects, and thus longest partial fission half-life at Z = 114, 120 or 126 and N = 172 or 184, is irrelevant concerning the longest 'total' half-life of SHEs. All regions for these SHEs are dominated by α decay. α–decay half-lives will only be modified by a factor of up to approximately 100, if the double shell closure is not located at Z = 114 and N = 184.

The line of reasoning is, however, different concerning the production cross section. The survival probability of the compound nucleus (CN) is determined among other factors significantly by the fission-barrier. Therefore, with respect to an efficient production yield, the knowledge of the location of minimal negative shell-correction energy is highly important. However, it may also turn out that shell effects in the region of SHEs are distributed across a number of subshell closures. In that case a wider region of less deep shell-correction energy would exist with corresponding modification of stability and production yield of SHEs.

5. Nuclear Reactions

The main features which determine the fusion process of heavy ions are (1) the fusion barrier and the related beam energy and excitation energy,

(2) the ratio of surface tension versus Coulomb repulsion which determines the fusion probability and which strongly depends on the asymmetry of the reaction partners (the product Z_1Z_2 at fixed $Z_1 + Z_2$),
(3) the impact parameter (centrality of collision) and related angular momentum, and
(4) the ratio of neutron evaporation and of γ emission versus the fission of the compound nucleus.

In fusion reactions towards SHEs the product Z_1Z_2 reaches extremely large values and the fission barrier extremely small values. In addition, the fission barrier itself is fragile, because it is solely built up from shell effects. For these reasons the fusion of SHEs is hampered twofold: (i) in the entrance channel by a high probability for reseparation and (ii) in the exit channel by a high probability for fission. In contrast, the fusion of lighter elements proceeds unhindered through the contracting effect of the surface tension and the evaporation of neutrons instead of fission.

The effect of Coulomb repulsion on the cross section starts to act severely for fusion reactions to produce elements beyond fermium. From there on a continuous decrease of cross section was measured from microbarns for the synthesis of nobelium down to picobarns for the synthesis of element 112. Data obtained in reactions with ^{208}Pb and ^{209}Bi for the 1n-evaporation channel at low excitation energies of about 10-15 MeV (therefore named *cold fusion*) and in reactions with actinide targets at excitation energies of 35-45 MeV (*hot fusion*) for the 4n channel are plotted in Figure 7a and b, respectively.

Some features which the data reveal are pointed out in the following:
(1) So far no data were measured below cross section values of about 0.5 pb. This is the limit presently set by experimental constraints. The experimental time necessary to observe one decay chain at a certain cross section is given on the right ordinate of Figure 7a. Considering the already long irradiation time to reach a cross section of 0.5 pb, it seems impractical to perform systematic studies on this cross section level or even below. Further improvement of the experimental conditions is mandatory. Note in this context that the experimental sensitivity increased by three orders of magnitude since the 1984 search experiment for element 116 using a hot fusion reaction [60].
(2) The cross sections for elements lighter than 113 decrease by factors of 4 and 10 per element in the case of cold and hot fusion, respectively. The decrease is explained as a combined effect of increasing probability for reseparation of projectile and target nucleus and fission of the compound nucleus. Theoretical consideration and empirical descriptions, see e.g. [61,62], suggest that the steep fall of cross sections for cold fusion reactions

may be strongly linked to increasing reseparation probability at high values of Z_1Z_2 while hot fusion cross sections mainly drop because of strong fission losses at high excitation energies. Extremely small values result from an extrapolating these data into the region of element 114 and above. However, strong shell effects for SHEs with spherical nuclear shapes could lead to an increase of the fission barrier and thus to an increase of the survival probability of the compound nucleus, see also discussion in Section 4. The relatively high values measured in Dubna for the synthesis of elements 114 and 116 would be in agreement with this argumentation. In the case of cold fusion only cross section limits are known for the synthesis of elements 116 and 118.

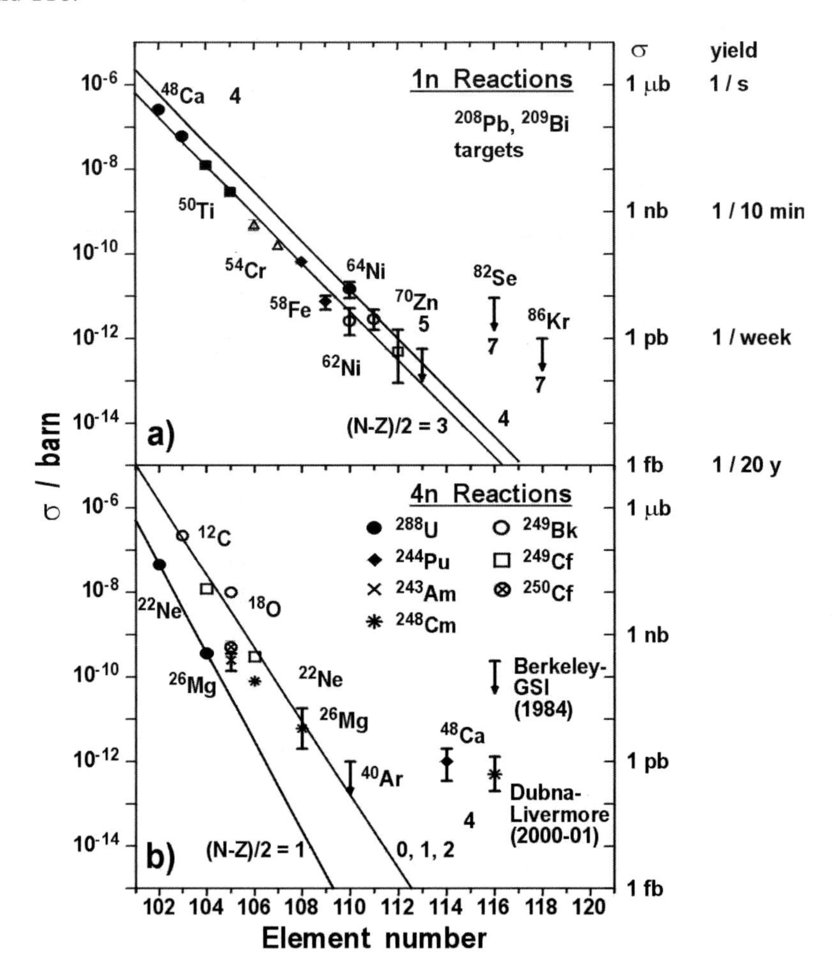

Fig. 7. Measured cross sections (a) for reactions with ^{208}Pb and ^{209}Bi targets and 1n evaporation and (b) for reactions with actinide targets and 4n evaporation. At the right ordinate cold fusion reaction yields are given which are obtained with the presently available technology. The values (N-Z)/2 denote the projectile isospin.

(3) Locally an increase of the cross section by a factor of 5.8 was measured for element 110 in cold fusion reactions when the beam was changed from ^{62}Ni to ^{64}Ni. It was speculated that this increase could be due to the increased value of the projectile isospin. However, the assumption could not be confirmed in the case of element 112 which was synthesized using the most neutron rich stable zinc isotope with mass number 70.

A number of excitation functions was measured for the synthesis of elements from No to 110 using Pb and Bi targets [15]. The maximum evaporation residue cross section (1n channel) was measured at beam energies well below a one dimensional fusion barrier [63]. At the optimum beam energy projectile and target are just reaching the contact configuration in a central collision. The relatively simple fusion barrier based on the Bass model [63] is too high and a tunneling process through this barrier cannot explain the measured cross section. Various processes are possible, and are discussed in the literature, which result in a lowering of the fusion barrier. Among these processes transfer of nucleons and an excitation of vibrational degrees of freedom are the most important [64,65].

Target nuclei of actinide targets are strongly deformed and the height of the Coulomb barrier depends on the orientation of the deformation axes. The reaction ^{48}Ca + ^{248}Cm, studied in Dubna, was performed at a beam energy resulting in an excitation energy of approximately 34 MeV [35]. The observed decay chains were assigned to the 4n-evaporation channel. An excitation function which could provide experimental evidence for an orientation effect on the fusion cross section is not yet measured.

It was pointed out in the literature [66] that closed shell projectile and target nuclei are favorable synthesizing SHEs. The reason is not only a low reaction Q-value and thus low excitation energy, but also that fusion of such systems is connected with a minimum of energy dissipation. The fusion path proceeds along cold fusion valleys, where the reaction partners maintain kinetic energy up to the closest possible distance. In this view the difference between *cold* and *hot* fusion is not only a result from gradually different values of excitation energy, but there exists a qualitative difference, which is on the one hand (cold fusion) based on a well ordered fusion process along paths of minimum dissipation of energy, and on the other hand (hot fusion) based on a process governed by the formation of a more or less energy equilibrated compound nucleus. Fusion of the doubly magic ^{48}Ca and actinide targets seems to proceed via an intermediate, sometimes also called "warm" fusion process, possibly along a fusion valley less pronounced than in the case of cold fusion. Triggered by the recent experimental success of heavy element synthesis, a number of theoretical studies are in progress aiming to obtain a detailed understanding of the processes involved [65,67-72].

Due to the great uncertainty concerning the influence of the various steps in the fusion of heavy elements, more precise experimental data are needed. It is especially important that various combinations of projectile and target be investigated, from very asymmetric systems to symmetric ones, and that excitation functions are measured. This provides information on how fast cross section decrease with increasing energy due to the fission of compound nuclei, and how fast cross section decrease on the low energy side due to the fusion barrier. From both slopes, information about the 'shape' of the fission and the fusion barriers can be obtained. At a high enough cross section, these measurements can be complemented by in-beam γ-ray spectroscopy using the recoil-decay tagging method in order to study the influence of angular momentum on the fusion and survival probability [73-75].

6. Summary and Outlook

The experimental work of the last two decades has shown that cross sections for the synthesis of the heaviest elements decrease almost continuously. However, recent data on the synthesis of element 114 and 116 in Dubna using hot fusion seem to break this trend when the region of spherical superheavy elements is reached. Therefore a confirmation is urgently needed that the region of spherical SHEs has finally been reached and that the exploration of the 'island' has started and can be performed even on a relatively high cross section level.

The progress towards the exploration of the island of spherical SHEs is difficult to predict. However, despite the exciting new results, many questions of more general character are still awaiting an answer. New developments will not only make it possible to perform experiments aimed at synthesizing new elements in reasonable measuring times, but will also allow for a number of various other investigations covering reaction physics and spectroscopy.

One can hope that, during the coming years, more data will be measured in order to promote a better understanding of the stability of the heaviest elements and the processes that lead to fusion. A microscopic description of the fusion process will be needed for an effective explanation of all measured phenomena in the case of low dissipative energies. Then, the relationships between fusion probability and stability of the fusion products may also become apparent.

An opportunity for the continuation of experiments in the region of SHEs at decreasing cross sections afford, among others, further accelerator developments. High current beams and radioactive beams are options for the future. At increased beam currents, values of tens of particle μA's may become accessible, the cross section level for the performance of experiments can be shifted down into the region of tens of femtobarns, and excitation functions can be measured on the level of tenths of picobarns. High currents, in turn, call for the development of new targets and separator improvements. Radioactive ion beams, not as intense as the ones with stable isotopes, will allow for approaching the closed neutron shell N = 184 already at lighter elements. Interesting will be the study of the fusion process using radioactive neutron rich beams.

The half-lives of spherical SHEs are expected to be relatively long. Based on nuclear models, which are effective predictors of half-lives in the region of the heaviest elements, values from microseconds to years have been calculated for various isotopes. This wide range of half-lives encourages the application of a wide variety of experimental methods in the investigation of SHEs, from the safe identification of short lived isotopes by recoil-separation techniques to atomic physics experiments on trapped ions, and to the investigation of chemical properties of SHEs using long-lived isotopes.

References

1. Hahn, O., Strassmann, F.: Naturwissenschaften **27**, (1939) 11.
2. Meitner, L., Frisch, O.R.: Nature **143**, (1939) 239.
3. Gamov, G.: Proc. R. Soc. London A **126**, (1930) 632.
4. von Weizsäcker, C.F.: Z. Phys. **96**, (1935) 431.
5. Göppert-Mayer, M.: Phys. Rev. **74**, (1948) 235.
6. Haxel, O., Jensen, J.H.D., Suess, H.D.: Phys. Rev. **75**, (1949) 1769.
7. Myers, W.D., Swiatecki, W.J.: Nucl. Phys. **81**, (1966) 1.
8. Meldner, H.: Ark. Fys. **36**, (1967) 593.
9. Nilsson, S.G., Nix, J.R., Sobiczewski, A., Szymanski, Z., Wycech, S., Gustafson, C., Möller, P.: Nucl. Phys. A **115**, (1968) 545.
10. Mosel, U., Greiner, W.: Z. Phys. A **222**, (1969) 261.
11. Fiset, E.O., Nix, J.R.: Nucl. Phys. A **193**, (1972) 647.
12. Randrup, J., Larsson, S.E., Möller, P., Nilsson, S.G., Pomorski, K., Sobiczewski, A.: Phys. Rev. C **13**, (1976) 229.
13. Münzenberg, G., Faust, W., Hofmann, S., Armbruster, P., Güttner, K., Ewald, H.: Nucl. Instr. and Meth. **161**, (1979) 65.
14. Hofmann, S., Faust, W., Münzenberg, G., Reisdorf, W., Armbruster, P., Güttner, K., Ewald, H.: Z. Phys. A **291**, (1979) 53.
15. Hofmann, S., Münzenberg, G.: Rev. Mod. Phys. **72**, (2000) 733.
16. Armbruster, P., Eidens, J., Grueter, J.W., Lawin, H., Roeckl, E., Sistemich, K.: Nucl. Instr. and Meth. **91**, (1971) 499.
17. Folger, H., Hartmann, W., Heßberger, F.P., Hofmann, S., Klemm, J., Münzenberg, G., Ninov, V., Thalheimer, W., Armbruster, P.: Nucl. Instr. and Meth. A **362**, (1995) 64.
18. Saro, S., Janik, R., Hofmann, S., Folger, H., Heßberger, F.P., Ninov, V., Schött, H.J., Andreyev, A.N., Kabachenko, A.P.,Popeko, A.G., Yeremin, A.V.: Nucl. Instr. and Meth. A **381**, (1996) 520.
19. Hofmann, S., Heßberger, F.P., Ackermann, D., Münzenberg, G., Antalic, S., Cagarda, P., Kindler, B., Kojouharova, J., Leino, M., Lommel, B., Mann, R., Popeko, A.G., Reshitko, S., Saro, S., Uusitalo, J., Yeremin, A.V.: Eur. Phys. J. A **14**, (2002) 147.
20. Münzenberg, G.: Rep. Prog. Phys. **51**, (1988) 57.
21. Hofmann, S.: Rep. Prog. Phys. **61**, (1998) 639.
22. Münzenberg, G., Hofmann S., Heßberger, F.P., Reisdorf, W., Schmidt, K.-H., Schneider, J.H.R., Armbruster, P.,Sahm, C.C., Thuma, B.: Z. Phys. A**300**, (1981) 107.
23. Münzenberg, G., Armbruster, P., Hofmann, S., Heßberger, F.G., Folger, H., Keller, J.G., Ninov, V., Poppensieker, K., Quint, A.B., Reisdorf, W., Schmidt, K.-H., Schneider, J.R.H., Schött, H.-J., Sümmerer, K., Zychor, I., Leino, M.E., Ackermann, D., Gollerthan, U., Hanelt, E., Morawek, W., Vermeulen, D., Fujita, Y., Schwab, T.: Z. Phys. A **333**, (1989) 163.
24. Hofmann, S., Heßberger, F.P., Ninov, V., Armbruster, P., Münzenberg, G., Stodel, C., Popeko, A.G., Yeremin, A.V., Saro, S., Leino, M.: Z. Phys. A **358**, (1997) 377.
25. Hofmann, S., Ninov, V., Heßberger, F.P., Armbruster, P., Folger, H., Münzenberg, G., Schött, H.J., Popeko, A.G., Yeremin, A.V., Andreyev, A.N., Saro, S., Janik, R., Leino, M.: Z. Phys. A **350**, (1995) 281.
26. Eichler, R., Brüchle, W., Dressler, R., Düllmann, Ch.E., Eichler, B., Gäggeler, H.W., Gregorich, K.E., Hoffman, D.C., Hübener, S., Jost, D.T., Kirbach, U.W., Laue, C.A., Lavanchy, V.M., Nitsche, H., Patin, J.B., Piquet, D., Schädel, M., Shaughnessy, D.A., Strellis, D.A., Taut, S., Tobler, L., Tsyganov, Y.S., Türler, A., Vahle, A., Wilk, P.A., Yakushev, A.B.: Nature **407**, (2000) 63.
27. Münzenberg, G., Armbruster, P., Folger, H., Heßberger, F.P., Hofmann, S., Keller, J., Poppensieker, K., Reisdorf, W., Schmidt, K.H., Schött, H.J., Leino, M.E., Hingmann, R.: Z. Phys. A **317**, (1984) 235.

28. Münzenberg, G., Armbruster, P., Berthes, G., Folger, H., Heßberger, F.P., Hofmann, S., Poppensieker, K., Reisdorf, W., Quint, B., Schmidt, K.H., Schött, H.J., Sümmerer, K., Zychor, I., Leino, M., Gollerthan, U., Hanelt, E.: Z. Phys., A **324**, (1986) 489.
29. Hofmann, S., Ninov, V., Heßberger, F.P., Armbruster, P., Folger, H., Münzenberg, G., Schött, H.J., Popeko, A.G., Yeremin, A.V., Andreyev, A.N., Saro, S., Janik, R., Leino, M.: Z. Phys. A **350**, (1995) 277.
30. Hofmann, S., Ninov, V., Heßberger, F.P., Armbruster, P., Folger, H., Münzenberg, G., Schött, H.J., Popeko, A.G., Yeremin, A.V., Saro, S., Janik, R., Leino, M.: Z. Phys., A **354**, (1996) 229.
31. Türler, A., Düllmann, Ch.E., Gäggeler, H.W., Kirbach, U.W., Yakushev, A., Schädel, M., Brüchle, W., Dressler, R., Eberhardt, K., Eichler, B., Eichler, R., Ginter, T.N., Glaus, F., Gregorich, K.E., Hofmann, D.C, Jäger, E., Jost, D.T., Lee, D.M., Nitsche, H., Patin, J.B., Pershina, V., Piquet, D., Qin, Z., Schausten, B., Schimpf, E., Schött, H.-J., Soverna, S., Sudowe, R., Thörle, P., Timokhin, S.N., Trautmann, N., Vahle, A., Wirth, G., Zielinski, P.: Eur. Phys. J. A, submitted 2002; and Düllmann, Ch.E. et al.: Nature **418** (2002) 859.
32. Münzenberg, G., Armbruster, P., Heßberger, F.P., Hofmann, S., Poppensieker, K., Reisdorf, W., Schneider, J.R.H., Schneider, W.F.W., Schmidt, K.H., Sahm, C.C., Vermeulen, D.: Z. Phys. A **309**, (1982) 89.
33. Münzenberg, G., Reisdorf, W., Hofmann, S., Agarwal, Y.K., Heßberger, F.P., Poppensieker, K., Schneider, J.R.H., Schneider, W.F.W., Schmidt, K.H., Schött, H.J., Armbruster, P., Sahm, C.C., Vermeulen, D.: Z. Phys., A **315**, (1984) 145.
34. Münzenberg, G., Hofmann, S., Heßberger, F.P., Folger, H., Ninov, V., Poppensieker, K., Quint, B., Reisdorf, W., Schött, H.J., Sümmerer, K., Armbruster, P., Leino, M.E., Ackermann, D., Gollerthan, U., Hanelt, E., Morawek, W., Fujita, Y., Schwab, T., Türler, A.: Z. Phys. A **330**, (1988) 435.
35. Oganessian, Yu.Ts., Utyonkov, V.K., Moody, K.J.: Physics of Atomic Nuclei **64**, (2001) 1349.
36. Ghiorso, A., Lee, D., Sommerville, L.P., Loveland, W., Nitschke, J.M., Ghiorso, W., Seaborg, G.T., Wilmarth, P., Leres, R., Wydler, A., Nurmia, M., Gregorich, K., Czerwinski, K., Gaylord, R., Hamilton, T., Hannink, N.J., Hoffman, D.C., Jarzynski, C., Kacher, C., Kadkhodayan, B., Kreek, S., Lane, M., Lyon, A., McMahan, M.A., Neu, M., Sikkeland, T., Swiatecki, W.J., Türler, A., Walton, J.T., Yashita, S.: Phys. Rev. C **51**, (1995) R2293.
37. Lazarev, Yu.A., Lobanov, Yu.V., Oganessian, Yu.Ts., Utyonkov, V.K., Abdullin, F.Sh., Polyakov, A.N., Rigol, J., Shirokovsky, I.V., Tsyganov, Yu.S., Iliev, S., Subbotin, V.G., Sukhov, A.M., Buklanov, G.V., Gikal, B.N., Kutner, V.B., Mezentsev, A.N., Subotic, K., Wild, J.F., Lougheed, R.W., Moody, K.J.: Phys. Rev. C **54**, (1996) 620.
38. Hofmann, S., Heßberger, F.P., Ackermann, D., Antalic, S., Cagarda, P., Cwiok, S., Kindler, B., Kojouharova, J., B., Lommel, B., Mann, R., Münzenberg, G., Popeko, A.G., Saro, S., Schött, H.J., Yeremin, A.V.: Eur. Phys. J. A **10**, (2001) 5.
39. Ninov, V., Gregorich, K.E., Loveland, W., Ghiorso, A., Hoffman, D.C., Lee, D.M., Nitsche, H., Swiatecki, W.J., Kirbach, U.W., Laue, C.A., Adams, J.L., Patin, J.B., Shaughnessy, D.A., Strellis, D.A.,Wilk, P.A.: Phys. Rev. Lett. **83**, (1999) 1104.
40. Ninov, V., Gregorich, K.E., Loveland, W., Ghiorso, A., Hoffman, D.C., Lee, D.M., Nitsche, H., Swiatecki, W.J., Kirbach, U.W., Laue, C.A., Adams, J.L., Patin, J.B., Shaughnessy, D.A., Strellis, D.A.,Wilk, P.A.: Phys. Rev. Lett. **89**, (2002) 039901.
41. Kutner, V.B., Bogomolov, S.L., Gulbekian, G.G., Efremov, A.A., Ivanov, G.N., Lebedev, A.N., Lebedev, V.Ya., Loginov, V.N., Oganessian, Yu.Ts.,Yakushev, A.B., Yazvitsky, N.Yu.: "Operation and recent development of ECR ion sources at the FLNR (JINR) cyclotrons". In: Proceedings of "15th International Conference on Cyclotrons and their Applications", Caen, France, 14-19 June 1998, pp. 405-408.

42. Yeremin, A.V., Andreyev, A.N., Bogdanov, D.D., Ter-Akopian, G.M., Chepigin, V.I., Gorshkov, V.A., Kabachenko, A.P., Malyshev, O.N., Popeko, A.G., Sagaidak, R.N., Sharo, S., Voronkov, E.N., Taranenko, A.V., Lavrentjev, A.Yu.: Nucl. Instr. and Meth. in Phys. Research A **350**, (1994) 608.
43. Yeremin, A.V., Bogdanov, D.D., Chepigin, V.I., Gorshkov, V.A., Kabachenko, A.P., Malyshev, O.N., Popeko, A.G., Sagaidak, R.N., Ter-Akopian, A.Yu., Lavrentjev, A.Yu.: Nucl. Instr. and Meth. in Phys. Research B **126**, (1997) 329.
44. Lazarev, Yu.A., Lobanov, Yu.V., Mezentsev, A.N., Oganessian, Yu.Ts., Subbotin, V.G., Utyonkov, V.K., Abdullin, F.Sh., Bechterev, V.V., Iliev, S., Kolesov, I.V., Polyakov, A.N., Sedykh, I.M., Shirokovsky, I.V., Sukhov, A.M., Tsyganov, Yu.S., Zhuchko, V.E.: "The Dubna gas-filled recoil separator: a facility for heavy element research". In: Proceedings of "The International School Seminar on Heavy Ion Physics", Dubna, Russia, 10-15 May 1993, Vol. II, pp. 497-502.
45. Oganessian, Yu.Ts., Yeremin, A.V., Gulbekian, G.G., Bogomolov, S.L., Chepigin, V.I., Gikal, B.N., Gorshkov, V.A., Itkis, M.G., Kabachenko, A.P., Kutner, V.B., Lavrentev, A.Yu., Malyshev, O.N., Popeko, A.G., Rohac, J., Sagaidak, R.N., Hofmann, S., Münzenberg, G., Veselsky, M., Saro, S., Iwasa, N., Morita, K.: Eur. Phys. J. A **5**, (1999) 63.
46. Oganessian, Yu.Ts., Yeremin, A.V., Popeko, A.G., Bogomolov, S.L., Buklanov, G.V., Chelnokov, M.L., Chepigin, V.I., Gikal, B.N., Gorshkov, V.A., Gulbekian, G.G., Itkis, M.G., Kabachenko, A.P., Lavrentev, A.Yu., Malyshev, O.N., Rohac, J., Sagaidak, R.N., Hofmann, S., Saro, S., Giardina, G., Morita, K.: Nature **400**, (1999) 242.
47. Oganessian, Yu.Ts., Utyonkov, V.K., Lobanov, Yu.V., Abdullin, F.Sh., Polyakov, A.N., Shirokovsky, I.V., Tsyganov, Yu.S., Gulbekian, G.G., Bogomolov, S.L., Gikal, B.N., Mezentsev, A.N., Iliev, S., Subbotin, V.G., Sukhov, A.M., Buklanov, G.V., Subotic, K., Itkis, M.G., Moody, K.J., Wild, J.F., Stoyer, N.J., Stoyer, M.A., Lougheed, R.W.: Phys. Rev. Lett. **83**, (1999) 3154.
48. Oganessian, Yu.Ts., Utyonkov, V.K., Lobanov, Yu.V., Abdullin, F.Sh., Polyakov, A.N., Shirokovsky, I.V., Tsyganov, Yu.S., Gulbekian, G.G., Bogomolov, S.L., Gikal, B.N., Mezentsev, A.N., Iliev, S., Subbotin, V.G., Sukhov, A.M., Ivanov, O.V., Buklanov, G.V., Subotic, K., Itkis, M.G., Moody, K.J., Wild, J.F., Stoyer, N.J., Stoyer, M.A., Lougheed, R.W.: Phys. Rev. C **62**, (2000) 041604.
49. Oganessian, Yu.Ts., Utyonkov, V.K., Lobanov, Yu.V., Abdullin, F.Sh., Polyakov, A.N., Shirokovsky, I.V., Tsyganov, Yu.S., Gulbekian, G.G., Bogomolov, S.L., Gikal, B.N., Mezentsev, A.N., Iliev, S., Subbotin, V.G., Sukhov, A.M., Ivanov, O.V., Buklanov, G.V., Subotic, K., Itkis, M.G., Moody, K.J., Wild, J.F., Stoyer, N.J., Stoyer, M.A., Lougheed, R.W., Laue, C.A., Karelin, Ye.A., Tatarinov, A.N.: Phys. Rev. C **63**, (2000) 011301.
50. Schädel, M.: J. Nucl. Radiochem. Sci. **3**, (2002) 113.
51. Kadkhodayan, B. Türler, A., Gregorich, K.E., Baisden, P.A., Czerwinski, K.R., Eichler, B., Gäggeler, H.W., Hamilton, T.M., Jost, D.T., Kacher, C.D., Kovacs, A., Kreek, S.A., Lane, M.R., Mohar, M., Neu, M.P., Stoyer, N.J., Sylwester, E.R., Lee, M.D., Nurmia, M.J., Seaborg, G.T., Hoffman, D.C.: Radiochim. Acta **72**, (1996) 169.
52. Kratz, J.V., Gober, M.K., Zimmermann, H.P., Schädel, M., Brüchle, W.,Schimpf, E., Gregorich, K.E., Türler, A., Hannink, N.J., Czerwinski, K.R., Kadkhodayan, B., Lee, D.M., Nurmia, M.J., Hoffman, D.C., Gäggeler, H., Jost, D., Kovacs, J., Scherer, U.W., Weber, A.: Phys. Rev. **C 45**, (1992) 1064.
53. Türler, A., Dressler, R., Eichler, B., Gäggeler, H.W., Jost, D.T., Schädel, M., Brüchle, W., Gregorich, K.E., Trautmann, N., Taut, S.: Phys. Rev. C **57**, (1998) 1648.
54. Wilk, P.A., Gregorich, K.E., Türler, A., Laue, C.A., Eichler, R., Ninov, V., Adams, J.L., Kirbach, U.W., Lane, M.R., Lee, D.M., Patin, J.B., Shaughnessy, D.A., Strellis, D.A., Nitsche, H., Hoffman, D.C.: Phys. Rev Lett. **85**, (2000) 2797.

55. Smolanczuk, R., Sobiczewski, A.: "Shell effects in the properties of heavy and superheavy nuclei". In: Proceedings of "The XV. Nuclear Physics Divisional Conference on Low Energy Nuclear Dynamics", St.Petersburg, Russia, 18-22 April 1995, pp. 313-320.
56. Cwiok, S., Dobaczewski, J., Heenen, P.H., Magierski, P., Nazarewicz, W.: Nucl. Phys. A **611**, (1996) 211.
57. Rutz, K., Bender, M., Bürvenich, T., Schilling, T., Reinhard, P.G., Maruhn, J.A., Greiner, W.: Phys. Rev. C **56**, (1997) 238.
58. Smolanczuk, R., Skalski, J., Sobiczewski, A.: Phys. Rev. C **52**, (1995) 1871.
59. Möller, P., Nix, J.R., Myers, W.D., Swiatecki, W.J.: Atomic Data and Nucl. Data Tables **59**, (1995) 185.
60. Armbruster, P., Agarwal, Y.K., Brüchle, M., Brügger, M., Dufour, J.P., Gäggeler, H., Heßberger, F.P., Hofmann, S., Lemmertz, P., Münzenberg, G., Poppensieker, K., Reisdorf, W., Schädel, M., Schmidt, K.-H., Schneider, J.H.R., Schneider, W.F.W., Sümmerer, K., Vermeulen, D., Wirth, G., Ghiorso, A., Gregorich, K.E., Lee, D., Leino, M., Moody, K.J., Seaborg, G.T., Welch, R.B., Wilmarth, P., Yashita, S., Frink, C., Greulich, N., Herrmann, G., Hickmann, U., Hildebrand, N., Kratz, J.V., Trautmann, N., Fowler, M.M., Hoffman, D.C., Daniels, W.R., von Gunten, H.R., Dornhöfer, H.: Phys. Rev. Lett. **54,** (1985) 406.
61. Reisdorf, W. Schädel, M.: Z. Phys. A **343**, (1992) 47.
62. Schädel, M., Hofmann, S.: J. Radioanal. Nucl. Chem. **203**, (1996) 283.
63. Bass, R.: Nucl. Phys. A **231**, (1974) 45.
64. von Oertzen, W.: Z. Phys. A **342**, (1992) 177.
65. Denisov, V.Yu., Hofmann, S.: Phys. Rev. C **61**, (2000) 034606.
66. Gupta, R.K., Parvulescu, C., Sandulescu, A. Greiner, W.: Z. Phys. A **283**, (1977) 217.
67. Denisov, V.Yu., Nörenberg, W.: Eur. Phys. J., submitted 2002.
68. Aritomo, Y., Wada, T., Ohta, M., Abe, Y.: Phys. Rev. C **59**, (1999) 796.
69. Zagrebaev, V.I.: Phys. Rev. C **64**, (2001) 034606.
70. Giardina, G., Hofmann, S., Muminov, A.I., Nasirov, A.K.: Eur. Phys. J. A **8**, (2000) 205.
71. Smolanczuk, R.: Phys. Rev. C **63**, (2001) 044607.
72. Adamian, G.G., Antonenko, N.V., Scheid, W.: Nucl. Phys. A **678**, (2000) 24.
73. Reiter, P., Khoo, T.L., Lister, C.J., Seweryniak, D., Ahmad, I., Alcorta, M., Carpenter, M.P., Cizewski, J.A., Davids, C.N., Gervais, G., Greene, J.P., Henning, W.F., Janssens, R.V.F., Lauritsen, T., Siem, S., Sonzogni, A.A., Sullivan, D., Uusitalo, J., Wiedenhöver, I., Amzal, N., Butler, P.A., Chewter, A.J., Ding, K.Y., Fotiades, N., Fox, J.D., Greenlees, P.T., Herzberg, R.D., Jones, G.D., Korten, W., Leino, M., Vetter, K.: Phys. Rev. Lett. **82**, (1999) 509.
74. Reiter, P., Khoo, T.L., Lauritsen, T., Lister, C.J., Seweryniak, D., Sonzogni, A.A., Ahmad, I., Amzal, N., Bhattacharyya, P., Butler, P.A., Carpenter, M.P., Chewter, A.J., Cizewski, J.A., Davids, C.N., Ding, K.Y., Fotiades, N., Greene, J.P., Greenlees, P.T., Heinz, A., Henning, W.F., Herzberg, R.-D., Janssens, R.V.F., Jones, G.D., Kankaanpää, H., Kondev, F.G., Korten, W., Leino, M., Siem, S., Uusitalo, J., Vetter, K., Wiedenhöver, I.: Phys. Rev. Lett. **84**, (2000) 3542.
75. Herzberg, R.-D., Amzal, N., Becker, F., Butler, P.A., Chewter, A.J.C., Cocks, J.F.C., Dorvaux, O., Eskola, K., Gerl, J., Greenlees, P.T., Hammond, N.J., Hauschild, K., Helariutta, K., Heßberger, F., Houry, M., Jones, G.D., Jones, P.M., Julin, R., Juutinen, S., Kankaanpää, H., Kettunen, H., Khoo, T.L., Korten, W., Kuusiniemi, P., Le Coz, Y., Leino, M., Lister, C.J., Lucas, R., Muikku, M., Nieminen, P., Page, R.D., Rahkila, P., Reiter, P., Schlegel, Ch., Scholey, C., Stezowski, O., Theisen, Ch., Trzaska, W.H., Uusitalo, J., Wollersheim, H.J.: Phys. Rev. C **65**, (2001) 014303.

Index

Chapter 2

Theoretical Chemistry of the Heaviest Elements

V. Pershina

Gesellschaft für Schwerionenforschung mbH, Planckstr. 1, D-64221 Darmstadt, Germany

1. Introduction

The last decade was marked with the discovery of five new members of the Periodic Table: The heaviest elements of the last transition element series 110 through 112 were identified in the Gesellschaft für Schwerionenforschung (GSI), Darmstadt [1-3] and some decay chains and fission products associated with production of even more heavy elements 116 and 114 were recently reported by the Joint Institute for Nuclear Research (JINR), Dubna [4]. This period of time was also very fruitful with studying chemical properties of the very heavy elements [5-9].

Conceptually, it is the atomic number and the electronic configuration of an element that define its position in the Periodic Table. Since they cannot be measured for the very heavy elements, information on its chemical behavior is often used to place an element in a chemical group. Unfortunately, with increasing nuclear charge the cross sections and the production rates drop so rapidly that such chemical information can be accessed only for elements with a half-life of the order of at least few seconds and longer. In this case, some fast chemistry techniques are used. They are based on the principle of chromatographic separations either in the gas phase exploiting the differences in volatility of heavy element compounds, or in the aqueous

M. Schädel (ed.), The Chemistry of Superheavy Elements, 31-94

phase by solvent extraction or ion exchange separations using differences in the complex formation (see the related chapters in this book). Chemistry of elements 104 through 108 has been successfully studied using these techniques [5-10]. Some first results were reported on gas-phase chemical experiments with element 112 [11].

Due to the very short half-lives, chemical information obtained from these experiments is limited to the knowledge of only a very few properties. It mostly answers the question about whether a new element behaves similarly or differently than its lighter congeners in the chromatographic separation processes: Due to very strong relativistic effects on the valence electronic shells of the heaviest elements, some deviations from trends known for the lighter homologues in the chemical groups are expected. The chemical composition of heavy element compounds cannot be experimentally established either and can only be assumed by comparing their behavior with those of lighter congeners. Presently, some important spectroscopic properties like ionization potentials (IP), electron affinities (EA), force constants (k_e), etc. cannot be measured at all. Thus, for the heaviest elements, theoretical studies become extremely important and are often the only source of useful chemical information. Moreover, theoretical predictions of experimentally studied properties are especially valuable in order to help design the required sophisticated and expensive experiments with single atoms.

For heavy elements, the use of the relativistic quantum theory, and molecular and atomic programs based on it, is mandatory [12-14]. It has been shown [15-17] that even trends in properties can be predicted erroneously by using non-relativistic approximations. Simple extrapolations of properties from those of the lighter elements are also unreliable and can lead to erroneous predictions. Relativistic quantum theory and quantum-chemical methods have been tremendously developed in the last two decades to accurately treat both relativistic and correlation effects for the many-electron problem. Density Functional Theory methods were improved with respect to the treatment of non-local (exchange-correlation) effects [18] and the *ab initio* Dirac-Fock theory has advanced due to the development of basis set techniques and treatment of electron correlation [19]. These improvements, along with the motivating experimental results, caused a boom in the molecular relativistic electronic structure studies of the heaviest elements as the number of publications in Figure 1 shows, and resulted in some remarkable achievements. The present publication is an overview of the results of the recent theoretical investigations in the area of the heaviest elements (Z=103 and higher) and it provides their comparative analysis. Special attention is paid to the predictive power of the theoretical studies for the chemical experiments. Problems and challenges of the theoretical research in this area related to the necessity of improving the used techniques and obtaining the desired chemical information are described.

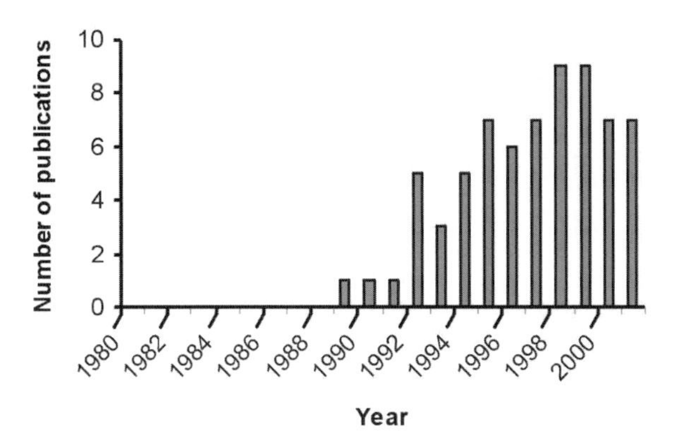

Fig. 1. Publications on molecular calculations for elements 104 and heavier.

2. Architecture of the Periodic Table

When G. T. Seaborg in 1944 introduced his ‘actinide’ concept the theory played not the last role in his decision to place newly discovered elements in a second series where the filling of the 5f-shell takes place, similarly to the ‘lanthanide’ series where the filling of the 4f-shell takes place. Thus, the filled-shell concept was in accord with the newly found periodicity in chemical properties and resulted in the discovery of all the heavy actinides at that time [20].

Since then, the theory advanced to such an extent that the Periodic Table, see Figure 2, is now predicted with sufficient accuracy up to very high Z numbers. That was possible due to the development of very accurate relativistic quantum chemical methods and programs, which could reliably calculate electronic configurations of heavy element atoms and ions. Ground states for the superheavy elements up to Z = 174 were predicted in the late sixties and early seventies by J.B. Mann [21], B. Fricke, J.T. Waber, W. Greiner [22] and J.P. Desclaux [23] using the Dirac-Fock (DF) and Dirac-Fock-Slater (DFS) methods. The results up to 1975 are summarized in the review of B. Fricke, see [12] and references therein. More accurate multiconfiguration Dirac-Fock (MCDF) [24-30] and coupled cluster (CC) calculations, see reviews [31,32], performed in the eighties and nineties basically confirmed those earlier predictions and furnished more reliable values of the electronic energy states.

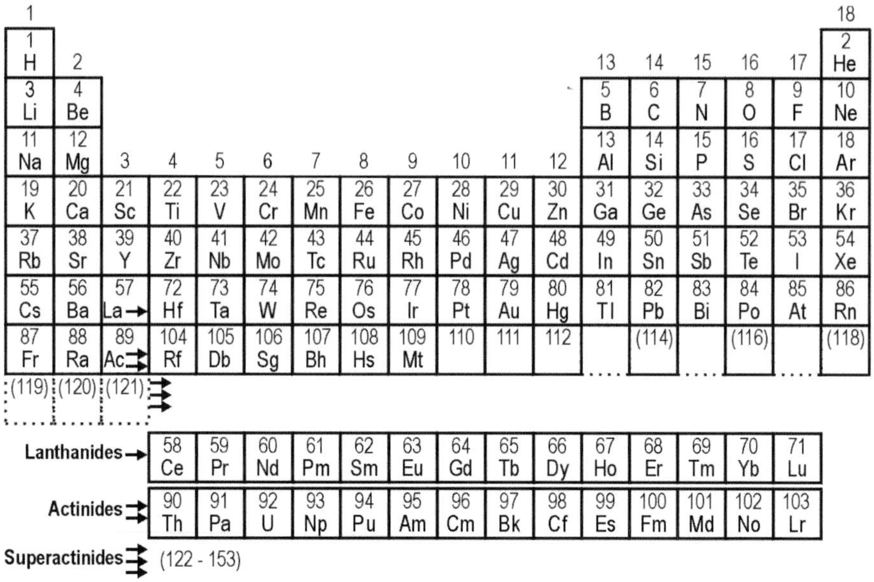

1	2	3	4	5	6	7	8	9	10	11	12	13	14	15	16	17	18
1 H																	2 He
3 Li	4 Be											5 B	6 C	7 N	8 O	9 F	10 Ne
11 Na	12 Mg											13 Al	14 Si	15 P	16 S	17 Cl	18 Ar
19 K	20 Ca	21 Sc	22 Ti	23 V	24 Cr	25 Mn	26 Fe	27 Co	28 Ni	29 Cu	30 Zn	31 Ga	32 Ge	33 As	34 Se	35 Br	36 Kr
37 Rb	38 Sr	39 Y	40 Zr	41 Nb	42 Mo	43 Tc	44 Ru	45 Rh	46 Pd	47 Ag	48 Cd	49 In	50 Sn	51 Sb	52 Te	53 I	54 Xe
55 Cs	56 Ba	57 La→	72 Hf	73 Ta	74 W	75 Re	76 Os	77 Ir	78 Pt	79 Au	80 Hg	81 Tl	82 Pb	83 Bi	84 Po	85 At	86 Rn
87 Fr	88 Ra	89 Ac→	104 Rf	105 Db	106 Sg	107 Bh	108 Hs	109 Mt	110	111	112		(114)		(116)		(118)
(119)	(120)	(121)→															

Lanthanides→	58 Ce	59 Pr	60 Nd	61 Pm	62 Sm	63 Eu	64 Gd	65 Tb	66 Dy	67 Ho	68 Er	69 Tm	70 Yb	71 Lu
Actinides→	90 Th	91 Pa	92 U	93 Np	94 Pu	95 Am	96 Cm	97 Bk	98 Cf	99 Es	100 Fm	101 Md	102 No	103 Lr
Superactinides→	(122 - 153)													

Fig. 2. Modern Periodic Table of the Elements

According to results of these calculations, in the first nine of the transactinide elements (Z = 104 to 112) the filling of the 6d shell takes place. Followed by them are the 7p elements 113 through 118, with element 118 falling into the group of noble gases. In elements 119 and 120, the filling of the 8s shell takes place, so that these elements will obviously be homologues of alkali elements in groups 1 and 2. The next element, 121, has a relativistically stabilized 8p electron in its ground state electronic configuration in contrast to what would be predicted by a simple extrapolation in the group. In the next element, Z=122, a 7d electron is added to the ground state, so that it is $8s^2 7d8p$ in contrast to the $7s^2 6d^2$ state of Th. This is the last element where accurate calculations [33] exist. Beyond element 122, the situation becomes more complicated: 7d, 6f and 5g levels, and further on 9s, $9p_{1/2}$ and $8p_{3/2}$ levels are located energetically so close to each other that in the Z = 160 region the usual classification on the basis of a simple electronic configuration may become invalid. The clear structures of the pure p, d, f and g blocks are not distinguishable any more, see the Periodic Table till Z=173 in [12]. Chemical properties of these elements influenced by all these mixed electronic shells will then be also so different to anything known before that any classification just of the basis of their knowledge will be impossible. It is interesting to note, however, that without relativistic effects, the chemistry of the heaviest superheavy elements would also be different to that of their lighter homologues due to very large shell structure effects. The modern Periodic Table up to element 153 is shown in Figure 2. Its structure cannot be understood without knowledge of the influence of relativistic effects on the electronic shells, which will be considered in the following section.

3. Relativistic and QED Effects on the Orbital Shells

With increasing Z of heavy elements causing a stronger attraction to the core an electron is moving faster, so that its mass increase is

$$m = m_0/[1 - (v/c)^2]^{1/2} \qquad (1)$$

where m_0 is the rest mass and v is the velocity of the electron. The effective Bohr radius

$$a_B = \frac{\hbar^2}{mc^2} = a_B{}^0\sqrt{1-(v/c)^2} \qquad (2)$$

decreases, as a consequence, for a hydrogen-like s and $p_{1/2}$ electrons. (The 1s electron of Sg, e.g., has v/c = 106/137 = 0.77, so that its radius shrinkage is 37%). The contraction and stabilization of the s and $p_{1/2}$ orbitals is known to be the direct relativistic effect and it was shown to originate from the inner K and L shells region [34]. This effect was originally thought to be large only for the "fast" electrons in inner core shells of heavy atoms. It was, however, shown that the direct relativistic stabilization is still large for the outer s and $p_{1/2}$ valence orbitals. Thus, e.g., 7s orbital for element 105, Db, is $\Delta_R\langle r\rangle_{7s}$ = 25% relativistically contracted, where $\Delta_R\langle r\rangle_{ns} = \langle r\rangle_{nr} - \langle r\rangle_{rel}/\langle r\rangle_{nr}$, see Figure 3. The contraction of the outer s and $p_{1/2}$ orbitals was recently explained as due to the admixing of higher bound and (partially) continuum orbitals due to relativistic perturbations [34].

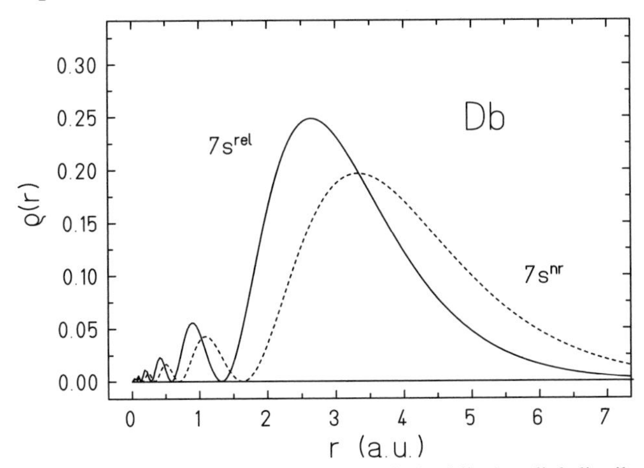

Fig. 3. Relativistic (solid line) and non-relativistic (dashed line) radial distribution of the 7s valence electrons in element 105, Db.

The effect of the ns orbital contraction reaches its maximum in the 6th row on Au (17.3%) and in the 7th row on element 112 (31%), the phenomenon being called the relativistic effects gold maximum and group 12 maximum, respectively. The same maximum is, consequently, observed with the relativistic stabilization of the 6s and 7s orbitals, see Figure 4. The shift of the maximum to element 112 in the 7th row in contrast to gold in the 6th row is due to the fact that in both the 111 and 112 elements the ground state electronic configuration is $d^q s^2$, while the electronic configuration changes from Au (d^9s^1) to Hg ($d^{10}s^2$).

The second (indirect) relativistic effect is the expansion of outer d and f orbitals: The relativistic contraction of the s and $p_{1/2}$ shells results in a more efficient screening of the nuclear charge, so that the outer orbitals which never come to the core become more expanded and energetically destabilized. While the direct relativistic effect originates in the immediate vicinity of the nucleus, the indirect relativistic effect is influenced by the outer core orbitals. It should be realized that though contracted s and $p_{1/2}$ core (innermore) orbitals cause indirect destabilization of the outer orbitals, relativistically expanded d and f orbitals cause the indirect stabilization of the valence s and p-orbitals. That partially explains the very large relativistic stabilization of the 6s and 7s orbitals in Au and element 112, respectively: Since d shells (it is also valid for the f shells) become fully populated at the end of the nd series, there will occur a maximum of the indirect stabilization of the valence s and p orbitals [34].

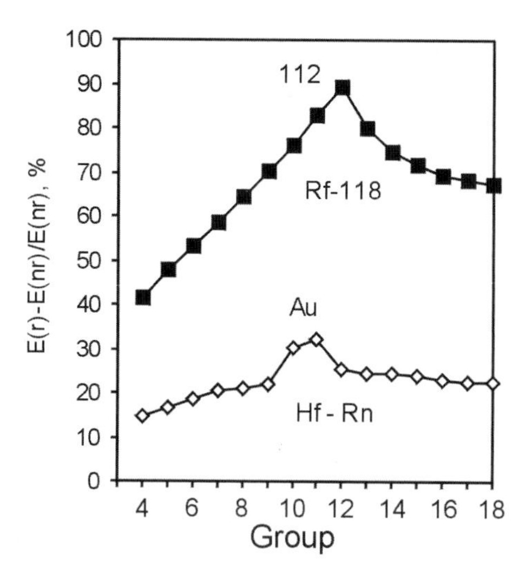

Fig. 4. The relativistic stabilization of the 6s and 7s orbitals in the 6th and 7th row of the Periodic Table. Re-drawn from the data of [17]. The relativistic DF data are from [23].

Figure 5 demonstrates the relativistic stabilization of the ns orbitals, as well as the destabilization of the (n-1)d orbitals for group-8 elements, as an example. One can see that trends in the relativistic and non-relativistic energies of the valence electrons are opposite from the 5d to the 6d elements. Thus, the non-relativistic description of the wave function would still give the right trend in properties from the 4d to the 5d elements, while it would result in the opposite and consequently wrong trend from the 5d to the 6d elements.

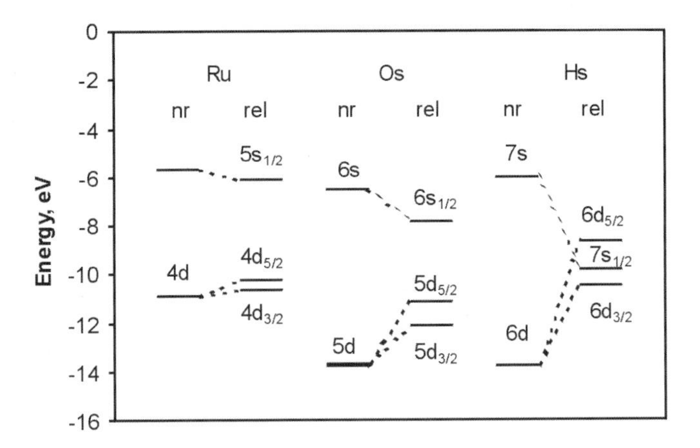

Fig. 5. Relativistic (DF) [23] and non-relativistic (HF) [17] energy levels of the valence ns and (n-1)d electrons for group-8 elements.

The third relativistic effect is the well-known spin-orbit (SO) splitting of levels with $l > 0$ (p, d, f, … electrons) into $j = l \pm ½$. It also originates from the inner region in the vicinity of nucleus. The SO splitting for the same l decreases with increasing number of subshells, i.e., it is much stronger for inner (core) shells than for outer shells. The SO splitting decreases with increasing l for the same principal quantum number, i.e., the $np_{1/2}$-$np_{3/2}$ splitting is larger than the $nd_{3/2}$-$nd_{5/2}$ and both are larger than the $nf_{5/2}$-$nf_{7/2}$. It is explained by the orbital densities in the vicinity of nucleus decreasing with increasing l. In transactinide compounds SO coupling becomes similar, or even larger, in size compared to typical bond energies. The SO splitting of the valence 7p electrons in element 118, e.g., is as large as 11.8 eV, see Figure 6. All the three relativistic effects are of the same order of magnitude and they grow roughly as Z^2.

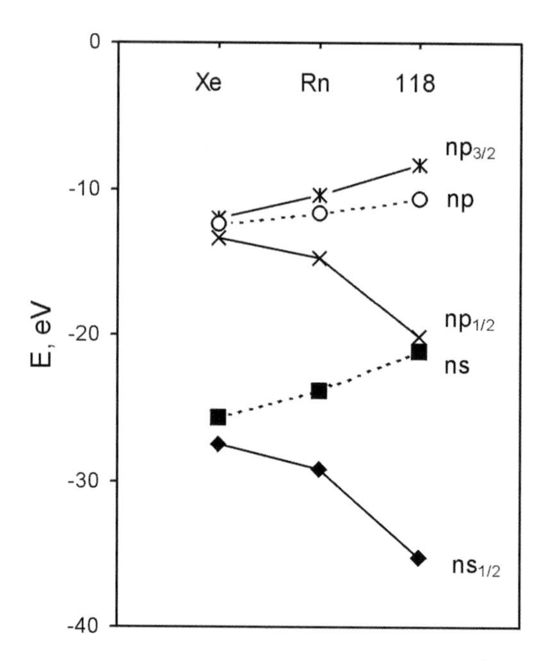

Fig. 6. Relativistic stabilization of the ns and $np_{1/2}$ orbitals and the spin-orbit splitting of the np orbitals for the noble gases Xe, Rn and element 118. The DF atomic energies are from [23] and the HF values are from [17].

Breit effects (accounting for magnetostatic interaction) on energies of the valence orbitals and on IP are usually small, e.g., 0.02 eV for element 121 [35]. They can, however, reach few % for the fine structure level splitting, e.g., 3.6 % in Tl ^2P splitting and are of the order of correlation effects there [14]. In element 121, they are of about 0.1 eV for transition energies between the states including f orbitals [35].

The quantum electrodynamic (QED) effects are known to be very important for inner-shells, e.g., in accurate calculations of X-ray spectra. For highly charged few electron atoms they were found to approach the Breit correction to the electron-electron interaction [36]. Similar effects were also found for valence ns electrons of neutral alkali-metal and coinage metal atoms [37]. Comparison of the valence ns Lamb shift (the vacuum polarization and self-energy) with the ns valence orbital energy and the relativistic, Breit, and nuclear volume contributions to it for coinage metals at the DF level is shown in Figure 7. The result for the valence ns electron is a destabilization, while for (n-1)d electron is an indirect stabilization. In the middle range (Z = 30–80) both the valence-shell Breit and the Lamb-shift terms behave similarly to the kinetic relativistic effects scaling as Z^2. For the highest Z values the increase is faster. The nuclear volume effects grow even faster with Z. Thus, the results show that valence ns-shell Lamb shifts are not negligible: They are of the order of 1-2% of the kinetic relativistic effects, which means that the existing studies of relativistic effects are up to 99% correct [37].

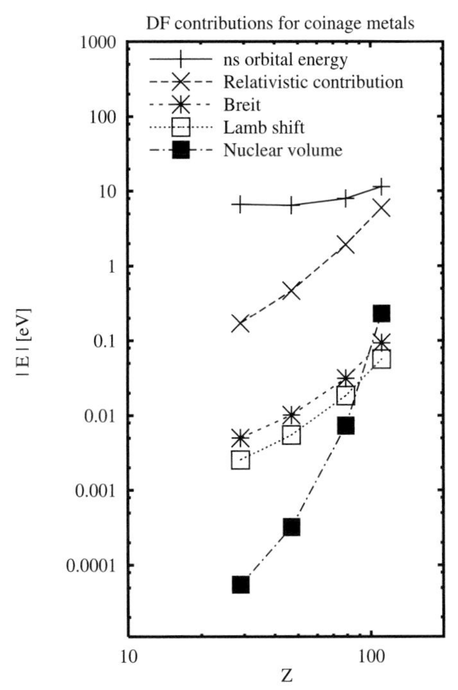

Fig. 7. Comparison of the valence ns Lamb shift with the orbital energy and the relativistic, Breit, and nuclear volume contributions to it for coinage metals. Reproduced from [37].

4. Relativistic Quantum Chemical Methods for Atoms and Molecules

Methods to calculate the electronic structures of very heavy element compounds are the same relativistic methods which can be applied to any relativistic systems. They were overviewed in application to transition elements [13], actinides [38], and transactinides [15-17]. They will, therefore, be only shortly described here with the accent put on those which were used for calculations of the transactinide systems.

4.1 AB INITIO SOLUTION OF THE DIRAC EQUATION

The Dirac-Coulomb-Breit (DCB) Hamiltonian

$$h_{DCB} = h_{DC} + \Sigma B_{ij} \tag{3}$$

is a satisfactory starting point for calculations of many-electron systems in the sense that it treats electrons relativistically and the most important part of the electron-electron interaction nonperturbatively. It contains the one-electron Dirac Hamiltonian h_{DC} including the nuclear potential, V^n, and the operator $V_{ij} = 1/r_{ij}$ for the instantaneous Coulomb interactions between electrons

$$h_{DC}=c\alpha_i p_i + (\beta - 1)c^2 + V^n + \Sigma 1/r_{ij.} \tag{4}$$

B_{ij} is the Breit term

$$B_{ij} = -1/2[(\alpha_i\alpha_j)r_{ij}^{-1} + (\alpha_i r_{ij})(\alpha_j r_{ij})r_{ij}^{-3}]. \tag{5}$$

The V^n includes the effect of the finite nuclear size, while some finer effect, like QED, can be added to the h_{DCB} perturbatively. The DCB Hamiltonian in this form contains all effects through the second order in α, the fine-structure constant.

Since the relativistic many-body Hamiltonian cannot be expressed in closed potential form, which means it is unbound, projection one- and two-electron operators are used to solve this problem [39]. The operator projects onto the space spanned by the positive-energy spectrum of the Dirac-Fock-Coulomb (DFC) operator. In this form, the “no-pair“ Hamiltonian [40] is restricted then to contributions from the positive-energy spectrum and puts Coulomb and Breit interactions on the same footing in the SCF calculations.

Since the Dirac equation is written for one electron, the real problem of *ab initio* methods for a many-electron system is an accurate treatment of the instantaneous electron-electron interaction, called electron correlation. The latter is of the order of magnitude of relativistic effects and may contribute to a very large extent to the binding energy and other properties. The DCB Hamiltonian (Equation 3) accounts for the correlation effects in the first order via the V_{ij} term. Some higher order of magnitude correlation effects are taken into account by the configuration interaction (CI), the many-body perturbation theory (MBPT) and by the presently most accurate coupled cluster (CC) technique.

4.1.1 *Atomic codes*

Earlier predictions of chemical properties of the heaviest elements were made on the basis of atomic single-configuration DF and DS calculations using approximations of eqs. (3-5) and numerical techniques [12,21-23]. Later, the MCDF method based on the CI technique was used. The advantage of this method is that it accounts for most of the correlation effects while retaining relatively small number of configurations. It can treat a large number of open shell configurations and can be applied to elements with any number of valence electrons. There are several modifications of this method: the codes of J.P. Desclaux [42], I.P. Grant [30,41] and S. Fröse-Fisher [43]. Calculations of IPs of the heaviest elements from Lr through element 108 [24-29] and elements 113 through 117 [30] were performed using the MCDF method. An average error for IP is about 1 eV. The limited accuracy of this method is due to omitting dynamic correlations, since excitations of the type (nj) → (n′j) cannot be handled, and some core polarization.

Another recently created correlated method is the no(virtual)pair DF (DCB) coupled-cluster technique of E. Eliav, U. Kaldor and I. Ishikawa [31,32,44]. It is based on the DCB Hamiltonian; Equation 3. Correlation effects are taken into account by action of the excitation operator

$$S = \sum_{m\geq 0}\sum_{n\geq 0}\left(\sum_{l\geq m+n} S_l^{(m,n)}\right) \tag{6}$$

defined in the Fock-space CC approach with respect to a closed-shell reference determinant. In addition to the traditional decomposition into terms with different total number of excited electrons (l), S is partitioned according to the number of valence particles (m) or holes (n). Presently, the method is limited to one or two (single-double excitations, CCSD, e.g. $(m,n) \leq 2$) particle valence sectors of the Fock space, i.e. it can treat the states which can be reached from a closed shell by adding or removing no more than two electrons. It was applied to the calculations of electronic energy states of a number of the heaviest elements up to Z = 122 [33]. The method is very accurate, with an average error of 0.1 eV for excitation energies, since it takes into account most of dynamic correlation effects, omitted from the MCDF method and a core polarization. Due to the mentioned limitation of the method, it can, however, presently not handle the elements of the middle of the 6d series (with the number of 6d electrons more than 2) leaving this area to the MCDF calculations.

4.1.2 *Molecular ab initio 'fully' relativistic methods*

Molecular fully relativistic *ab initio* methods use the same DFC or DCB Hamiltonians, Equation 3. Based on them molecular (DF)-LCAO codes inclu-ding correlation effects are still under development [19] and calculations are restricted to molecules with very few atoms. Algebraic solution of the molecular Dirac equation encounters difficulties due to the fact that the Dirac operator is unbound. Some special techniques like the projection operator technique are used to avoid connected with it the variational collapse. In contrast to non-relativistic calculations, large basis sets are needed to describe accurately the inner shell region where relativistic pertur-bation operators are dominant. The condition of the kinetic balance should be observed relating the large and small components of the four-component wave-function [41]. (Kinetically balanced Gaussian type wave functions with a Gaussian distribution for the nuclear potential are presently best suited.) Calculations of two-electron integrals require large disk space and computational time. Correlation effects are taken into account by using the same techniques as those of the atomic codes: CI [45] or MBPT (e.g., the second order Møller-Plesset, MP2 [46]) and, recently, the CCSD [47].

The following calculations for heavy and superheavy elements should be mentioned: PtH_2 (DF–KRCCSD) [47], $(113)_2$ (DFC) [48], 111H, 117H, (113)(117) (DF) [49-51] and $114H_4$ (DF-MP2) [52]. The aim of those calculations was mainly to test relativistic and correlation effects on model systems. The calculations for the molecules, where experimental data are available, are very promising, see Table 1.

4.2 RELATIVISTIC PSEUDO-HAMILTONIANS

An efficient way to solve a many-electron problem is to apply relativistic effective core potentials (RECP). According to this approximation, frozen inner shells are omitted and replaced in the Hamiltonian h_{nl} by an additional term, a pseudopotential (U^{REP})

$$h_{nl} = -\nabla^2/2 - Z/r + U^{REP} + V^c_{nl} - V^{ex}_n \quad (7)$$

$$U^{REP} = U^{AREP} + U^{SO} \quad (8)$$

$$U^{AREP} = \Sigma\, U_l(r)\,|lm_l\rangle\langle lm_l| \quad (9)$$

where AREP stands for a spin-averaged or scalar relativistic effective core potential, and V^c and V^{ex} are the Coulomb and exchange operators. The SO splitting is obtained using the effective SO operator, U^{SO}. As a result, the number of basis functions is drastically diminished and, hence the number of two-electron integrals. Then, the one-electron integrals are solved for the valence basis functions and this additional term by applying *ab initio* schemes at the SCF level or with electron correlation (CI, MP2, CCSD(T), where T stands for 'triple') included.

The U^{AREP} is usually fitted to one-component or two-component all-electron DF or DHF relativistic atomic wave functions [53]. Such potentials for elements Am through 118 were generated by C.S. Nash *et al.* [54]. The calculations are still very sophisticated and expensive. The RECP have recently been further improved by introducing a special technique (of nonvariational restoration of the electronic structure in the core) to correct an error introduced by a smoothing of the orbitals in the core region (nodeless wavefunction), which may be essential for dissociation and transition energies (as high as 2000 cm^{-1}). The method is called the generalized RECP, GRECP / RCC [55]. Calculations for ground and excited states of HgH [56] show very good agreement with experiment for binding and excitation energies.

Another type of methods belonging to this group is the energy-adjusted pseudo potentials (PP) which uses atomic spectra (energies) for generating the U^{REC} [57-59]. To achieve a high accuracy, a large number of correlated states has to be taken into account in the fitting procedure, which technically can be too demanding and expensive. SO effects can be included in average relativistic (AR)PP calculations by adding a SO operator, so that a SOPP are obtained. These PP were generated for the transactinides [60]. Several RECP and PP calculations were performed for gas-phase compounds of the heaviest elements [17].

The quasirelativistic (QR) PP of Hay and Wadt [61] use two-component wave functions, but the Hamiltonian includes the Darwin and mass-velocity terms and omits the spin-orbit effects. The latter are then included via the perturbation operator after the wave functions have been obtained. The advantage of the method is the possibility to calculate quite economically rather large systems. The method is implemented in the commercial system "Gaussian 98". It has extensively been applied to calculations of transition-element and actinide systems [62].

The accuracy of the various relativistic, non-relativistic, correlated and non-correlated methods in comparison with experimental results is shown in Table 1 for AuH, a sort of a test molecule (see also [63,64]). The data of Table 1 demonstrate the importance of relativistic and electron correlation effects. Thus, relativistic effects diminish the equilibrium bond length (R_e) by 0.26 Å (the HF-DF difference without correlation) or by 0.21 Å (the HF+MP2 - DF+MP2 difference with correlation), and enlarge the binding energy (D_e) by 0.70 eV (the HF-DF difference without correlation) or by 2.21 eV (the HF+MP2 - DF+MP2 difference with correlation). Correlation diminishes R_e on the DF level by 0.07 eV, but enhances D_e by 1.34 eV. Thus, even for AuH correlation amounts almost to 50% of the chemical bond strength. No additivity of correlation and relativistic effects was shown.

4.3 DENSITY FUNCTIONAL THEORY

Density Functional Theory (DFT) is primarily a theory of electronic ground state structure expressed in terms of the electronic density, see [18,65] and references therein,

$$\rho(\vec{r}) = \sum_i n_i \phi_i^+(\vec{r}) \phi_i(\vec{r}) \tag{10}$$

Due to its relative simplicity, the DFT became extremely useful in the application to large heavy-element molecules, clusters, solutions and solids. Systems with the large number of atoms can be treated with sufficient accuracy. The computing time in the DFT for a system of many atoms grows as N_{at}^2 or N_{at}^3, while in traditional methods as $\exp(N_{at})$. The present upper limit is of $N_{at} \sim 200$. The modern DFT is in principle exact and the accuracy depends on the adequate knowledge of the exchange-correlation energy functional $V^{ex}[\rho(r)]$ [18]. Based on it DFT methods are alternative and complementary, both quantitatively and conceptually, to the traditional methods where the many-electron wave function $\Psi(r_1 \ldots r_N)$ is used. The Hamiltonian used here is the same as given in Equation 3 plus an additional exchange-correlation term $V^{ex}[\rho(r)]$. Thus, the single-particle equations are

$$\left[t + V^n + V^c + V^{ex}\right]|\phi_i\rangle = \varepsilon_i |\phi_i\rangle \tag{11}$$

$$V^C = \int \frac{\rho(\vec{r}')}{|\vec{r} - \vec{r}'|} d\vec{r} \tag{12}$$

$$V^{ex}{}_{GGA} = f\left[\rho(\vec{r}), |\nabla\rho(\vec{r})|\right] \tag{13}$$

$V^{ex}{}_{GGA}$ or $V^{ex}{}_{RGGA}$, the (relativistic) generalized gradient approximation (RGGA), is the highest level of approximation for the exchange-correlation potential taking into account the non-local effects [66]. This approximation reduces the failure of the (relativistic) local density approximation, (R)LDA (the one level below approximation), describing the asymptotic behavior of the exchange and correlation potential in a wrong way, since it falls off exponentially and not as $1/r$.

There are several methods based on eqs. (10-13). The so-called 'fully relativistic' DFT methods use four-component (numerical or combined with some Slater type) wave functions ϕ_i, which means that the spin-orbit is included explicitly from the beginning. A predecessor of this group of the methods was the intrinsically approximate Dirac Slater discrete variational (DS-DV) method of Ellis and Rosen [67], where the Slater exchange-correlation term $V^{ex} = -3C[3\rho(r)/8\pi]^{1/3}$ was used. The DS-DV method was extensively applied to calculations of various gas-phase and aqueous phase compounds of the transactinides [15,16]. The introduction of the RGGA for V^{ex} [66] (Equation 13) and of an accurate integration scheme, the minimization of an error in the total energy by an additional fitting procedure and the use of optimized basis sets have resulted in a drastic reduction of errors in the binding energy calculations [68-70] which was previously a main drawback of the DS-DV method. Another error in D_e introduced by the single-particle approximation could be corrected by using the MCDF method for atomic calculations when calculating dissociation energies [70]. The method is presently used by us for optimization of geometry and binding energy calculations of heavy element compounds. The results will be discussed in the following chapters.

Another four-component 'fully relativistic' DFT program is the Beijing density functional (BDF) method [71]. Four-component numerical atomic spinors obtained by finite-difference atomic calculations are used for cores, while basis sets for valence spinors are a combination of numerical atomic spinors and kinetically balances Slater-type functions. The non-relativistic GGA for V^{ex} is used. Calculations of small transactinide systems, like 111X and 114X (X = H, F, Cl, Br, I, O) were reported [72,73].

One-component quasirelativistic density-functional method extensively used in the calculations of the electronic structure of transition element and actinide compounds, known as the Amsterdam (ADF) method [74], is the successor of the quasirelativistic Hartree-Fock-Slater (QR-HFS) one [75]. An earlier version of the method used a QR Hamiltonian (Darwin and mass-velocity terms) and the GGA for V^{ex}. The spin-orbit effects were taken into account by perturbation operators. In the modern version of the method the Hamiltonian contains relativistic corrections already in the zeroth-order and is called the zeroth-order regular approximation (ZORA) [76]. Recently, the spin operator was also included in the ZORA Fock operator [77]. The ZORA method uses analytical basis functions, and gives reliable optimization of geometry and bonding. For elements with a very large SO splitting, like 114, ZORA can deviate from the 4-component DFT results due to improper description of the $p_{1/2}$ spinors [73]. The basis functions for the elements beyond 103 have, however, not yet been generated. Another one-component quasirelativistic scheme [78] applied to the calculations of dimers of elements 111 and 114 [72,73] is a modification of the ZORA method. Some other relativistic DFT codes like those of N. Rösch [79,80], T. Ziegler [81], and of D.A. Case and C.Y. Young [82] should also be mentioned here. They were, however, not applied to the heaviest elements. The accuracy of some of the DFT methods for AuH is shown in Table 1.

Table 1. Accuracy of different molecular methods for AuH showing importance of relativistic and correlation effects. R_e is the equilibrium bond length and D_e is the binding energy

Method	R_e, Å	D_e, eV	Ref.
HF	1.831	1.08	46
HF + MP2	1.711	1.90	46
DF	1.570	1.78	46
DF + MP2	1.497	3.11	46
PP + CI	1.504	2.86	58
RECP + MP2	1.512	3.08	53
RECP + KRCCSD(T)	1.510	3.31	63
DKH	1.539	3.33	79
ZORA (MP)	1.537	3.33	72
BDF	1.537	3.34	72
Experiment	1.524	3.36	83

4.4 OTHER METHODS

There is quite a number of other theoretical methods appropriate for calculations of the electronic structure of heavy elements with some limitations, like, e.g., expansion techniques. One should mention here the Douglas-Kroll-Hess (DKH) [84] method, which can be applied if SO effects are small (it was also used for calculations of the electronic structure of 111H [50]). For other methods see [17].

5. Atomic Properties of the Transactinides

Electronic configurations, ionization potentials (IP), atomic/ionic radii (AR/IR), polarizabilities (α) and stability of oxidation states are important atomic properties. From knowledge of trends in these properties one can assess the similarity of the heaviest elements to their lighter homologues in the chemical groups. Differences in the chemical behavior of the transactinides and their lighter homologues can result from larger ionic radii, and from different energies and radial extension of the valence orbitals.

5.1 ELECTRONIC CONFIGURATIONS

Lr was the first element where a strong relativistic stabilization of the $7p_{1/2}$ electron was supposed to result in an unusual ground state electronic configuration, $7s^27p_{1/2}$ [85], in contrast to Lu($7s^26d$). The MCDF calculations [24] taking into account correlation effects by the configuration mixing have, indeed, shown that the strong relativistic stabilization of the 7p orbitals results in the $7s^27p_{1/2}$ ground state in contrast to the $7s^26d$ ground state shown by single-configuration DF calculations [23]. More accurate CCSD calculations [86] confirmed later this result. The next excited state, $7s^26d_{3/2}$ ($^2D_{3/2}$), was found to be about 0.16 eV higher in energy in good agreement with the "corrected" value of 0.186 eV of the MCDF calculations [24].

For Rf, the MCDF calculations [24,25] have shown again a different electronic configuration, $7s^26d7p$, than the one of Hf ($6s^25d^2$). This state was shown to be J=2 (80% of $7s^26d7p_{1/2}$), with the first excited state $7s^26d^2$ (95%) lying 0.5 eV [24] or 0.24 eV [25] higher in energy. The $7s^27p^2$ configuration proposed in [87] as the ground state turned out to be 2.9 eV above the $7s^26d7p_{1/2}$ state. The $7s^26d7p$ ground state of Rf was, however, not confirmed by more accurate CCSD calculations [88]. Inclusion of correlation effects of higher orders (f-electrons) in the CCSD calculations resulted in the inversion of the $7s^26d7p$ and $7s^26d^2$ configurations with the latter being more stable.

For elements 105 through 108, the MCDF calculations confirmed the $7s^26d^q$ ground states [27-29]. The CCSD calculations also confirmed the $7s^26d^q$ electronic configurations of elements 111 and 112 [89,90], as well as the

$7s^27p^q$ ground states for elements 113, 114 and 115 [91-93] obtained by the earlier DF and DS calculations. The MCDF calculations for elements 113 through 118 [30] have also given the $7s^27p^q$ state as the ground. Thus, the relativistic stabilization of the 7s electrons of the transactinide elements results in the stability of the $7s^2$ electron pair in the ground and first ionized states over the entire 7th row of the Periodic Table, see Table 2, which is different from the states of some elements of the 6th row like Pt($5d^96s$) and Au($5d^{10}6s$) or from the 1+ ionized states, d^qs, of Ta, W, Os or Hg. (The non-relativistic configuration of element 111 is, for example, $6d^{10}7s$ [89]).

The relativistic stabilization of the 7p electrons manifests itself in some excited states different than those of the lighter homologues, e.g., $6d7s^27p(^3D_2)$ of Rf lying 0.3 eV higher in energy in contrast to the $6d^27s^2(^3F_3)$ state of Hf.

For some time, the relativistic stabilization of the $7s^2$ and $7p_{1/2}$ electrons was a reason to suggest an enhanced stability of lower oxidation states like Lr^+, Db^{3+} or Sg^{4+}, and to design gas-phase experiments to detect a p-character of Lr and Rf taking into account different volatilities of d- and p-element compounds. The MCDF calculations of multiple IPs for elements 104 through 108 [26-29] and estimates of redox potentials (see Section 7.1) did not confirm the stability of lower oxidation states. Neither was the character of Rf confirmed by molecular calculations, see [15,16].

Recently, accurate CCSD calculations were reported for elements heavier than 118 [94,35,33]. For element 121, the CCSD calculations have given the $8s^28p$ configuration [35], as was predicted by the DF [12] and the DFT [95] calculations. This configuration is a result of the relativistic stabilization of the $8p_{1/2}$ electrons and is lower than $8s^27d$ (which is the ground state for the lighter group-3 homologues) by 0.4 eV. The lowest state of element 121 anion is $8s^28p^2$ compared to the $7s^27p6d$ state of Ac^- [35]. For element 122, the CCSD calculations [33] have given the $8s^27d8p$ ground state in contrast to the $7s^26d^2$ state for Th, the nearest homologue: the relativistic stabilization of the 8p orbitals is responsible for such a change.

5.2 ORBITAL ENERGIES, IONIZATION POTENTIALS AND OXIDATION STATES

The non-relativistic (HF) and relativistic (DF) orbital energies [23] for elements 104 through 118 are shown in Figure 8. IPs, experimental in the case of the 5d elements and calculated for the 6d elements (at the CCSD and MCDF levels), are shown in Figure 9. Single IP are listed in Tables 2 and 3.

Table 2. Electronic configurations, stable oxidation states and single ionization potentials (IP) for elements 104 through 112

Element	104[b]	105[c]	106[c]	107[c]	108[c]	109[d]	110[d]	111[b]	112[b]
Chemical group	4	5	6	7	8	9	10	11	12
Stable oxidation states[a]	**4**,3	**5**,4,3	**6**,4	**7**,5,4,3	**8**,6,4,3	3,6,1	6,4,2,0	**5**,**3**,-1	**4**,2,**0**
M	$6d^{2}7s^{2}$	$6d^{3}7s^{2}$	$6d^{4}7s^{2}$	$6d^{5}7s^{2}$	$6d^{6}7s^{2}$	$6d^{7}7s^{2}$	$6d^{8}7s^{2}$	$6d^{9}7s^{2}$	$6d^{10}7s^{2}$
IP_1, eV	6.01	6.9	7.8	7.7	7.6	8.7	9.6	10.6	11.97
M^{+}	$6d7s^{2}$	$6d^{2}7s^{2}$	$6d^{3}7s^{2}$	$6d^{4}7s^{2}$	$6d^{5}7s^{2}$	$6d^{6}7s^{2}$	$6d^{7}7s^{2}$	$6d^{8}7s^{2}$	$6d^{9}7s^{2}$
IP_2, eV	14.37	16.0	17.9	17.5	18.2	(18.9)	(19.6)	(21.5)	22.49
M^{2+}	$7s^{2}$	$6d^{3}$	$6d^{3}7s$	$6d^{3}7s^{2}$	$6d^{5}7s$	?	?	?	$6d^{8}7s^{2}$
IP_3, eV	23.8[c]	24.6	25.7	26.6	29.3	(30.1)	(31.4)	(31.9)	(32.8)
M^{3+}	7s	$6d^{2}$	$6d^{3}$	$6d^{2}7s^{2}$	$6d^{3}7s^{2}$	?	?	?	?
IP_4, eV	31.9[c]	34.2	35.4	37.3	37.7	(40)	(41)	(42)	(44)
M^{4+}	[Rn]	6d	$6d^{2}$	$6d^{2}7s$	$6d^{2}s^{2}$	$(6d^{5})$	$(6d^{6})$	$(6d^{7})$	$(6d^{8})$
IP_5, eV		44.6	47.3	49.0	51.2	(51)	(53)	(55)	(57)
M^{5+}		[Rn]	6d	$6d^{2}$	$6d^{3}$				
IP_6, eV			59.2	62.1	64.0				
M^{6+}			[Rn]	6d	$6d^{2}$				
IP_7, eV				74.9	78.1				
M^{7+}				[Rn]	6d				
IP_8, eV					91.8				

[a] bold = most stable states in the gas phase, underlined = most stable in solutions; [b] CCSD calculations: 104 [88]; 111 [89]; 112 [90]; [c] MCDF calculations 104 [25,26]; 105 [38], 106[28], 107 and 108 [29]; [d] DF calculation [23]. The values of the IP in the parentheses are extrapolations, see [12].

Table 3. Ionization potentials (in eV) for elements 113 through 118

Calc.	113	114[a]	115	116	117	118	Ref.
DF/DS	7.4	8.5	5.5	6.6	7.7	8.7	12
MCDF	7.11	8.04 (8.51[b])	4.65	5.96	6.51	7.74 (7.6[b])	30
CCSD	7.306	8.539	5.58	-	-		91-93

[a] 8.36 as DFC CCSD [96]; 8.45 as ECP-CCSD(T); 8.26 eV as BDF and 8.00 as ZORA [73]; 8.41 as RECO+SOCI [97]; [b] Relativistic CI version of RECP [98].

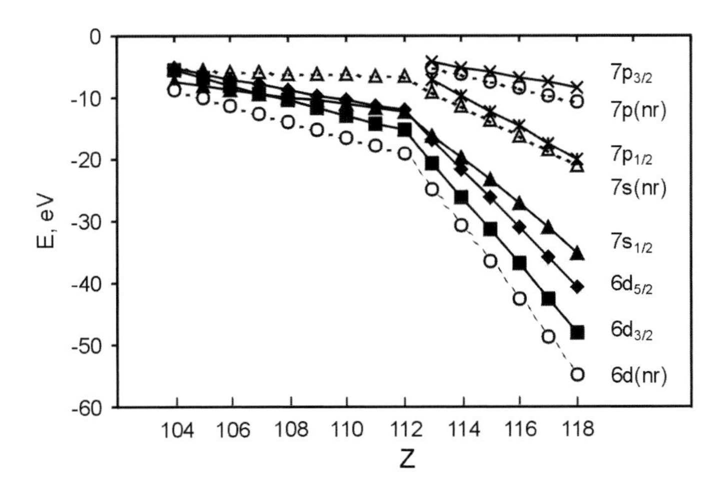

Fig. 8. Relativistic (DF, solid lines) and non-relativistic (HF, dashed lines) orbital energies for elements 104 through 118. The data are from [23].

The shape of the lines in Figure 9 reflects the energy level structures shown in Figure 8 and the ionization processes. The IP of Sg and Bh are very similar to those of W and Re. This is due to the fact that the ionized states are different in W ($s^2d^4 \rightarrow sd^4$) and Sg ($s^2d^4 \rightarrow s^2d^3$), and Re ($s^2d^5 \rightarrow sd^5$) and Bh ($s^2d^5 \rightarrow s^2d^4$), respectively, in contrast to Hf ($s^2d^2 \rightarrow s^2d$) and Rf ($s^2d^2 \rightarrow s^2d$), or Ta ($s^2d^3 \rightarrow s^2d^2$) and Db ($s^2d^3 \rightarrow s^2d^2$). Since the energies of the 6s orbitals in W and Re and of the 6d orbitals in Sg and Bh, respectively, are very similar, the ionization energies from those orbitals are also similar.

Multiple MCDF IPs for elements 104 through 108 are given in [26-29], and some CCSD values for elements 112-115, 121 and 122 are given in [90-93,35,33]. They are useful in estimates of stabilities of higher oxidation states. The calculations [26-29] have shown the multiple IP to decrease within the transition element groups (Figure 10). The reason for that is the proximity of the relativistic valence 7s and 6d orbitals (Figures 6 and 8). That makes the excitation energies of the 6d elements smaller than those of the lighter 4d and 5d homologues, resulting in an enhanced stability of the maximum oxidation states. Due to the same reason lower oxidation states at the beginning of the 6d series will be unstable: the step-wise ionization process shown in Figure 11 for group-6 elements, as an example, results in the d^2 and not in the $7s^2$ configurations for Db^{3+} or Sg^{4+} [99]. Since the 6d orbitals of the 6d elements are more destabilized than the (n-1)d orbitals of the 4d and 5d elements, the Db^{3+} and Sg^{4+} will even be less stable than Ta^{3+} and W^{4+}.

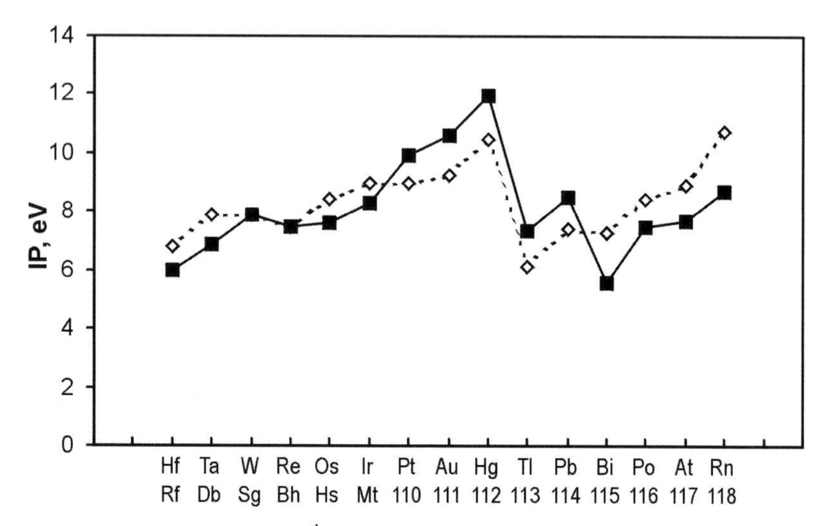

Fig. 9. Ionization potentials for the 6th row elements (dashed line, experimental values) and the 7th row (solid line, calculated values). The CCSD calculations are for Rf and elements 111- 115 [88-92] and the MCDF [26-30] calculations are for the other transactinide elements.

The destabilization of the 6d orbitals at the end of the transactinide series is also the reason for the 6d electrons to be chemically active. As a consequence, an increase in the stability of the highest oxidation states can be expected, e.g. of the 3+ and 5+ states of element 111. The 1+ oxidation state was predicted to be very unstable [12]. Due to a relatively high electron affinity of element 111, 1- oxidation state can then be stable with appropriate ligands.

The IP of element 112 of 11.97 eV is the largest among those of all the considered elements [90]; for comparison, the IP of Hg is 10.43 eV. Together with its closed shell configuration this is the reason why element 112 is considered to be the most inert out of all the elements of the 7th row, though less inert than Xe. The EA of element 112 is 0. The 6d orbitals should, however, be active in element 112. The first ionized electron of element 112 is a 6d electron resulting in the $6d^97s^2$ excited state in contrast to the $5d^{10}6s$ state of Hg. Thus, the 6d electrons of element 112 should play an important role. That may mean that higher oxidation states, 4+, or d-d metal interactions should be more active in element 112 compounds than in those of Hg.

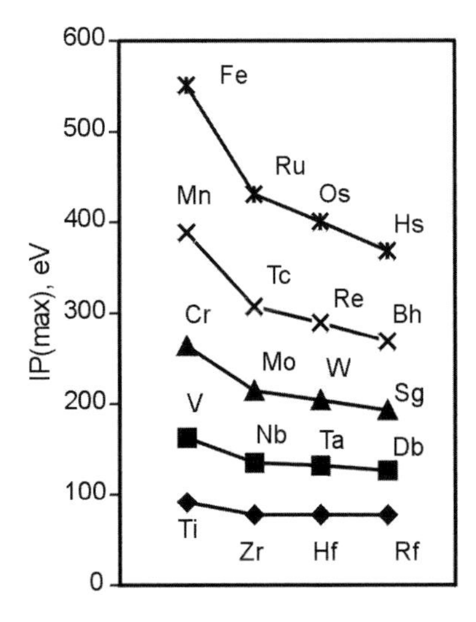

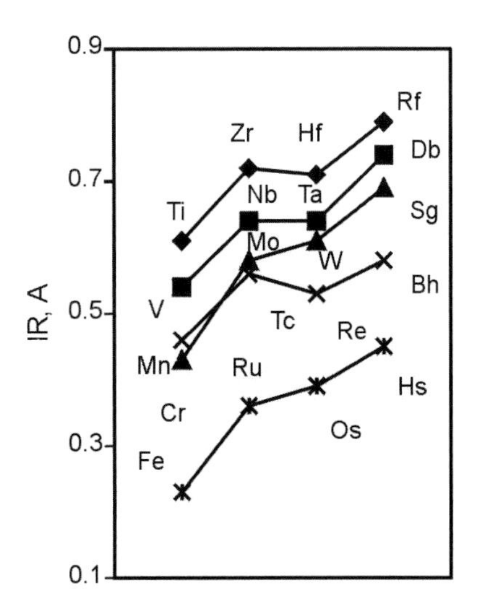

Fig. 10. Multiple ionization potentials (IP_{max}) and ionic radii (IR) for elements 104 through 108 in their maximum oxidation states obtained from the MCDF calculations [26-29].

The large relativistic stabilization of the $7s^2$ electrons and, hence, a large 7s-7p gap hindering the hybridization, see Figure 8, is the reason for an enhanced stability of lower oxidation states at the beginning of the 7p-series. Due to the inertness of the $7s^2$ electrons, the 1+ oxidation state will be more important for element 113 than the 3+ state, though the first IP is still very high: IP(113) = 7.306 V is the largest in group 13 as compared to IP(Tl) = 6.108 eV. The increased stability of the $7p_{1/2}$ orbital is also the reason for the highest EA(113) = 0.68 eV [91] in the group as compared to EA(Tl) = 0.4 eV. However, the relativistic stabilization of the 7s orbital results in a dramatic reduction in the $d^{10}s \rightarrow d^9s^2$ excitation energy to 0.1 eV for element 113^{2+} as compared to 8 eV for Tl^{2+}. Thus, the 2+ or 3+ oxidation states were thought to be probable for element 113. The 6d orbitals should be still accessible for hybridization at the beginning of the 7p series (elements 113 and 114) and should take part in bonding leading to the formation of higher oxidation states. The 5+ state was, therefore, suggested as stable in $113F_5$ [100].

The very large SO splitting of the 7p electrons, see Figure 8, results in the closed-shell configuration of element 114 ($7s^27p_{1/2}^2$) and a little participation of the $7p_{1/2}$ electrons in bonding, thus implying its relative inertness. That suggests that the 2+ state should predominate over the 4+ state to a greater extent than in the case of Pb. The first IP of element 114 was calculated by quite a number of methods, see Table 3. The DCB CCSD result of 8.539 eV is the most accurate [92]. The trend to an increase in the IP in the group is continued with element 114 due to the relativistic stabilization of the $7p_{1/2}$

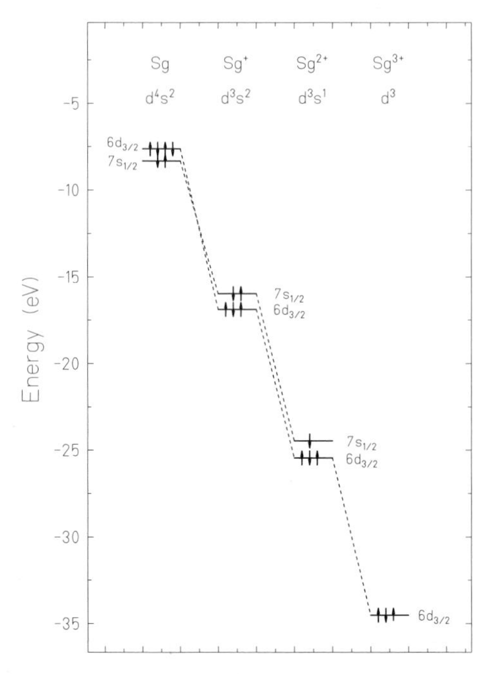

Fig. 11. MCDF orbital energies of the neutral through the third ionized state of Sg. Reproduced from [99].

electrons. For comparison, the IP of Sn and Pb are 7.34 eV and 7.42 eV, respectively. The EA is predicted to be zero [96] within a given accuracy of the DFC CCSD method due to limited basis sets, as compared to EA(Pb) = 0.36 eV. Because of the relatively destabilized 6d shell, element 114 should form coordination compounds like the hexafluoride $114F_6$ [12].

The plot in Figure 9 shows a sharp decrease in IP from element 114 to 115 due to a very large $7p_{1/2}$-$7p_{3/2}$ SO splitting, which is three times larger than the $6p_{1/2}$-$6p_{3/2}$ splitting of the 6p elements. In element 115, 1+ oxidation state should, therefore, be important, since an electron is added to the destabilized $7p_{3/2}$ orbital. Consequently, element 116 should be stable in the 2+ oxidation state. Element 117 has one electron missing at the $7p_{3/2}$ shell. Due to the relativistic stabilization of the $7p_{1/2}$ shell it should, therefore, be more stable in the 1+ and 3+ oxidation states than the lighter homologues, but less stable in the 5+ and 7+ oxidation states. The 1- oxidation state becomes less important in group 17 due to the destabilization of the $7p_{3/2}$ orbital (the EA of element 117 is the smallest in the group [12]).

For element 118, the calculations [98] show its IP to be about that of element 114 and less than IP(Rn) = 10.74 eV and IP(112) = 11.97 eV, see Table 3. Due to the smallest IP in group 18, element 118 is expected to be the least inert and the most electropositive out of all noble gases. It was predicted to form compounds with F and even Cl. The oxidation states 2+ and 4+ will be more important than the 6+ state because of the relativistically stabilized $7p_{1/2}$ electrons, see Figure 6. The outer 8s orbital of element 118 is relativistically stabilized to give the atom a positive EA = 0.056 eV according to the DCB CCSD calculations [101]. The inclusion of both relativistic and correlation effects was required to obtain this result. Similar calculations did not give a ^2S bound state for Rn^-.

The IP of element 119 is relativistically increased from 3.31 eV to 4.53 eV, as the DK CCSD calculations show [102]. Due to the relativistic stabilization of the 8s orbital, EA(119) = 662 meV, being the highest in group 1, according to the DCB CCSD calculations [94]. The calculated CCSD IP(121) = 4.45 eV and EA(121) = 0.57 eV, the highest in group 3 [35]. The CCSD IP of element 122 is 5.59 eV as compared to the IP of 6.52 eV for Th [33]: this relative decrease is due to the ionized $8p_{1/2}$ electron of element 121. The DF and DFS IPs of heavier elements can be found in [12].

5.3 RADIAL CHARGE DISTRIBUTIONS, ATOMIC/IONIC RADII AND POLARIZABILITY

An ionic radius of an element is defined by the maximum of the radial charge density, r_{max}, of an outer valence orbital, or by its expectation value $\langle r_{nj} \rangle$. The DF $\langle r_{nj} \rangle$ values for elements up to Z = 120 are tabulated by Desclaux [23]. The MCDF r_{max} for elements 104 through 108 are given in [26-29]. As discussed earlier, the relativistic ns orbitals contract within the chemical groups, while the (n-1) orbitals expand. This means that bond lengths in molecules with the predominant contribution of the 7s orbitals should relativistically decrease, while those with the predominant contribution of the (n-1)d orbitals should increase. The latter case is expected for compounds of the transactinides in higher oxidation states where the 6d orbitals are involved. The situation is more complex when both types of the orbitals participate to an equal extent in bonding (see further discussion for 111H). The smallest r_{max} of the 7s orbitals of element 112 suggests that the 112-X bond length should also be the smallest. For the 7p elements, except for element 113 (where contracted $7p_{1/2}$ orbital contributes mostly to bonding), bond lengths are expected to increase within the groups due to the participation of the expanded $7p_{3/2}$ orbitals.

The IR of elements 104 through 108 in various oxidation states were estimated on the basis of a linear correlation between r_{max} obtained from the atomic MCDF calculations [26-29] and experimentally known IR for their lighter homologues [103]. They are summarized in Table 4 and Figure 10 for these elements in the highest oxidation states.

Table 4. Estimated IR (in Å) (for the coordination number CN=6) of elements 104 through 108 in the maximum oxidation states [26-29]. Experimental data [103] are given for the lighter elements.

Group 4		Group 5		Group 6		Group 7		Group 8[a]	
Ti^{4+}	0.61	V^{5+}	0.54	Cr^{6+}	0.44	Mn^{7+}	0.46	Fe^{8+}	0.23
Zr^{4+}	0.72	Nb^{5+}	0.64	Mo^{6+}	0.59	Tc^{7+}	0.57	Ru^{8+}	0.36
Hf^{4+}	0.71	Ta^{5+}	0.64	W^{6+}	0.60	Re^{7+}	0.53	Os^{8+}	0.39
Rf^{4+}	0.79[b]	Db^{5+}	0.74[b]	Sg^{6+}	0.65[b]	Bh^{7+}	0.58	Hs^{8+}	0.45

[a] For CN=4; [b] a correlation of IR with the $\langle \bar{r}_{nl} \rangle$ of the 6p orbitals gives IR = 0.74 Å for Rf^{4+}, 0.66 Å for Db^{5+}, and 0.63 Å for Sg^{6+} [104]. More realistic values obtained from the geometry optimization of molecular compounds are IR = 0.76 Å for Rf^{4+} and 0.69 Å for Db^{5+} (see further).

The data of Table 4 and Figure 10 show that the IR of the 4d and 5d elements are almost equal due to the lanthanide contraction which is 86% a non-relativistic effect, while the IR of the transactinides are about 0.05 Å larger than the IR of the 5d elements due to an orbital expansion of the $6p_{3/2}$ orbitals being the outer orbitals for the maximum oxidation state. The IR of the lighter 6d elements are however smaller than the IR of the actinides since the latter undergo the actinide contraction of 0.030 Å which is mostly a relativistic effect [13,105].

Relativistically decreased atomic sizes and increased IPs of the 6d elements are related to relativistically decreased dipole polarizabilities, α. (The latter were shown to change linearly with IP^{-2} [106,102]). Relativistic effects on α change roughly as Z^2. For element 112, α should be the smallest in the group and in the 7th row, and it is relativistically decreased from 74.66 a.u. to 25.82 a.u., as was shown by the PP CCSD(T) calculations [107,108]. As a consequence, element 112 should form the weakest van der Waals bond and should be extremely volatile. For element 119, α is also relativistically decreased from 693.94 a.u. to 184.83 a.u., as was calculated at the CCSD(T) level [102]. The polarizability of 118 is expected to be the highest in the group suggesting its highest reactivity.

6. The Electronic Structure and Properties of Gas-Phase Compounds. Role of Relativistic Effects

Transactinides are now known to form volatile halides and oxyhalides as do their lighter homologues in respective chemical groups. These were the first gas-phase compounds studied experimentally. Therefore, there was widespread interest in their electronic structures and fundamental properties such as ionicity or covalence, bonding and thermochemical stability, as well as trends within the groups. The question of the magnitude of the influence of relativistic effects on those properties was also of great interest.

6.1 ELEMENTS 104 THROUGH 108

6.1.1 *Hydrides of elements 104 and 106*

The group-4 and 6 hydrides, MH_4 (M = Ti, Zr, Hf and Rf) and MH_6 (M = Cr, Mo, W and Sg) are the simplest systems which were used quite a time ago as models to study the influence of relativistic effects on molecular properties. The DF one-center expansion (DF-OCE) calculations [109] showed relativistic effects to decrease the bond length of RfH_4 and SgH_6, so that $R_e(RfH_4)$ is only 0.03 Å larger than $R_e(HfH_4)$, and $R_e(SgH_6)$ is 0.06 Å larger than $R_e(WH_6)$. The relativistic contraction of orbitals and the relativistic contraction of bond lengths were shown to be two parallel, but largely independent effects. The calculations showed a decrease in the dissociation energy of RfH_4 as compared to that of HfH_4 and a slight increase in it of SgH_6 as compared to that of WH_6.

6.1.2 *Halides and oxyhalides of elements 104, 105 and 106*

Using density-functional methods, mostly the DS-DV, a large series of calculations was performed for halides and oxyhalides of elements 104 through 106: MCl_4 (M = Zr, Hf, and Rf), MCl_5 and MBr_5, $MOCl_3$ and $MOBr_3$ (M = V, Nb, Ta and Db), MF_6 and MCl_6, $MOCl_4$, MO_2Cl_2, MO_4^{2-}, $M(CO)_6$ (M = Mo, W and Sg). Results until 1999 are overviewed in [15,16,110]. Various electronic structure properties such as IP, EA, electron transition energies, charge density distribution and bonding, as well as their trends have been predicted for the transactinides and their lighter homologues. The calculations of binding energies and optimization of geometries were performed for MCl_4 (M = Ti, Zr, Hf and Rf) [111], MCl_6 (M = Mo, W and Sg) [112], $MOCl_3$ (M = Tc, Re and Bh) [113], and MO_4 (M = Ru, Os and Hs) [114] using the RGGA DFT method. Recently, CCSD RECP calculations were also performed for halides and oxyhalides of elements 104 through 106, such as $RfCl_4$, MCl_6, $MOCl_4$, MO_2Cl_2, MO_3 (M = W and Sg), MCl_5 and MBr_5 (M = Ta and Db) [115]. Results of the *ab initio* DF calculations were reported for $RfCl_4$ [116].

The influence of relativistic effects on the electronic structure and properties of the 6d transactinides was analyzed in detail on the example of MCl_5 (M = V, Nb, Ta and Db) [117]. Opposite trends in the relativistic and non-relativistic energies of the valence orbitals from the 5d to the 6d elements were shown to result in opposite trends in molecular orbital (MO) energies, see Figure 12. Thus, the highest occupied MO (HOMO) of the 3p(Cl) character, responsible for the IP, is relativistically stabilized in each compound and becomes more bound from $TaCl_5$ to $DbCl_5$. Consequently, molecular IP increase in the group due to relativistic effects. Non-relativistically, the trend would be opposite from $TaCl_5$ to $DbCl_5$. The lowest unoccupied MO (LUMO) of predominantly (n-1)d character, responsible for EA, is, in turn, destabilized in each compound and becomes more destabilized from the Ta to Db molecule due to the relativistic destabilization of the (n-1)d orbitals. Consequently, the EA decrease in the group as a result of relativity. A relativistic increase in the energy gap ΔE between the HOMO and LUMO and, hence, in the energy of the lowest charge-transfer transitions $E_{3p(Cl)\rightarrow d}$ associated with the reduction of the metal is indicative of the increasing stability of the maximum oxidation state. This is shown by a correlation between $E_{3p(Cl)\rightarrow d}$ and reduction potentials E^0(V-IV) for MCl_5 (M = V, Nb, Ta and Db) in Figure 13. Thus, non-relativistically, Db^{5+} would have been even less stable than Nb^{5+}.

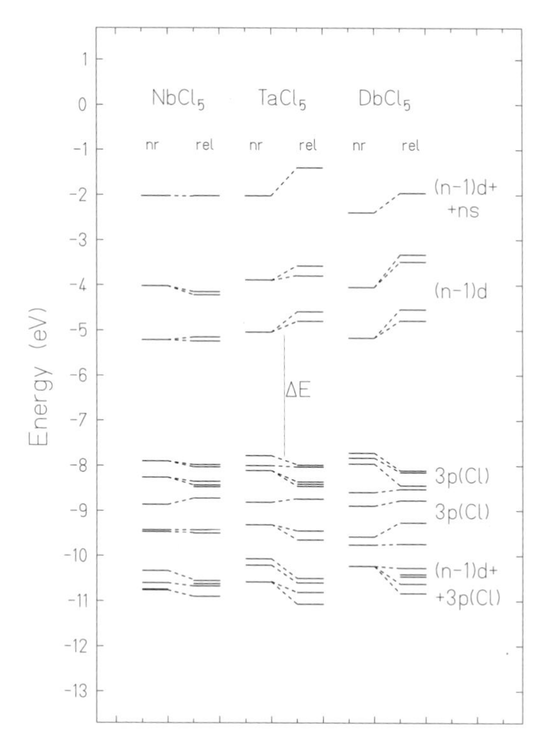

Fig. 12. Energy of relativistic (rel) and non-relativistic (nr) energy levels for MCl_5 (M = Nb, Ta, and Db) obtained from the DS-DV and HFS calculations. ΔE indicates the HOMO-LUMO energy difference. Reproduced from [117].

The Mulliken analysis [122] of the valence orbital populations has shown the bonding in the 6d-elements to be dominated by a large contribution of both the $6d_{3/2}$ and $6d_{5/2}$ orbitals, e.g., 70% in $DbCl_5$, typical of d-element compounds [117]. The 7s, as well as both the $6p_{1/2}$ and $6p_{3/2}$ orbitals, each contribute about 15 % to the bonding. Effective charges (Q_M) were shown to decrease in the group and overlap populations (OP) to increase, see Figure 14. (The OP is an amount of the electronic density located on the bond between atoms in a molecule and is a direct counterpart of the covalent contribution to the binding energy [122]). A comparison of the relativistic with non-relativistic calculations shows this increase to be a purely relativistic effect due to the increasing contribution of the relativistically stabilized and contracted 7s and $7p_{1/2}$ AO, as well as of the expanded 6d AO in bonding, see Figure 15. Non-relativistically, the trends would be just opposite from Ta to Db due to the opposite radial distributions of the atomic wavefunctions and their energies.

Thus, relativistic effects define continuation of the trends in the IP, EA, bonding and stabilities of oxidation states in going over to the 6d elements, while the non-relativistic description of these properties would give opposite and, therefore, wrong trends.

The calculated R_e and D_e for $RfCl_4$, MCl_5 (M = Ta and Db), MCl_6, $MOCl_4$, MO_2Cl_2, MO_3 (M = W and Sg) using both the DFT [111-114] and RECP [115] methods are summarized in Tables 5 and 6. Some earlier estimates of D_e obtained via a correlation between OP and covalent part of the binding energy (deduced from the experimental dissociation enthalpies minus the calculated Coulomb contribution into it) are also given there [123].

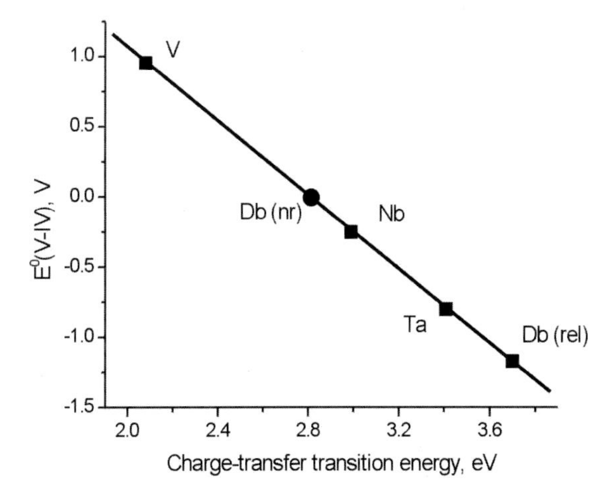

Fig. 13. Correlation between reduction potentials E^0(V-IV) and energies of the lowest charge transfer transitions in MCl_5 (M = V, Nb, Ta and Db). The non-relativistic value for Db is shown with a filled circle. Reproduced from [16].

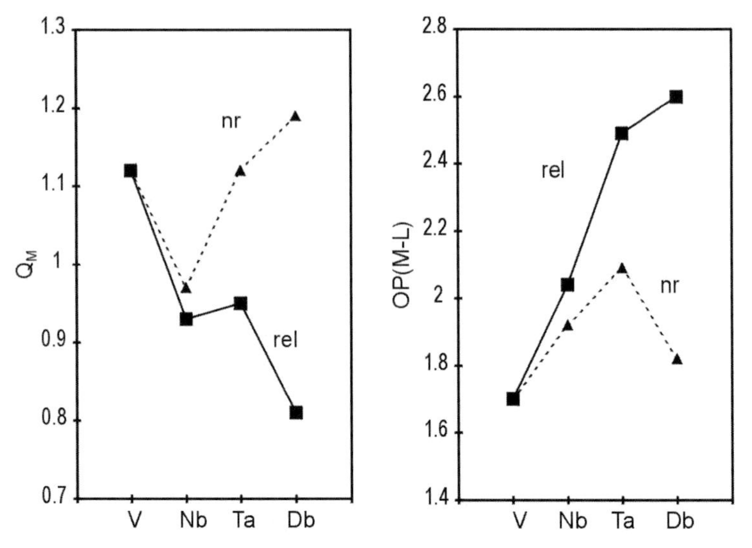

Fig. 14. Relativistic (rel) and non-relativistic (nr) effective charges, Q_M, and overlap populations, OP, for MCl_5 (M = V, Nb, Ta and Db)). L denotes the ligand. The data are from [117].

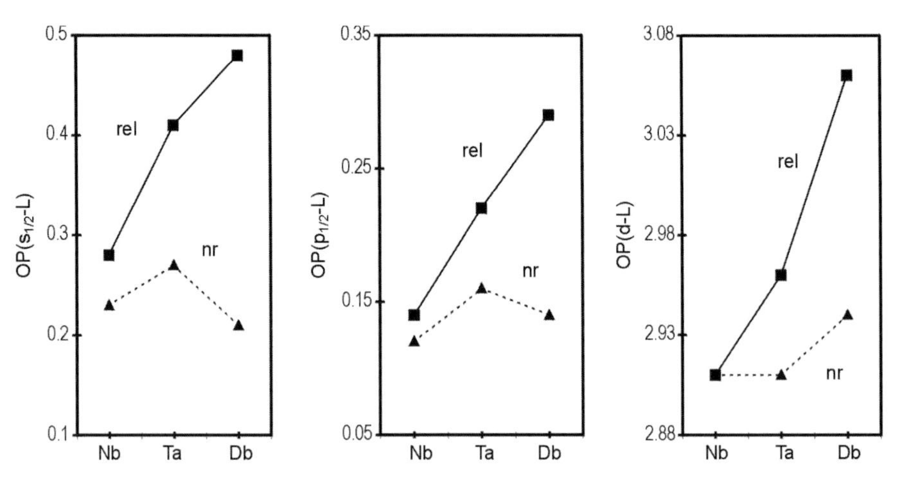

Fig. 15. Relativistic (rel) and non-relativistic (nr) partial overlap populations in MCl_5 (M = Nb, Ta and Db). L denotes valence orbitals of the ligand. The data are from [117].

Table 5. Optimized R_e (in Å) for MCl_4 (M = Hf and Rf), MCl_6, $MOCl_4$ and MO_2Cl_2 (M = W and Sg) as a result of the RECP-CCSD(T) [115] and DFT [111,112] calculations

Molecule	RECP-CCSD(T)		DFT		Experiment		
	M-O	M-Cl	M-Cl		M-O	M-Cl	Ref.
			RLDA	RGGA			
$HfCl_4$		-	2.307	2.344		2.318	118
$RfCl_4$		2.381	2.360	2.402		-	-
WCl_6		2.319	2.36[a]	-		2.26	119
$SgCl_6$		2.359	2.45[a]	-		-	-
$WOCl_4$	1.67	2.317	-	-	1.685	2.28	120
$SgOCl_4$	1.72	2.364	-	-	-	-	-
WO_2Cl_2	1.700	2.282	-	-	1.72	2.27	121
SgO_2Cl_2	1.749	2.339	-	-	-	-	-
WO_3	1.735	-	-	-	-	-	-
SgO_3	1.777	-	-	-	-	-	-

[a] DFT without the RGGA approximation [112]

The data of Table 6 show good agreement for D_e of $RfCl_4$ between the RECP and DFT calculations. It is also notable that earlier estimates of D_e [123] made by using ionic and covalent contributions to the binding energy deviate very little from the RECP values, thus indicating that the former procedure is justified. What is more important, a trend to a decrease in the D_e values from the W to Sg compounds found in [123] was confirmed by the RECP calculations [115]. Both types of the data also agree with an increase in the stability of compounds with increasing number of oxygen atoms from MCl_6 to $MOCl_4$ and to MO_2Cl_2 (M = Mo, W and Sg), as is experimentally known for the lighter elements. Thus, SgO_2Cl_2 was recommended [123] for the high-temperature gas phase experiments as the most stable 106 compound and was, indeed, experimentally observed [124,125]. $SgCl_6$ and $SgOCl_4$ were shown to be unstable with respect to the loss of Cl and to transform into compounds of Sg^V [112].

Table 6. Atomization energies (D_e in eV) for MCl_4 (M = Hf and Rf), MCl_6, $MOCl_4$ and MO_2Cl_2 (M = W and Sg) as a result of the RECP-CCSD(T) and DFT calculations

Molecule	RECP-CCSD(T)[a]	DFT	Experiment[f]
$HfCl_4$	-	21.14[b]	20.61
$RfCl_4$	18.8	19.5[b] (18.97)[c]	-
WCl_6	19.9	22.18[d]	21.7
$SgCl_6$	19.9	20.05[d]	-
$WOCl_4$	21.5	(23.0)[e]	23.0
$SgOCl_4$	21.0	21.2[e]	-
WO_2Cl_2	22.2	(23.5)[e]	23.5
SgO_2Cl_2	21.0	21.8[e]	-
WO_3	18.9	-	-
SgO_3	17.8	-	-

[a] [115]; [b] [111]; [c] corrected by the differences between calculated and experimental values for the lighter homologues; [d] the DFT without RGGA [112]; [e] estimated from the DS-DV calculations [123]; [f] calculated via a Born-Haber cycle.

The authors of [115] evaluated contributions of spin-orbit and correlation effects into bonding for these compounds. The SO interaction causes a decrease in D_e of 1-1.5 eV for the transactinides molecules at the CCSD(T) level. The effects are larger for the Sg compounds than for the Rf due to an increasing $d_{3/2}$-$d_{5/2}$ splitting. As a consequence, the D_e of SgO_2Cl_2 becomes 1.2 eV smaller than the D_e of WO_2Cl_2. Electron correlation effects on D_e were found to be larger in the W compounds than in the Sg compounds and they become more significant as the number of oxygen atoms increases.

The Mulliken population analysis data (Q_M and OP), as well as dipole moments, μ, for group-6 oxyhalides are given in Table 7. Both the DFT and RECP results show a decrease in Q_M and an increase in OP from the W to Sg compounds, with the values being very similar. The RECP μ are somewhere larger than the DS-DV ones, but both values show an increase in μ from the W to Sg oxyhalide.

Table 7. The Mulliken population analysis data (Q_M and OP) and dipole moments (μ) for MO_2Cl_2 (M = W and Sg)

	Molecule	RECP		DS-DV[c]
		MP2[a]	CCSD(T)[b]	
Q_M	WO_2Cl_2	1.46	1.71	1.08
	SgO_2Cl_2	1.32	1.52	0.97
OP	WO_2Cl_2	2.07	2.03	2.23
	SgO_2Cl_2	2.53	2.55	2.34
μ, D	WO_2Cl_2	0.92	1.51	1.35
	SgO_2Cl_2	1.90	2.39	1.83

[a] [115], correlation at the MP2 level; [b] [115], the correlation at the quadratic CI singles and doubles (QCISD) geometries; [c] [123].

The PP calculations were performed for DbO along with NbO and TaO. Relativistic effects were shown to stabilize the $^2\Delta_{3/2}$ ground state electronic configuration in DbO, as that in TaO, in contrast to the $^4\Sigma^-$ state of NbO [57].

6.1.3 *Oxyhalides of group-7 elements Tc, Re and Bh*

The first chemical experiment for element 107, Bh, and its homologues Tc and Re, was conducted in the O_2/HCl media, so that the trioxyhalides MO_3Cl (M = Tc, Re, and Bh) were formed [9]. TcO_3Cl and ReO_3Cl are known to be the most stable and most volatile species among oxychlorides of these elements. To predict the outcome of the experiment, the electronic structures of MO_3Cl (M = Tc, Re and Bh) were calculated using the RGGA DFT method [113]. The obtained MO energies for BhO_3Cl are very similar to those of TcO_3Cl and ReO_3Cl indicating close analogy in properties between the compounds. The optimized geometrical structures are given in Table 8 showing a small deviation of the calculated R_e of 0.05 Å from the experimental values for ReO_3Cl. In Table 9, Q_M, OP, μ, α, IP and D_e are presented as a result of the calculations. The RGGA D_e for ReO_3Cl agrees well with the thermochemical ΔH_{diss} with the difference of 0.54 eV.

Table 8. Optimized geometrical parameters, bond lengths (R_e) and angles (∠), for MO_3Cl (M = Tc, Re, and Bh) in a C_{3v} symmetry

	R_e(M=O), Å		R_e(M-Cl), Å		∠ClMO	
	Calc.	Exp.	Calc.	Exp.	Calc.	Exp.
TcO_3Cl	1.69	(1.75)[a]	2.30	(2.22)[a]	107° 40′	(106° 20′)[c]
ReO_3Cl	1.71	1.761[b]	2.28	2.23[b]	109° 40′	108° 20′ [b]
BhO_3Cl	1.77	(1.82)[c]	2.37	(2.30)[c]	114° 20′	(113° 00′)[c]

[a] Values estimated on the basis of experimental IR [103]; [b] [126]; [c] "normalized" values (corrected by the difference with experiment for ReO_3Cl).

The Mulliken analysis data of Table 9 show BhO_3Cl to be more covalent (a larger OP) than TcO_3Cl and ReO_3Cl. Such an increase in covalence is typical of halides and oxyhalides of the elements at the beginning of the transactinide series (Rf, Db and Sg) and this is a relativistic effect as discussed earlier. The dipole moments increase from the Tc to the Bh compound

Table 9. Effective charges (Q_M), total overlap populations (OP), dipole moments (μ), dielectric dipole polarizabilities (α), ionization potentials (IP) and atomization energies (D_e) for MO_3Cl (M = Tc, Re, and Bh). From [113].

Property	TcO_3Cl	ReO_3Cl	BhO_3Cl
Q_M	1.28	1.21	1.13
OP	1.93	2.20	2.31
μ, D	0.93	1.29	1.95
α, 10^{24} cm^3	4.94	5.91	7.50
IP, eV	12.25	12.71	13.01
D_e (calc.), eV	23.12	24.30	22.30
ΔH_{diss} (exp.), eV[a]	-	23.76	-

[a] Calculated via a Born-Haber cycle

similarly to the group-6 oxyhalides. The polarizabilities also increase in this direction. The data of Table 9 were used to predict adsorption enthalpies of MO_3Cl on the surface of the chromatography column for the gas-phase chromatography experiments [9] (see Section 6.1.6).

6.1.4 *Tetroxides of group-8 elements Ru, Os and Hs*

Element 108, Hs, was expected to be a member of group 8 of the Periodic Table and, thus, to form a very volatile tetroxide, HsO_4, by analogy with Ru and Os. To predict the outcome of the first gas-phase experiment on Hs [10], the electronic structure calculations for the group-8 gas-phase tetroxides, RuO_4, OsO_4, and HsO_4, were performed using the DFT method [114]. Optimized R_e are shown in Table 10. The RLDA values are closer to the experimental values than the RGGA ones. The "experimental" R_e for HsO_4 was predicted by correcting the calculated value by the difference between the calculated and experimental values for the lighter homologues.

The calculated MO energies of HsO_4 are very similar to those of RuO_4 and OsO_4 indicating their close analogy. An increasing value of the energy gap between HOMO and LUMO is indicative of an increase in the stability of the 8+ oxidation state from Ru to Os, and further to Hs in the tetroxides. This is a relativistic effect caused by the relativistic destabilization of the 6d orbitals. The calculated Q_M, OP, α, IP and D_e are given in Table 10 in very good agreement with the experiment for the IP. The IP of HsO_4 is almost equal to the IP of OsO_4 according to the transition-state calculations, and slightly lower according to the total energy calculations. The calculated α show a "saw-tooth" behavior in the group. As expected, HsO_4 is more covalent than RuO_4 and OsO_4. Such an increase in covalence has the same

Table 10. Effective charges (Q_M), total overlap populations (OP), dielectric dipole polarizabilities (α), ionization potentials (IP) and atomization energies (D_e) for MO_4 (M = Ru, Os, and Hs). From [114].

Property	RuO_4	OsO_4	HsO_4	Ref. (exp.)
R_e (calc.), Å	1.730	1.751	1.810	
R_e (exp.), Å	1.706	1.711	(1.775)	127
Q_M	1.45	1.46	1.39	
OP	1.92	1.94	2.17	
α (calc), 10^{24} cm^3	6.48	5.96	6.26	
α (exp.), 10^{24} cm^3	-	6.30	-	128
IP (calc), eV[a]	12.25	12.35	12.28	
IP (calc), eV[b]	12.32	12.26	12.27	
IP (exp.), eV	12.19	12.35	-	129
D_e (calc.), eV	27.48	27.71	28.44	
ΔH_{diss} (exp.), eV[c]	19.11	21.97	-	-

[a] Via the difference in the total energies; [b] via the transition-state procedure; [c] calculated via a Born-Haber cycle.

relativistic origin as that for all other group-4 through 7 compounds, as discussed earlier. The D_e values indicate that HsO_4 should be more stable than OsO_4. The calculated molecular properties (Table 10) were used to predict adsorption enthalpies of the group-8 tetroxides on the surface of the chromatography column for the gas-phase chromatography experiments [10] (see Section 6.1.6).

6.1.5 *Comparison of properties of group-4 through 8 halides and oxyhalides*

The DFT and RECP calculations predicted similar trends in the electronic structure and related properties for oxides, halides and oxyhalides of group-4 through 8 elements. In these transactinide compounds, bonding is defined by a strong participation of both the $6d_{3/2}$ and $6d_{5/2}$ orbitals, with a large admixture of the 7s and $7p_{1/2}$ orbitals. The covalence increases within the groups, see Figure 16, and this increase is a result of the relativistic stabilization and contraction of the valence 7s and $7p_{1/2}$ orbitals and destabilization and expansion of the 6d orbitals.

Trends in D_e and R_e for various types of compounds are summarized in Figure 17. To put the various calculated data on the same footing, the experimental ΔH_{diss} for the 4d and 5d compounds and "extrapolated" (corrected by the difference between the calculated and experimental values for the lighter compounds) D_e for the transactinides were used. The data of Figure 17 show a decrease in D_e from 5d to the 6d elements for almost all the compounds except for MO_4 and MCl_6, though the DFT calculations have found a similar decrease in D_e from WCl_6 to $SgCl_6$ [112]. In all the groups, relativistic effects steadily enhance bonding with increasing Z, though increasing SO splitting of the d orbitals partially diminishes it reaching about 1 - 1.5 eV at the 6d compounds [115]. This is one of the reasons for most of the transactinide compounds having atomization energies lower than those of the 5d homologues. The other reason is a decrease in the ionic contribution to bonding which is also a relativistic effect.

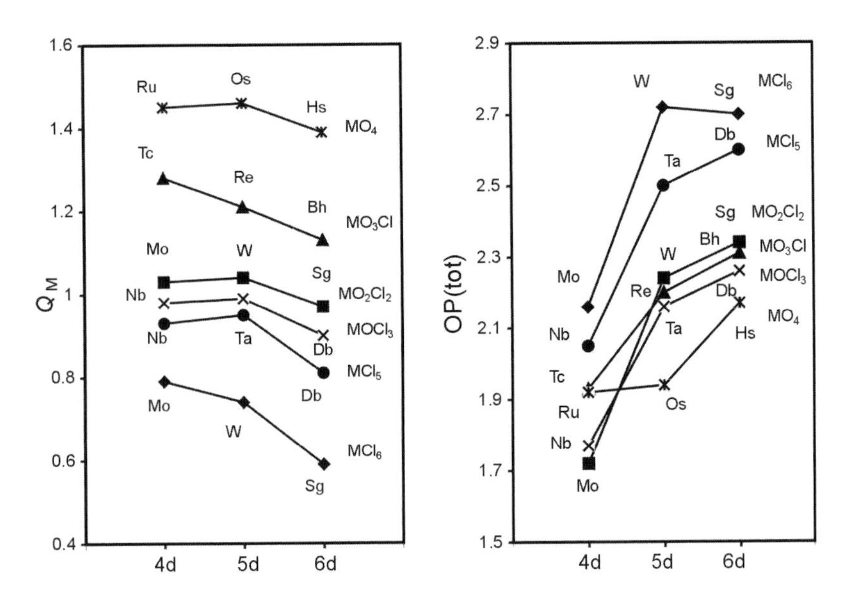

Fig. 16. Effective charges (Q_M) and total overlap populations (OP) for group-4 through 8 halides and oxyhalides obtained as a result of the Mulliken population analysis in the DS-DV and DFT calculations [15,113-114].

Among other important trends, one should mention a decrease in the metal-ligand bond strength of the halides with increasing group number, and from the 5d to the 6d compounds within the same group. Thus, $SgCl_6$ was shown to be unstable [15]. Consequently, $BhCl_7$ should not exist. This is also connected with a decrease in the relative stability of the maximum oxidation state along the transactinide series, see Figure 13 in [15]. The R_e values, see Figure 17, reflect the experimentally known similarity of the bond lengths for the 4d and 5d compounds (the lanthanide contraction effect) and an increase in R_e of about 0.05 Å in the 6d compounds in comparison with R_e of the 5d compounds.

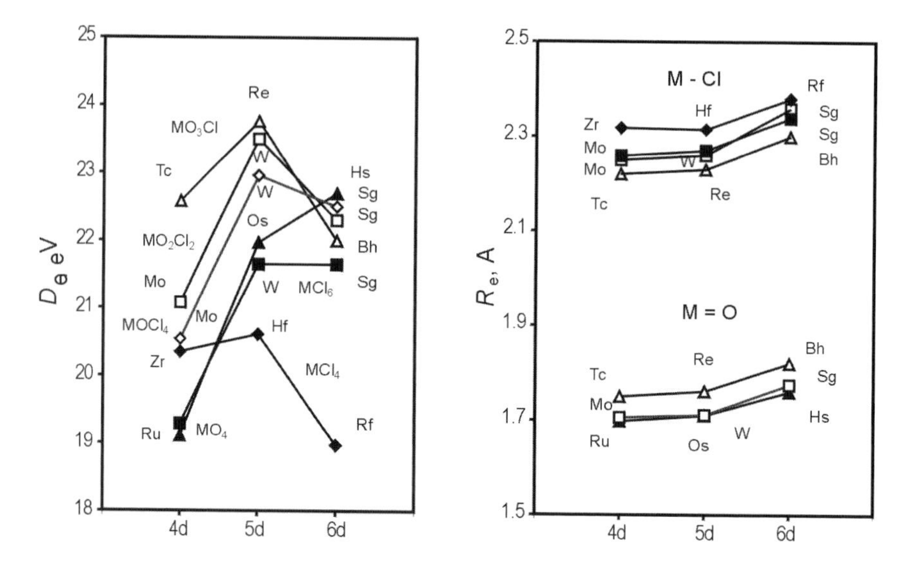

Fig. 17. Atomization energies, D_e, (experimental for the 4d and 5d elements and extrapolated from the DFT and RECP calculations for the 6d elements) and optimized bond lengths, R_e, for various halides, oxides and oxyhalides of group-4 through 8 elements [113-115,122-123].

An increase in μ of the low symmetry molecules and an increase in α within the groups are observed for all the considered compounds. One should note here that μ and α are relativistically decreased due to the relativistic increase in IP, decrease in the molecular size and increase in the covalence of the compounds.

6.1.6 *Predictions of volatility for group-4 through 8 compounds*

Identification of properties of the transactinides by studying their volatility is a difficult task due to the fact that there are several quantities associated with this phenomenon. Available experimental data are often contradictory and hardly correlate with a single electronic structure parameter.

In the gas-phase chromatography experiments, the measure of volatility is assumed to be either a deposition temperature in a thermochromatography column or the temperature in an (isothermal) column at which 50% ($T_{50\%}$) of the desired product (50% "chemical yield") are passing through the gas chromatography column [8]. The values are then correlated with the adsorption enthalpies (ΔH_{ads}). The latter is then related to the sublimation enthalpy (ΔH_{sub}), a property of the macro-amount using some models, see

Chapter 6. The most accepted measure of volatility in macro-chemistry is, however, the equilibrium vapor pressure over a substance, P_{mm}. Boiling points basically correlate with P_{mm}.

A larger series of the DS-DV calculations [15,16] yielded data that permitted establishment of some relationships between molecular properties such as covalence or dipole moment and volatility. Thus, the following relationships have been established for the group-4 through 8 highest halides and oxyhalides:

- higher OP of the 5d halides compared to those of the 4d halides correspond to their higher volatility;
- higher OP of the group-5 halides compared to those of the group-4 halides are consistent with their higher volatility;
- smaller OP of oxyhalides compared to those of the corresponding pure halides are related to their lower volatility;
- much lower OP of halides of the actinides compared to those of halides of the d elements are in line with the lower volatility of the actinides;
- increasing dipole moment s of the low symmetry compounds of the group-5 $MOCl_3$, group-6 $MOCl_4$, MO_2Cl_2, or group-7 MO_3Cl are the reason for the observed decrease in volatility within those groups.

These observations led to conclusions that the higher covalence (larger OP) of high symmetry halides or oxides of the transactinides should result in their higher volatility compared to their 4d and 5d homologues. The largest μ of lower symmetry transactinide compounds should result in their lowest volatility as compared to those of the lighter homologues within the groups. Experimentalists were also advised to use the equilibrium vapor pressure of the lighter homologues as a benchmark for the "one-atom-at-a-time" measurements of the adsorption temperatures or $T_{50\%}$ of the chemical yield. Some hypothetical P_{mm} was predicted for transactinide compounds [130]. As a result of these theoretical considerations it was predicted that volatility should change as $RfCl_4$ > $HfCl_4$, $DbCl_5$ > DbO_3Cl, MoO_2Cl_2 > WO_2Cl_2 > SgO_2Cl_2 or TcO_3Cl > ReO_3Cl > BhO_3Cl [16,123,131,113]; see Section 6.1.7. This was, indeed, confirmed experimentally by the observation of higher volatility of $RfCl_4$ than that of $HfCl_4$ [132] and lower volatility of DbO_3Cl, SgO_2Cl_2, or BhO_3Cl in comparison with those of their lighter homologues [133,124,125,9]; see the Chapter 7.

To determine ΔH_{ads} of a very heavy molecule on a surface often of a vaguely known structure, e.g., the quartz can be covered with a reactive material, is presently a formidable task for quantum chemical calculations. To find a way out, some models accounting for physical adsorption with the use of the calculated molecular electronic structure properties were applied to predictions of ΔH_{ads} [113,114]. The models are based on the principle of intermolecular interactions subdivided into usual types for the long-range forces: the dipole-dipole, dipole-polarizability and the van der Waals (dispersion) interactions [134].

In the experiments on studying volatility of group-7 elements [9] a mixture of O_2 and HCl was used to produce the volatile MO_3Cl (M = Tc, Re and Bh). Thus, the quartz chromatography column was supposed to be covered with adsorbed HCl, so that the interaction of MO_3Cl should take place via an outer Cl atom carrying some effective charge. The long-range forces equations were modifies accordingly [113], so that the interaction energy

$$E(x) = -\frac{2Qe\mu_{mol}^{2}}{x^2} - \frac{Q^2e^2\alpha_{mol}}{2x^4} - \frac{3}{2}\frac{\alpha_{mol}\alpha_{slab}}{\left(\frac{1}{h\nu_{mol}} + \frac{1}{h\nu_{slab}}\right)} \tag{14}$$

where μ, $h\nu_{mol}$ (roughly IP) and α_{mol} belong to the molecule and those with index "slab" to the surface. The calculated contributions to *E(x)* for a typical value of Q(Cl) = -0.4 are given in Table 11. The data there are indicative of an increase in the energies of all three types of interactions from the Tc to the Bh compound. The interaction distance x for ReO_3Cl was deduced by setting the calculated *E(x)* to the experimental $\Delta H_{ads}(ReO_3Cl)$ = -61 kJ/mol. Assuming then the proportionality between x and molecular size $\Delta H_{ads}(TcO_3Cl)$ = -48.2 kJ/mol and ΔH_{ads} (BhO_3Cl) = -78.5 kJ/mol were determined using the calculated *E(x)* of Table 9. The obtained adsorption enthalpies show that volatility changes as $TcO_3Cl > ReO_3Cl > BhO_3Cl$. The predicted sequence and the ΔH_{ads} are in excellent agreement with experimental values of –51 ± 2 kJ/mol and –77 ± 8 kJ/mol for TcO_3Cl and BhO_3Cl, respectively [9].

Table 11. Contributions to the interaction energy *E(x)* between neutral MO_3Cl molecules (M = Tc, Re, and Bh) and Cl^Q (surface) for Q = -0.4. From [113].

Molecule	μ-*Qe*	α-*Qe*	α-α(Cl)
	$E10^{16}x^2$, eV cm^2	$E10^{32}x^4$, eV cm^3	$E10^{48}x^6$, eV cm^6
TcO_3Cl	2.23	5.69	379.0 6
ReO_3Cl	3.10	6.81	460.65
BhO_3Cl	4.67	8.64	591.17

Experimental volatility studies of MO_4 (M = Ru, Os and Hs) were performed in a chromatography column with a pure quartz surface [10]. For these highly symmetric molecules having no dipole moments, the interaction should be of a pure van der Waals type. With the use of a model of the molecule-slab interaction and the relation between the dielectric constant of a molecular solid and its polarizability, the interaction energy is expressed in the following way [114]

$$E(x) = -\frac{3}{16}\left(\frac{\varepsilon-1}{\varepsilon+2}\right)\frac{\alpha_{mol}}{\left(\frac{1}{h\nu_{slab}}+\frac{1}{h\nu_{mol}}\right)x^3} \tag{15}$$

where $h\nu$ denotes roughly IP of the molecule and the slab, and ε is the dielectric constant of quartz. Similar expressions were obtained for the cases where the surface is covered with molecular oxygen, and when the oxygen has an effective charge. The calculated contributions to *E(x)* between MO_4 (M = Ru, Os, and Hs) and the various types of surfaces are shown in Table 12. Using the same technique to define *x* as previously used in the case of the group-7 oxychlorides, i.e. by setting the interaction energies from Table 12 equal to the experimentally known ΔH_{ads} (OsO_4) = -38 ±1.5 kJ/mol, $\Delta H_{ads}(RuO_4)$ = -40.4 ±1.5 kJ/mol and $\Delta H_{ads}(HsO_4)$ = -36.7 ±1.5 kJ/mol were

Table 12. Contributions to the interaction energies *E(x)* between MO_4 (M = Ru, Os, and Hs) and a) pure quartz surface; b) surface covered with O_2; c) surface with an effective charge *Qe* (*Q* = -0.4). From [114].

Molecule			α-α(SiO_2)	α-α(O_2)	α-*Qe*
	$10^{24}\alpha$, cm^3	$h\nu_0$, eV	$E10^{24}x^3$, eV cm^3	$E10^{24}x^3$, eV cm^3	$E10^{32}x^4$, eV cm^4
RuO_4	6.48	12.15	4.73	5.31	10.01
OsO_4	5.96	12.35	4.48	4.75	9.41
HsO_4	6.26	12.28	4.63	4.90	9.73

obtained. Thus, the volatility of group-8 tetroxides in the particular experiments was expected to change as $RuO_4 < OsO_4 \leq HsO_4$. This sequence would be in accord with an idea of increasing volatility with increasing covalence of the compounds. Equal volatilities of HsO_4 and OsO_4 are also expected from the extrapolation within the group [135].

In experiments with Os and Hs, performed under similar conditions, it was, however, found that HsO_4 condenses at a higher temperature than OsO_4 indicating that the Hs oxide is less volatile [10]. Because so few events were detected it is premature to say whether or not the experimental data are in disagreement with the predictions. It would be very interesting to discuss the influence of relativistic effects on the ΔH_{ads} of these molecules. This can be done after non-relativistic quantum-chemical calculations are performed.

6.1.7 *Summary of the volatility studies*

Observed trends in volatility of group-4 through 8 compounds compared to the theoretical predictions are summarized in Table 13.

Table 13. Trends in volatility of the heaviest element compounds and their lighter homologues in chemical groups

Group	Compounds	Theoretically predicted	Ref.	Experimentally observed	Ref.
4	MCl_4, MBr_4	Hf < Rf	16	Hf < Rf	132
5	ML_5 (L = Cl, Br)	Nb < Ta < Db	130	(DbO_3Br)	133
		$DbCl_5 > DbOCl_3$	131	$DbCl_5 > DbOCl_3$	133
6	MO_2Cl_2	Mo > W > Sg	123	Mo > W > Sg	124,125
7	MO_3Cl	Tc > Re > Bh	113	Tc > Re > Bh	9
8	MO_4	Ru < Os ≤ Hs	114	Os > Hs	10

6.2 ELEMENTS 109 AND 110

Molecular calculations [136] were performed only for $110F_6$ using the DS-DV method. The calculations have shown $110F_6$ to be very similar to PtF_6, with very close values of IP. Relativistic effects were shown to be as large as crystal-field splitting.

6.3 ELEMENTS 111 AND 112

The special interest in the chemistry of elements 111 and 112 originates from the expectation of anomalous properties of their compounds as compared to homologues in groups 11 and 12 due to the predicted maximum of relativistic effects on the 7s electron shell; see Section 3. The large relativistic destabilization and expansion of the 6d orbitals are also expected to influence their chemistry.

6.3.1 *Dimers and fluorides of element 111*

The relativistic contraction of the 7s orbitals ("Group 11 maximum") results in the atomic size of element 111 being similar to that of gold and smaller than that of silver. For compounds, where the 7s orbitals contribute predominantly to bonding, the smallest size of the element 111 species in the group is expected.

The electronic structure of the simplest molecule 111H, a sort of a test system like AuH, was studied in detail. The properties of 111H predicted by various calculations (HF, DF, DK, PP, PP CCSD(T), DFT, etc.) are compared in Table 14. A more extended table can be found in [17,50]. The data demonstrate importance of relativistic and correlation effects. A comparison of the relativistic (DF or ARPP) with the non-relativistic (HF or NRPP) calculations shows that the bonding is considerably increased by relativistic effects doubling the dissociation energy, though the SO splitting diminishes it by 0.7 eV (the ARPP CCSD – SOPP CCSD difference). Thus, the trend to an increase in D_e from AgH to AuH turned out to be inversed from AuH to 111H, see the figure in [50]. This is in agreement with the BDF calculations [72]. The PP CCSD calculations have shown that R_e is substantially shortened by relativity, ΔR_e = -0.4 Å, and it is the smallest in the series AgH, AuH and 111H, so that the trend to a decrease in R_e is continued with 111H [50]. The BDF calculations, however, show R_e(111H) to be slightly longer than R_e(AuH). The different trends in R_e obtained in these two types of the calculations are obviously connected with a different contribution of the contracted 7s and expanded 6d orbitals to bonding (though the 6d contribution was found to de predominant in both cases). Both types of the calculations show the trend to an increase in k_e to be continued with 111H having the largest value of all known diatomic molecules. The μ was shown to be relativistically decreased from AgH to AuH and to 111H indicating that 111H is more covalent and element 111(I) is more electronegative than Au(I) [50,72]. In [72], much larger SO effects were found on R_e (SO increased) and k_e (SO decreased).

Table 14. Bond length (R_e), dissociation energy (D_e) and force constant (k_e) for 111H calculated using various approximations

Method	R_e, Å	D_e, eV	k_e, mdyn Å^{-1}	Ref.
HF	2.015	0.90	1.01	50
NRPP CCSD(T)	1.924	1.83	1.11	50
DF	1.520	1.56	4.66	50
ARPP	1.505	2.32	4.98	50
ARPP CCSD(T)	1.498	3.79	4.77	50
SOPP CCSD(T)	1.503	3.05	4.72	50
BDP	1.546	2.77	3.66	72
ZORA(MP)	1.530	2.87[a]	4.26	72

[a] ZORA(MP) + SO

The DFT calculations for other dimers AuX and 111X (X = F, Cl, Br, O, Au, 111) show that except for the oxide, the 111 compounds have slightly longer bond lengths (relativistically increased due to the radial extension of the $6d_{5/2}$ orbitals), lower D_e, but higher k_e in comparison with those of AuX [72]. Relativistic effects follow a similar pattern as that in 111H except for 111F and 111O where the SO splitting increases D_e. The PP [107] and BDF [72] calculations show, however, different trends in R_e and D_e from AuF to 111F, so that the question about the right trends remains open.

To study the stability of higher oxidation states, energies of the following decomposition reactions $MF_6^- \rightarrow MF_4^- + F_2$ and $MF_4^- \rightarrow MF_2^- + F_2$ were calculated on the PP MP2 and CCSD levels [107,137]. The results confirmed the fact that relativistic effects stabilize higher oxidation states due to a larger involvement of the 6d orbitals in bonding. $111F_6^-$ was shown to be the most stable in the group. The SO coupling stabilizes the molecules in the following order: $MF_6^- > MF_4^- > MF_2$. This order is consistent with the relative involvement of the (n-1)d electrons in bonding for each type of molecules.

6.3.2 *Dimers and fluorides of element 112*

The relativistic effects maximum on the 7s electronic shell in element 112 was confirmed by molecular calculations. The PP calculations [108] have, indeed, shown that the relativistic bond length contraction in $112H^+$ is similar to that in 111H, and that $R_e(112H^+)$ is the shortest among CdH^+, HgH^+ and $112H^+$ and is similar to $R_e(ZnH^+)$. Due to relativistic effects $112H^+$ should be the most stable in the group-12 hydrides.

The Mulliken population analysis for MF_2 and MF_4 (M = Hg and 112) [108] suggests that the 6d orbitals of element 112 should be involved in bonding to a larger extent than the 5d orbitals of Hg. The calculated energies of the decomposition reactions $MF_4 \rightarrow MF_2 + F_2$ and $MF_2 \rightarrow M + F_2$ have confirmed the conclusion that the 4+ oxidation state of element 112 will be accessible and the 2+ state not [12]. The neutral state of element 112 should also be more preferred than that of Hg.

Since the most exciting property of element 112 is its predicted high volatility [12,138-140], experiments are planned to detect this element in its elemental state by gas-phase chromatography separations. Element 112 and its nearest homologue Hg are supposed to be sorbed on Au and Pd metal detectors [11]. Element 112 is expected to have adsorption enthalpies on the Au and Pd surfaces between those of Hg and Rn, as semi-empirical estimates of ΔH_{ads} show [138,141]. To predict the outcome of those experiments on the quantum-mechanical level, calculations of adsorption of Hg and element 112 on metal surfaces of interest are highly desirable. As a first step, the ability of element 112 to form intermetallic compounds was tested by the DFT calculations for 112M, where M = Cu, Pd, Pt, Ag and Au [142]. The calculated D_e are shown in Figure 18 together with experimental ΔH_{ads} of Hg on the corresponding metal surfaces [143]. The calculations revealed an increase in R_e of about 0.06 Å and a decrease in D_e of about 15 – 20 kJ/mol from HgX to 112X. The Mulliken population analysis shows this decrease to be a result of the drastic relativistic stabilization and, therefore, inertness of the $7s^2$ shell leading to a large decrease in the 7s(112)-6s(Au) overlap compared to the 6s(Hg)-6s(Au) one. The contribution of the other valence orbitals is almost unchanged.

The plots in Figure 18 show good agreement for the trends between the two types of data, the experimental ΔH_{ads} for Hg and the calculated D_e for HgM and 112M, with the interaction of Hg and element 112 with Pd being the strongest. Thus, the DFT calculations describe the bonding in these systems correctly, so that the difference in D_e between HgX and 112X might be related to the difference in ΔH_{ads} of Hg and element 112 on the metal surfaces. Thus, element 112 is expected to be more weakly adsorbed than Hg, though not as weakly sorbed as was expected earlier [138]. Calculations of adsorption of Hg and element 112 on the metal surfaces are in progress using the embedded cluster model [144]

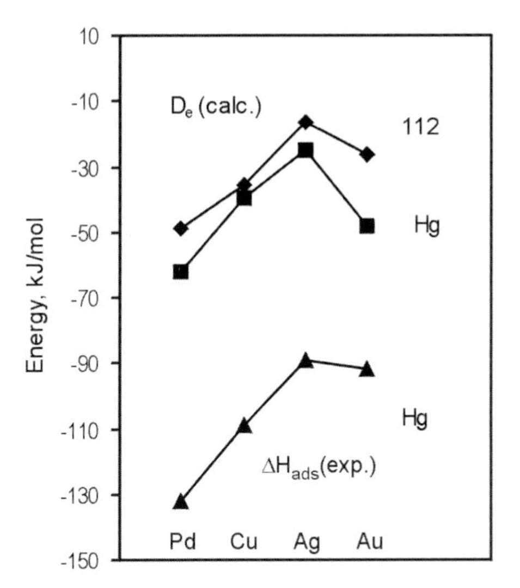

Fig. 18. Calculated binding energies for HgX and 112X (X = Pd, Cu, Ag and Au) and measured ΔH_{ads} on the corresponding metal surfaces [143]. Reproduced from [142].

For the case of adsorption of element 112 on the quartz surface, simple calculation [145] of the van der Waals interaction of Hg and 112 using the slab model (see Section 6.1) and the calculated α and IP of element 112 [108,90] have given $\Delta H_{ads}(112) = 27 \pm 1$ kJ/mol, as compared to $\Delta H_{ads}(Hg) = 34$ kJ/mol and $\Delta H_{ads}(Rn) = 21$ kJ/mol. Thus, the adsorption properties of element 112 are expected to be between those of Hg and Rn.

6.4 ELEMENTS 113-117

The very large SO splitting of the 7p orbitals is expected to influence properties of compounds of elements 113 through 117. The RECP CCSD calculations [100,146-147] were performed for simple MH (M = 113-118) to study the influence of relativistic effects. The results are summarized in Table 15. In 113H, the 6d and 7s orbitals were shown to participate little in bonding and all the effects are defined by the large participation of the $7p_{1/2}$ shell. A large spin-orbit contraction of the $7p_{1/2}$ orbital results in the large SO contraction of the 113-H bond: $\Delta R_e(SO) = -0.206$ Å according to the RECP CCSD(T) calculations [146] and $\Delta R_e(SO) = -0.16$ Å according to the DFC CCSD(T) and PP SO CCSD(T) calculations [100]. This also agrees with the RECP CI calculations [148]. The R_e of 113H is the only case of the bond contraction among all the MH (M = elements 113 – 118). For 114H through 118H, both the contracted $7p_{1/2}$ and expanded $7p_{3/2}$ orbitals take part in bonding, with the former being gradually less available for bonding.

Table 15. Bond lengths (R_e), dissociation energies (D_e) and the SO effects on them, Δ(SO), for MH (M = 113 - 117) [146, 73]

Molecule	R_e, Å	ΔR_e(SO), Å	D_e, eV	ΔD_e(SO), eV
TlH	1.927	-0.021	1.98	-0.47
113H	1.759	-0.206	1.46	-0.93
PbH	1.884	0.001	1.61	-0.71
114H	1.972	0.068	0.43	-2.18
BiH	1.836	0.019	2.24	0.08
115H	2.084	0.206	1.82	-0.23
PoH	1.784	0.031	2.27	-0.29
116H	1.988	0.171	1.81	-0.63
AtH	1.742	0.032	2.31	-0.68
117H	1.949	0.171	1.79	-1.04

In the series of the group-13 hydrides, a reversal of the trend to an increase in R_e and μ was predicted from TlH to 113H [100,146]. Thus, element 113 was found to be more electronegative than Ga, In, Tl and even Al. The binding energy was shown to be destabilized by the SO effects, so that RECP ΔD_e(SO) = -0.93 eV [146] and PP ΔD_e(SO) = -0.97 eV [100] in good agreement with each other. The reason for that is the large atomic SO destabilization of element 113. Decreasing D_e and k_e from BH to 113H were predicted. The $(113)_2$, being of only academic interest, should also be weakly bound, as the DF calculations show [149]: the $7p_{1/2}$ electron yields a weak bond having 2/3π bonding and ½σ antibonding character.

The PP and DCB calculations [100,107,147] for MF (M stands for all group 13 metals) have shown increasing R_e and μ from TlF to 113F, in contrast to decreasing values from TlH to 113H [146]. These different trends in R_e and μ for the MF compounds as compared to MH are explained by a more ionic nature of the MF molecules. For a dimer (113)(117), the SO interaction should change the sign of the dipole moment, as the results of the DF calculations show [49].

Participation of the 6d electrons of element 113 in bonding was confirmed by the PP calculations [100] for $113H_3$, $113F_3$ and $113Cl_3$ and RECP for $113H_3$ and $113F_3$ [147]. As a consequence of the involvement of the 6d orbitals, a T-shape geometric configuration rather than trigonal planar was predicted for these molecules. The stability of the high-coordination compound $113F_6^-$ with the metal in the 5+ oxidation state is foreseen. $113F_5$ will probably be unstable since the energy of the reaction $MF_5 \rightarrow MF_3 + F_2$ is less than -100 kJ/mol [100]. The calculated energies of the decomposition reaction $MX_3 \rightarrow MX + X_2$ (M = B, Al, Ga, In, Tl and element 113) suggest a decrease in the stability of the 3+ oxidation state in the group.

A relative inertness of element 114 due to the closed-shell configuration was confirmed by calculations for some of its compounds. The RECP calculations for 114H have found a SO bond length expansion ΔR_e(SO) of 0.068 Å [146] (0.030 Å [100]) in agreement with the BDF value of 0.094 Å [73], see Table 15. Relatively small ΔR_e(SO) for PbH and 114H in comparison with those for TiH and 113H were interpreted as the participation of both contracted $np_{1/2}$ and expanded $np_{3/2}$ orbitals in bonding. This is also, obviously, the reason why R_e(114H) = 1.96 Å [146] is larger than R_e(PbH) = 1.88 Å. The SO bond strength decrease ΔD_e(SO) was found to be very large in 114H: -2.18 eV in RECP [146], –2.07 eV in RECP CI [100] and –2.02 in BDF calculations [73], much larger than that of PbH (-0.71 eV) and the largest in the series of MH (M = 113 through 117). Thus, D_e(114-H) = 0.43 eV shown by the RECP calculations (0.40 eV according to BDF [73] and 0.5 eV according to RECP CI calculations [100]) is the smallest in the series. The inertness of element 114 is proven to be a combined result of both the SO splitting and the double occupancy of the $7p_{1/2}$ spinor. The electronic structures of 114X (X = F, Cl, Br, I, O, O_2) were calculated using the RECP, ZORA and BDF methods [73]. Similar effects on the bond strengths like those on MH were found for these 114 compounds. $114O_2$ was predicted to be thermodynamically unstable in contrast to PbO_2.

Energies of the decomposition reactions $MX_4 \rightarrow MX_2 + X_2$ and $MX_2 \rightarrow M + X_2$ (X = H, F and Cl) for group-14 elements were calculated at the PP CCSD(T) and DHF levels [96,52]. The results show a decreasing trend in the stability of the 4+ oxidation state in the group in agreement with the predictions of [150] based on atomic calculations and simple models of bonding. The instability was shown to be a relativistic effect. The neutral state was found to be more stable for element 114 than that for Pb. Thus, element 114 is expected to be less reactive than Pb, and as reactive as Hg. The possibility of the existence of $114F_6^{2-}$ is considered [96].

Estimates of formation enthalpies of MX_2 and MX_4 (X = F, Cl, Br, I, SO_4^{2-}, CO_3^{2-}, NO_3^- and PO_4^{3-}) for Po and element 116 [150] made on the basis of the MCDF atomic calculations confirmed the instability of the 4+ state of element 116. The chemistry of element 116 is expected to be mainly cationic: an ease of formation of the divalent compounds should approach that of Be or Mn, and tetravalent compounds should be formed with the most electronegative atoms, e.g., $116F_4$.

The trend to a decrease in the $np_{1/2}$ orbital involvement in bonding in group 17 was found to be continued further with element 117 [49,146,151]. The bonding in 117H, a closed shell molecule, was shown to be formed by the $7p_{3/2}$ orbital and is, therefore, 2/3 of the bonding without the SO splitting. Thus, the weakening and lengthening of the chemical bond is predicted by the DHF (ΔR_e=0.13 Å) [49] and RECP (ΔR_e=0.17 Å) [146] calculations. In the row HI, HAt and H117, an increasing trend in R_e and a decreasing trend in D_e are continued with H117. Analogously to the lighter homologues, element 117 should form dimers X_2. The DCB CCSD(T) calculations for X_2 (X = F through At) [152] found a considerable antibonding σ character of the HOMO of At_2 due to the SO coupling (without the SO coupling, it is an antibonding π orbital). Thus, the bonding in $(117)_2$ is predicted to have considerable π character [17]. The 117Cl is also predicted to be bound by a single π bond and have the relativistically (SO) increased bond length [153].

6.5 ELEMENT 118

In group 18, noble gases become less inert with increasing Z. A relatively high reactivity of element 118 was foreseen by earlier considerations [154,12]. Element 118 was predicted to be the most electropositive in the group and to be able to form a 118-Cl bond.

The RECP calculations for the reactions $M + F_2 \rightarrow MF_2$ and $MF_2 + F_2 \rightarrow MF_4$, where M = Xe, Rn and element 118 [155] confirmed an increase in the stabilities of the 2+ and 4+ oxidation states. The SO effects stabilize $118F_4$ by a significant amount of about 2 eV, though they elongate R_e by 0.05 Å. Thus, increasing trends in R_e and D_e are continued with element 118. The influence of the SO interaction on the geometry of MF_4 was investigated by the RECP-SOCI [151] and RECP CCSD calculations [155]. In both papers it was shown that a D_{4h} geometrical configuration for XeF_4 (calculated in agreement with experiment) and for RnF_4 (calculated) becomes slightly unstable for $118F_4$. A T_d configuration was shown to be more stable than the D_{4h} in $118F_4$ by 0.25 eV [151] and 0.17 eV [155] in agreement with each other. The reason for that was the availability of only stereochemically active $7p_{3/2}$ electrons for bonding. Thus, a SO modification of the valence shell electron pair repulsion (VSERP) theory was suggested in [151]. Han *et al* [155], however, claim that this modification should be applied with caution, since no non-linear $118F_2$ structure has been detected as a minimum at the HF level of theory. An important observation was made that the fluorides of element 118 will most probably be ionic rather than covalent, as in the case of Xe. This prediction might be useful for future gas-phase chromatography experiments. RECP calculations [146] for 118H have shown the van der Waals bond to be stabilized by about 2.0 meV by the SO effects and $\Delta R_e(SO)$ = -0.019 Å. Trends in molecular stability were predicted as follows: RnH << HgH < PbH, 118H << 114H < 112H for the hydrides, and $RnF_2 < HgF_2 < PbF_2$ and $112F_2 < 114F_2 < 118F_2$ for the difluorides. So far, elements heavier than 118 were not studied at the MO level. Some general aspects of their chemistry can be found in [12,156].

7. Aqueous Chemistry of the Transactinides

7.1 REDOX POTENTIALS OF THE TRANSACTINIDES

The knowledge of the relative stability of oxidation states, i.e., redox potentials, is very important for a chemical application. Trends in the stability of various oxidation states of the very heavy elements were predicted earlier on the basis of atomic relativistic DF and DS calculations in combination with some models based on a Born-Haber cycle (see [12]). The conclusions were, however, not always unanimous and varied depending on the model. Later, this topic received a more detailed consideration [157,158,99] using results of the atomic MCDF calculations.

For an oxidation-reduction reaction

$$M^{z+n} + ne \leftrightarrow M^{z+} \quad (16)$$

the redox potential $E°$ is defined as

$$E° = -\Delta G°/nF, \quad (17)$$

where $\Delta G°$ is the free energy of reaction (16) and F is the Faraday number. $E°$ could then be estimated using a correlation between $E°$ and IP, since

$$\Delta G° = -(\mathrm{IP} + \Delta G°_{\mathrm{hydr}}), \quad (18)$$

where $\Delta G°_{\mathrm{hydr}}$ is a free energy of hydration being a smooth function of atomic number and it can easily be evaluated. Thus, using the calculated MCDF IP [26-29] and experimentally known $E°$ [159], unknown values of redox potentials for the transactinides and some of their lighter homologues were determined for group-4, 5 and 6 elements in acidic solutions [157,158,99]. One of those correlations for group-6 species is shown in Figure 19, as an example. The redox potentials for the transactinides are summarized in Table 16.

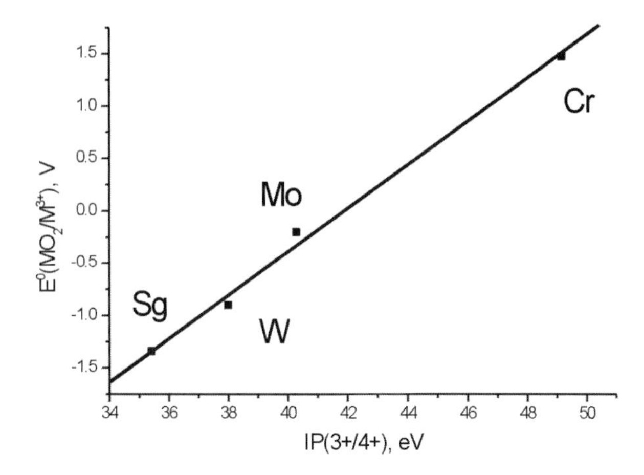

Fig. 19. Correlation between IP(3+/4+) and standard potentials $E°(MO_2/M^{3+})$, where M = Cr, Mo, W, and Sg. Reproduced from [99].

Results of these estimates show the following trends in the groups and along the 7th period: The stability of the maximum oxidation state increases in groups 4 through 6, while that of lower oxidation states decreases. Along the 7th row, the stability of the maximum oxidation state decreases: $E°(Lr^{3+}/Lr^{2+})$ < $E°(Rf^{IV}/Rf^{3+})$ < $E°(Db^{V}/Db^{IV})$ < $E°(Sg^{VI}/Sg^{V})$, or a similar trend for $E°(M^{max}/M)$, see Table 16. (The roman numbers denote the valence states in complex compounds). The increasing stability of the maximum oxidation state is a relativistic effect, as is discussed in Section 5.2. The estimates of redox potentials confirm that the 3+ and 4+ states for Db and Sg, respectively, will be unstable, see Figure 19 for Sg. The reasons for that are discussed in Section 5.2. Redox potentials for Rf, Db and Sg in comparison with the lighter homologues can be found in [99,157,158].

Table 16. Redox potentials of Lr, Rf, Db and Sg in aqueous acidic solutions

Potential	Lr[a]	Rf	Db[f]	Sg[h]
$E°(M^{VI}/M^{V})$	-	-	-	-0.046 (MO_3/M_2O_5) -0.05 (M^{VI}, H^+/M)
$E°(M^{V}/M^{IV})$	-	-	-1.0 (M_2O_5/MO_2) -1.13 (MO_2^+/MO^{2+})	-0.11 (M_2O_5/MO_2) -0.35 (M^V, H^+/M^{IV},H^+)
$E°(M^{IV}/M^{3+})$	8.1	-1.5 (M^{4+}/M^{3+})[c]	-1.38 (MO_2/M^{3+})	-1.34 (MO_2/M^{3+}) -0.98 ($M(OH)_2^{2+}/M^{3+}$)
$E°(M^{3+}/M^{2+})$	-2.6	-1.7 (M^{3+}/M^{2+})[c]	-1.20	-0.11
$E°(M^{3+}/M)$	-1.96[b]	-1.97 (M^{3+}/M)[d]	-0.56	0.27
$E°(M^{IV}/M)$	-	-1.85 (M^{4+}/M)[e] -1.95 (MO_2/M)[e]	-0.87 (MO_2/M)[d]	-0.134 (MO_2/M) -0.035 ($M(OH)_2^{2+}/M$)
$E°(M^{V}/M)$	-	-	-0.81 (M_2O_5/M)	-0.13 (M_2O_5/M)[d]
$E°(M^{VI}/M)$	-	-	-	-0.12 (MO_3/M) -0.09 (M^{VI}, H^+/M)

[a] [160]; [b] [159]; [c] [161]; [d] roughly estimated from the other E°; [e] [157]; [f] [158]; [h] [99]

7.2 HYDROLYSIS AND COMPLEX FORMATION

Complex formation is studied experimentally by liquid-liquid extractions, or anion exchange separations. For a simple complex ML_n, the cumulative complex formation constant

$$\beta_n = [ML_n][M]^{-1}[L]^{-n} \tag{19}$$

is a measure of its stability. For step-wise processes, consecutive constants K_i are used. If one molecular complex, ML_i, among many others is extracted into an organic phase, its distribution coefficient, K_d, dependent on β_i, is approximately a measure of its stability. Consequently, a sequence in the K_d values for a studied series, i.e. for elements of one group, reflects a sequence in the stability of their complexes [162]. Thus, by calculating $\log\beta_i$ one can predict and understand sequences in the extraction of complexes into an organic phase or their sorption on cation or anion exchange resins.

The complex formation is known to increase in the transition element groups. In aqueous solutions it is, however, competing with hydrolysis [163]. This may change trends in the stabilities of complexes and, finally, in their extraction into an organic phase.

One should distinguish between hydrolysis of cations and hydrolysis of complexes. The former process is described as a process of a successive loss of protons

$$M(H_2O)_n^{z+} \leftrightarrow MOH(H_2O)_{n-1}^{(z-1)+} + H^+ \tag{20}$$

In acidic solutions, hydrolysis involves either the cation, anion, or both, and is competing with the complex formation described by the following equilibrium

$$xM(H_2O)_{w^\circ}^{z+} + yOH^- + iL^- \leftrightarrow M_xO_u(OH)_{z-2u}(H_2O)_wL_a^{(xz-y-i)+} + (xw^\circ + u - w)H_2O \tag{21}$$

7.2.1 *Theoretical model for the prediction of complex formation*

For a reaction like Equation 20, a consecutive formation constant

$$\log K_i = -\Delta G^r/2.3RT, \tag{22}$$

where ΔG^r is the free energy change of the reaction. To define it in a straightforward way binding or total energies of species in the left and right parts of equation (20) should be calculated. Since it is almost impossible to do it with sufficient accuracy for very large, highly coordinated aqueous species, the following model was suggested by V. Pershina [164,165].

In a fashion analogous to that of A. Kassiakoff and D. Harker [163,170], the following expression for the free energy of formation of the $M_xO_u(OH)_v(H_2O)_w{}^{(xz-2u-v)+}$ species from the elements was adopted

$$-\Delta G^f(u,v,w)/2.3RT = \sum a_i + \sum a_{ij} + \log P - \log(u!v!w!2^w) + (2u+v+1)\log 55.5 \qquad (23)$$

The first term on the right hand side of Equation 23, $\sum a_i$, is the non-electrostatic contribution from M, O, OH, and H_2O, which is related to the overlap population, OP. For a reaction,

$$\Delta \sum a_i = \Delta E^{OP} = k\,\Delta OP, \qquad (24)$$

where k is an empirical coefficient. The next term, $\sum a_{ij}$, is a sum of each pairwise electrostatic (Coulomb) interaction:

$$E^C = \sum a_{ij} = -B \sum_{ij} Q_i Q_j / d_{ij} \qquad (25)$$

where d_{ij} is the distance between moieties i and j; Q_i and Q_j are their effective charges and $B = 2.3RTe^2/\varepsilon$, where ε is a dielectric constant. For a reaction, ΔE^C is the difference in E^C for the species in the left and right parts. P in Equation 23 is the partition function representing the contribution of structural isomers if there are any. The last two terms are statistical: one is a correction for the indistinguishable configurations of the species, and the other is a conversion to the molar scale of concentration for the entropy. Σa_{ij} and Σa_i for each compound are then calculated directly via Mulliken numbers implemented in the DS-DV or DFT methods; see Section 2.2. To predict $\log K_i$ or $\log \beta_i$ for transactinide complexes, coefficients k and B should be then defined by fitting $\log K_i$ to experimental values for the lighter homologues, as it is shown in [167]. Using this model, hydrolysis and complex formation constants were predicted for a large number of aqueous compounds of group-4 through 6 elements [164-170] in excellent agreement with experiment; see Section 4.3. Results of these calculations and comparison with experimental data reveal that a change in the electrostatic metal-ligand interaction energy (ΔE^C) contributes predominantly to ΔG^r. Thus, just by calculating the ΔE^C all the trends in the complex formation can correctly be predicted.

7.2.2 *Element 104*

Element 104 has extensively been studied experimentally. The accumulated experimental results have revealed quite a number of surprises and disagreements. Thus, it was not clear whether Rf is more or less hydrolyzed than Hf, and there were often disagreements in the sequences of the K_d

values of group-4 elements obtained by their ion exchange separations from HF and HCl solutions, see [6] and Chapter 5.

Hydrolysis of group-4 elements and their complex formation were studied theoretically [168] using the model described in the previous section. The following reactions were considered: the first hydrolysis step

$$M(H_2O)_8^{4+} \leftrightarrow MOH(H_2O)_7^{3+}, \tag{26}$$

the step-wise fluorination process

$$M(H_2O)_8^{4+} \leftrightarrow MF(H_2O)_7^{3+} \ldots \leftrightarrow \ldots MF_3(H_2O)_5^{+} \ldots \leftrightarrow \tag{27}$$

$$MF_4(H_2O)_2 \leftrightarrow \ldots MF_5(H_2O)^{-} \leftrightarrow MF_6^{2-}, \tag{28}$$

and the chlorination process

$$M(H_2O)_8^{4+} + 6HCl \leftrightarrow MCl_6^{2-}. \tag{29}$$

The calculated ΔE^C for reactions of eqs. (26-29) are given in Table 17. The ΔE^C data, see the first line of Table 17, show the following trend in hydrolysis for the group-4 elements (reaction 27): Zr > Hf > Rf. The first hydrolysis constant $\log K_{11}(\mathrm{Rf}) \approx -4$ was then determined in good agreement with the experimental values of -2.6 ± 0.7 [171], as compared to $\log K_{11}(\mathrm{Zr}) = 0.3$ and $\log K_{11}(\mathrm{Hf}) = -0.25$ [163]. The predicted trend is also in agreement with the experimental data for Zr and Hf [163]. One should note here that a simple model of hydrolysis [163] implying that the hydrolysis changes with the ratio of a cation charge to its size would give the opposite and hence a wrong trend from Zr to Hf, since $IR(Zr^{4+}) > IR(Hf^{4+})$.

Table.17. Calculated Coulomb parts of the free energies, ΔE^C (in eV), of some typical hydrolysis and complex formation reactions (26-29) for Zr, Hf and Rf. From [168].

Reactions	Zr	Hf	Rf
$M(H_2O)_8^{4+} \leftrightarrow MOH(H_2O)_7^{3+}$	-4.69	-4.61	-4.11
$M(H_2O)_8^{4+} \leftrightarrow MF(H_2O)_7^{3+}$	0.002	0.015	0.395
$M(H_2O)_8^{4+} \leftrightarrow MF_4(H_2O)_4$	15.86	15.84	16.52
$M(H_2O)_8^{4+} \leftrightarrow MF_6^{2-}$	50.76	50.91	51.10
$MF_2(H_2O)_6^{2+} \leftrightarrow MF_3(H_2O)_5^{+}$	4.96	4.94	5.06
$MF_2(H_2O)_6^{2+} \leftrightarrow MF_4(H_2O)_4$	12.90	12.89	12.97
$MF_2(H_2O)_6^{2+} \leftrightarrow MF_6^{2-}$	47.81	47.96	47.61
$M(H_2O)_8^{4+} \leftrightarrow MCl_4$	47.99	47.82	47.65
$M(H_2O)_8^{4+} \leftrightarrow MCl_6^{2-}$	52.15	52.50	53.06
$MOH(H_2O)_7^{3+} \leftrightarrow MCl_4$	52.68	52.44	51.76
$MOH(H_2O)_7^{3+} \leftrightarrow MCl_6^{2-}$	56.84	57.11	57.17

The following information can be deduced from the calculated ΔE^C. For the cation exchange separations (CIX) performed at < 0.1 M HF (no hydrolysis), i.e., for the extraction of the positively charged complexes, the K_d values will have the following trend in group-4: $Zr \leq Hf < Rf$. This is caused by the decreasing trend in the formation of the positively charged complexes according to Equation 27: $Zr \geq Hf > Rf$. (In the case of the formation of complexes with a lower positive charge from complexes with a higher positive charge (Equation 27) a sequence in the K_d values is opposite to a sequence in the complex formation, since complexes with a higher charge are better sorbed on the CIX resin than those with a lower charge). This trend was, indeed, observed in the experiments on the CIX separations of group-4 elements at low HF concentrations [172]. For the formation of anionic complexes separated by the anion exchange (AIX) resins, the trend becomes more complicated depending on pH, i.e., depending on whether the fluorination starts from hydrated or hydrolyzed species. Thus, for experiments conducted in 10^{-3} - 10^{-1} M HF (where some hydrolyzed or partially fluorinated species are present), the trend for the formation of MF_6^{2-} (Equation 28) is reversed within group 4: $Rf \geq Zr > Hf$. Such a trend was observed in the experiments on the AIX separations of group-4 elements from mixed 0.02 M HF and 0.3-0.4 M HCl solutions [173]. The weaker sorption of Rf from HF solutions of > 10^{-3} M on the AIX column was, however, found in [172]. This was explained (and also shown by additional experiments) by a strong competition between NO_3^- and Rf complexes for adsorption on the active resin sites.

For the AIX separations at 4-8 M HCl, where no hydrolysis should occur at such high acidities, the data of Table 17 suggest that the trend in the complex formation and K_d values should definitely be continued with Rf: $Zr > Hf > Rf$. The AIX separations [174] of group-4 elements from aqueous 4-8 M HCl solutions have, however, shown an inverse sequence in K_d values: $Rf > Zr > Hf$. Taking into account the above mentioned arguments, this result cannot find its theoretical explanation.

The TBP extraction of group-4 elements from 8 M HCl showed the K_d of Rf in between those of Zr and Hf: $Zr > Rf > Hf$ [175]. Such an inversion of the trend is consistent with the theoretical trend for the formation of the MCl_4 species, see Table 17, though some further calculations for the $MCl_4(TBP)_2$ complexes should be performed.

7.2.3 *Element 105*

Hydrolysis of group-5 cations and Pa was studied theoretically [164] for the reaction

$$M(H_2O)_6^{5+} \leftrightarrow M(OH)_6^- + 6H^+ \quad (30)$$

where M = Nb, Ta, Db, and Pa. The calculated relative ΔE^C of reaction (30) are given in Table 18. The ΔE^C data are indicative of the following trend in hydrolysis of group-5 cations: Nb > Ta > Db >> Pa.

Table 18. E^C and ΔE^C (in eV) for reaction $M(H_2O)_6^{5+} \leftrightarrow M(OH)_6^-$, where M = Nb, Ta, Db and Pa. From [164]

Complex	Nb	Ta	Db	Pa
$M(OH)_6^-$	-21.74	-23.33	-21.48	-19.53
$M(H_2O)_6^{5+}$	-21.92	-25.38	-25.37	-29.71
ΔE^C	0.18	2.05	3.89	9.18

This sequence is in agreement with experiments on hydrolysis of Nb, Ta, and Pa [163]. A simple model of hydrolysis [163] does not reproduce the difference between Nb and Ta having the same IR. The present model based on the real (relativistic) distribution of the electronic density does reproduce this difference.

The complex formation of group-5 elements in HF, HCl and HBr solutions was studied theoretically [165,166]. A motivation for this study was the unexpected behavior of Db (Ha at that time) in the extractions into triisooctyl amine (TIOA) from mixed HCl/HF solutions [176]: Db was extracted similarly to Pa and not to Ta.

In HCl solutions, a large variety of complexes, such as $M(OH)_2Cl_4^-$, $MOCl_4^-$, $MOCl_5^{2-}$ and MCl_6^- (M = Nb, Ta, Db and Pa) can be formed with different degrees of hydrolysis according to the following equilibrium

$$M(OH)_6^- + iL^- \leftrightarrow MO_u(OH)_{z-2u}L_i^{(6-i)-} \tag{31}$$

To predict stability of those complexes, the DFT calculations of the electronic structures were performed for the above mentioned complexes of Nb, Ta, Db and Pa. As a result, the ΔE^C for reaction (31) are given in Table 19.

Table 19. ΔE^C (in eV) for reaction $M(OH)_6^- \leftrightarrow M(OH)_nCl_m^-$, where M = Nb, Ta, Db, and Pa. From [165,166]

Metal	$M(OH)_2Cl_4$	$MOCl_4$	MCl_6
Nb	13.56	18.40	19.57
Ta	14.32	19.80	20.78
Db	14.29	19.67	20.46
Pa	11.68	16.29	17.67

The data of Table 19 show the following trend in the complex formation of group-5 elements: Pa >> Nb > Db > Ta. Taking into account the work of transfer of the complexes between the phases, the following trend was predicted for the extraction of group-5 anions by an anion exchanger

$$Pa >> Nb \geq Db > Ta. \tag{32}$$

Thus, complexes of Pa should be formed in much more dilute HCl solutions, while much higher acid concentrations are needed to form complexes of Ta. The calculations also predicted the following sequence in the formation of various types of complexes as a function of the acid concentration: $M(OH)_2Cl_4^- > MOCl_4^- > MCl_6^-$ in full agreement with experimental results for Nb, Ta and Pa. The calculations also reproduced the sequence F > Cl > Br in the formation of ML_6^- (L = F, Cl, and Br) as a function of the ligand L, as the data of Table 20 show.

Table 20. ΔE^C (in eV) for reaction $M(OH)_6^- \leftrightarrow ML_6^-$, where M = Nb, Ta, Db and Pa, and L = F, Cl and Br. From [166]

Complex	F	Cl	Br
NbL_6^-	12.20	19.57	21.40
TaL_6^-	12.69	20.78	22.63
DbL_6^-	12.38	20.46	22.11
PaL_6^-	12.19	17.67	19.91

The theoretical investigations have shown that the trend in the complex formation and extraction (Equation 32) known for Nb, Ta and Pa turned out to be *reversed* in going to Db. This could not be predicted by any extrapolation of this property within the group, which would have given a wrong trend, but came as a result of considering real chemical equilibria and calculating relativistically the electronic structure of the complexes.

According to these results, a recommendation was made to conduct the AIX separations in pure HCl or HF solutions to try to observe the predicted sequence, Equation 32. Accordingly, the amine separations of the group-5 elements were systematically redone by W. Paulus *et al.* [177]. The reversed extraction sequence Pa > Nb ≥ Db > Ta has been established exactly as theoretically predicted. That was the first time when predictions of extraction behavior of the heaviest elements based on quantum-chemical calculations were made, and also confirmed by specially designed experiments.

7.2.4 *Element 106*

Experiments on the CIX separations of element 106 from 0.1 M HNO_3 solutions showed that Sg did not elute from the CIX column, in contrast to Mo and W [178]. This non-tungsten-like behavior of Sg was tentatively attributed to its lower tendency to hydrolyze (deprotonate) compared to that of W. To interpret the behavior of Sg in these experiments and to predict its hydrolysis at various pH values, free energies of the following protonation reactions for Mo, W and Sg

$$MO_4^{2-} \leftrightarrow MO_3(OH)^- \leftrightarrow MO_2(OH)_2(H_2O)_2 \leftrightarrow MO(OH)_3(H_2O)_2^+ \leftrightarrow M(OH)_4(H_2O)_2^{2+} \leftrightarrow \ldots \leftrightarrow M(H_2O)_6^{6+} \quad (33)$$

were considered theoretically. The ΔE^C for the consecutive protonation steps were calculated using the DFT method [167]. The results shown in Table 21 indicate that for the first two protonation steps, the trend in group-6 is reversed: Mo < Sg < W. For the further protonation process, the trend is continued with Sg: Mo < W < Sg

Table 21. ΔE^C (in eV) for the stepwise protonation of MO_4^{2-} (M = Mo, W and Sg). From [167]

Reaction	ΔE^C		
	Mo	W	Sg
$MO_4^{2-} + H^+ \leftrightarrow MO_3(OH)^-$	-12.28	-13.13	-12.96
$MO_3(OH)^- + H^+ + 2H_2O \leftrightarrow MO_2(OH)_2(H_2O)_2$	-21.43	-22.08	-21.61
$MO_2(OH)_2(H_2O)_2 + H^+ \leftrightarrow MO(OH)_3(H_2O)_2^+$	-5.84	-6.35	-6.65
$MO(OH)_3(H_2O)_2^+ + H^+ \leftrightarrow M(OH)_4(H_2O)_2^{2+}$	-0.43	-0.76	-1.23
$M(OH)_4(H_2O)_2^{2+} + 4H^+ \leftrightarrow \ldots M(H_2O)_6^{6+}$	41.97	38.71	37.11

Thus, the same reversal in the trend is predicted for the protonation of oxyanions of the group-6 elements as that for the complex formation of the group-4 and 5 elements. The predicted trends in the complex formation are in agreement with experiments for Mo and W at various pHs [163]. For the protonation/hydrolysis of positively charged complexes, the predicted trend Mo < W < Sg is in line with the experimental observations for Sg [178]. Using the procedure described in Section 7.2.1, log*K* were determined for Sg, as given in Table 22 [167].

Table 22. log*K* for the step-wise protonation of MO_4^{2-} (M = Mo, W and Sg). From [167]

Reaction	$\log K_n$		
	Mo	W	Sg
$MO_4^{2-} + H^+ \leftrightarrow MO_3(OH)^-$	3.7	3.8	3.74
$MO_3(OH)^- + H^+ + 2H_2O \leftrightarrow MO_2(OH)_2(H_2O)_2$	3.8	4.3	4.1± 0.2
$MO_4^{2-} + 2H^+ + 2H_2O \leftrightarrow MO_2(OH)_2(H_2O)_2$	7.50	8.1	8.9 ± 0.1
$MO_2(OH)_2(H_2O)_2 + H^+ \leftrightarrow MO(OH)_3(H_2O)_2^+$	0.93	0.98	1.02

Complex formation of Mo, W, and Sg in HF solutions was studied theoretically [179] by considering the following step-wise fluorination process

$$MO_4^{2-}\ [\text{or } MO_3(OH)^-] + HF \leftrightarrow MO_3F^- \leftrightarrow$$
$$MO_2F_2(H_2O)_2 \leftrightarrow MO_2F_3(H_2O)^- \leftrightarrow MOF_5 \qquad (34)$$

The calculated ΔE^C indicate a very complicated dependence of the complex formation on pH and the HF concentration. At low HF concentrations (<~0.1 M HF), a reversal of trends occurs, while at high HF molarities (> ~0.1 M HF), the trend in continued with Sg: Mo < W < Sg. The obtained sequences are in agreement with experiment for Mo and W [180]. Future experiments on the AIX separations of group-6 elements from HF solutions will clarify the position of Sg in this row.

7.2.5 *Summary of the aqueous chemistry studies*

Predicted trends in hydrolysis, complex formation and extraction of complexes at various experimental conditions in comparison with experimental results are summarized in Table 23.

Table 23. Trends in hydrolysis and complex formation of the heaviest element compounds and their lighter homologues in chemical groups

Group	Complexes	Experimental conditions	Theoretically predicted[a]	Experimentally observed	Ref. exp.
4	Hydrolysis of M^{4+}	pH ≤ 2	Zr > Hf > Rf	Zr > Hf > Rf	163
	$MF_x(H_2O)^{z-x}_{8-x}$ (x ≤ 4)	[HF] < 10^{-1} M	Zr > Hf > Rf	Zr > Hf > Rf	172
	MF_6^-	[HF] > 10^{-3} M	Rf ≥ Zr > Hf	Rf ≥ Zr > Hf	173
	MCl_6^-	4-8 M [HCl]	Zr > Hf > Rf	Rf > Zr > Hf	174
5	Hydrolysis of M^{5+}	all pH	Nb > Ta > Db	Nb > Ta	163
	$M(OH)_2Cl_4^-$ $MOCl_4^-$, MCl_6^-	all [HCl]	Pa >> Nb ≥ Db > Ta	Pa >> Nb ≥ Db >Ta	177
6	Hydrolysis of M^{6+}	pH = 0 ÷1	Mo > W > Sg	Mo > W > Sg	178
	Hydrolysis of $MO_2(OH)_2$	pH > 1	Mo > Sg > W	Mo > W	163
	MO_3F^-, MO_2F_2	< 0.1 M [HF], pH > 1	Mo > Sg > W	Mo > W	180 181
	$MO_2F_3^-$	> 0.1 M [HF]	Mo < Sg < W	Mo < W	180
	MOF_5^-	>> 0.1 M[HF]	Mo < W < Sg	Mo < W	180

[a] Refs. [164-168]

7.4 PROSPECTS FOR AQUEOUS CHEMISTRY STUDIES

From element 107 on, the maximum oxidation state is expected to be relatively unstable in solutions. It would, therefore, be interesting to conduct experiments probing the stability of lower oxidation states. The stability of Bh^{VII} relative to Bh^{IV} could be established by AIX separations of group-7 elements in acidic solutions. In HCl solutions Tc and Re undergo the complexation reaction $MO_4^- + HCl \leftrightarrow MCl_6^{2-}$ simultaneously with reduction. The K_d curves for Tc and Re show peaks at about 7-8 M HCl associated with the reduction [182]. The peak for Tc is at lower HCl concentrations than that of Re indicating an earlier reduction of Tc, which means that Tc is less stable in the 7+ oxidation state (or more stable in the 4+ state) than Re. The position of the peak on the K_d curve for Bh would, therefore, be indicative of the relative stability of its 7+ oxidation state. It would also be interesting to conduct similar reduction experiments with Hs. Its homologues are known to have the following reduction potentials: $RuO_4 + nHCl \rightarrow RuCl_5OH^{2-} + Cl_2 + nH_2O$ of E° = 0.14 V and $OsO_4 + 8HCl \rightarrow H_2OsCl_6 + Cl_2 + 4H_2O$ of E° = -0.36 V.

Element 112 should have an extensive complex ion chemistry, like other elements at the second half of the 6d transition series. A tendency to form stronger bonding with "soft" ligands is foreseen in [12] by analogy with Hg showing increasing stability of the aqueous complexes from F to Cl to Br and to I (Table 24). The increasing stability constants in this row has, however, another reason, namely the decreasing hydrolysis $HgX_2 + H_2O \leftrightarrow Hg(OH)X$: 1.4% (Cl), 0.08% (Br), 0% (I) [183]. The stability of the gas and crystal phase compounds of Hg decreases from F to Cl to Br and to I, see Table 24.

Table 24. Formation enthalpies (in kcal/mol) for some compounds of Hg [183]

Compound	Phase	F	Cl	Br	J
HgX_2	gas	-70.2	-35.0	-20.4	-3.84
HgX_2	aqueous	-	-6.9	-10.7	-15.0
HgX_3^-	aqueous	-	-2.2	-3.0	-3.6
HgX_4^{2-}	aqueous	-	0.1	-4.1	-4.0

The formation enthalpy of $112F_2$ was calculated at the PP level as -75.33 kcal/mol as compared to the calculated -88.4 kcal/mol for HgF_2 [108]. Thus, taking into account the decreasing stability of the 2+ oxidation state in group 12, experiments could be conducted with probably only $112I^+$ and $112I_2$. The possibility of formation of $112F_5^-$ and $112F_3^-$ was also considered [108] (by analogy with Hg where the addition of an F^- to HgF_2 or HgF_4 was found energetically favorable), though these compounds will undergo strong hydrolysis in aqueous solutions and will not be stable. Thus, the only possibility would be the formation of $112Br_5^-$ or $112I_5^-$.

Element 114 should also have a greater tendency to form complexes in solutions than Pb. Since the stability of the 2+ state increases within group 14, element 114 would probably form $M^{2+} + X_2 \leftrightarrow MX^+$ (X = Cl, Br and I) and $M^{2+} + X_2 \leftrightarrow MX_2$ or $MX_2 + X_2 \leftrightarrow MX_3^-$ or MX_4^{2-} by analogy with Pb. (In 11 M HCl $PbCl_6^{4-}$ is known). As in group 12, the stability of the gas phase compounds of Pb decreases from F to Cl to Br and to I, while in aqueous solutions it is the other way around due to a decreasing hydrolysis from F to Cl to Br and to I (fluoride complexes are not known) according to the reaction $MX_2 \leftrightarrow M(OH)X$, $M(OH)_2$ or $M(OH)_3^-$ [184]. Since the 2+ oxidation state of element 114 should be more stable than Pb^{2+}, element 114 can be extracted as MBr_3^- or MI_3^-. In [107], the existence of $114F_6^-$ was suggested, though in solutions this compound will undergo strong hydrolysis. The stability of various complexes of element 114 versus stability of the hydrolysis products could be a subject of further theoretical investigations.

8. Conclusions and Outlook

Advances in relativistic quantum theory and computational methods made it possible to predict properties of the heaviest element compounds by performing accurate calculations of their electronic structures. Relativistic atomic and molecular calculations in combination with various models were useful in helping to design sophisticated and expensive chemical experiments. Experimental results, in turn, were helpful in defining the scope of the theoretical problems and provided an important input. The synergism between the theoretical and experimental research in the last decade led to better understanding the chemistry of these exotic species.

Extensive DFT and PP calculations have permitted the establishment of important trends in chemical bonding, stabilities of oxidation states, crystal-field and SO effects, complexing ability and other properties of the heaviest elements, as well as the role and magnitude of relativistic effects. It was shown that relativistic effects play a dominant role in the electronic structures of the elements of the 7th row and heavier, so that relativistic calculations in the region of the heaviest elements are indispensable. Straight-forward extrapolations of properties from lighter congeners may result in erroneous predictions. The molecular DFT calculations in combination with some physico-chemical models were successful in the application to systems and processes studied experimentally such as adsorption and extraction. For theoretical studies of adsorption processes on the quantum-mechanical level, embedded cluster calculations are under way. RECP were mostly applied to open-shell compounds at the end of the 6d series and the 7p series. Very accurate fully relativistic DFB *ab initio* methods were used for calculations of the electronic structures of model systems to study relativistic and correlation effects. These methods still need further development, as well as powerful supercomputers to be applied to heavy element systems in a routine manner. Presently, the RECP and DFT methods and their combination are the best way to study the theoretical chemistry of the heaviest elements.

Though substantial progress has been reached in understanding the chemistry of the heaviest elements, there is still a number of open questions needed to be answered from both the experimental and theoretical points of view. For the superactinides, investigations of the chemical properties will be even more exciting, since the resemblance of their properties to those of their lighter homologues will be much less pronounced.

9. Conversion Factors

1 a.u. = 0.5291770644 Å, 1Å = 10^{-10} m
1 eV = 96.4905 kJ/mol = 23.0618 kcal/mol

References

1. Hofmann, S., Ninov, V., Hessberger, F.P., Armbruster, P., Folger, H., Münzenberg, G., Schött, H.J., Popeko, A.G., Yeremin, A.V., Andreev, A.N., Saro, S., Janik, R., Leino, M.: Z. Phys. A **350**, (1995) 277.
2. Hofmann, S., Ninov, V., Hessberger, F.P., Armbruster, P., Folger, H., Münzenberg, G., Schött, H.J., Popeko, A.G., Yeremin, A.V., Andreev, A.N., Saro, S., Janik, R., Leino, M.: Z. Phys. A **350**, (1995) 281.
3. Hofmann, S., Ninov, V., Hessberger, F.P., Armbruster, P., Folger, H., Münzenberg, G., Schött, H.J., Popeko, A.G., Yeremin, A.V., Saro, S., Janik, R., Leino, M.: Z. Phys. A **354**, (1996) 229.
4. Oganessian, Yu. Ts., Utyonkov, V.K., Lobanov, Yu. V., Abdulin, F. Sh., Polyakov, A.N., Shirokovsky, I.V., Tsyganov, Yu. S., Gulbekian, G.G., Bogomolov, S.L., Gikal, B.N., Mezentsev, A.N., Iliev, S., Subbotin, V.G., Sukhov, A.M., Ivanov, O.V., Buklanov, G.V., Subotic, K., Itkis, M.G., Moody, K.J., Wild, J.F., Stoyer, N.J., Stoyer, M.A., Lougheed, R.W., Laue, C.A., Karelin, Ye. A., Tatarinov, A.N.: Phys. Rev. C **63**, (2000) 011301 (R).
5. Schädel, M.: Radiochim. Acta, **70/71**, (1995), 207; ibid. **89**, (2001) 721.
6. Kratz, J.V.: "Fast Chemical Separation Procedures for Transactinides". In: *Heavy Elements and Related New Phenomena,* Eds. Greiner, W., Gupta, R.K., World Scientific, Singapore, (1999) 43-63.
7. Hoffman, D.C.: Chem. Eng. News, May 24 (1994); Hoffman, D. C.: J. Chem. Ed. **76**, (1999) 331.
8. Gäggeler, H.W.: "Chemistry of the Transactinide Elements". In: Proceedings of "The Robert A. Welch Foundation Conference on Chemical Research XXXIV. Fifty Years with Transuranium Elements", Houston, Texas, 22-23 October 1990, pp. 255-276.
9. Eichler, R., Brüchle, W., Dressler, C.E., Düllman, C.E., Eichler, B., Gäggeler, H.W., Gregorich, K.E., Hoffman, D.C., Hübener, S., Jost, D.T., Kirbach, U.W., Laue, C.A., Lavanchy, V.M., Nitsche, H., Patin, J.B., Piguet, D., Schädel, M., Shaughnessy, D.A., Strellis, D.A., Taut, S., Tobler, L., Tsyganov, Y. S., Türler, A., Vahle, A., Wilk, P.A., Yakushev, A.B.: Nature **407**, (2000) 63.
10. Düllman, Ch., Brüchle, W., Dressler, R., Eberhardt, K., Eichler, B., Eichler, R., Gäggeler, H. W., Ginter, T. N., Glaus, F., Gregorich, K. E., Hoffman, D. C., Jäger, E., Jost, D. T., Kirbach, U. W., Lee, D. E., Nitsche, H., Patin, J. B., Pershina, V., Piguet, D., Qin, Z., Schädel, M., Schausten, B., Schimpf, E., Schött, H.-J., Soverna, S., Sudowe, R., Thörle, P., Timokhin, S. N., Trautmann, N., Türler, A., Vahle, A., Wirth, G., Yakushev, A. B., Zielinski, P. M.: Nature **418**, (2002) 859.
11. Yakushev, A.B. , Buklanov, G.V., Chelnokov, M.L., Chepigin, V.I., Dmitriev, S.N., Gorshkov, V.A., Lebedev, V.Ya, Malyshev, O.N., Oganessian, Yu.Ts., Popeko, A.G., Sokol, E.A., Timokhin, S.N., Vasko, V.M., Yeremin, A.V., Zvara, I. In: Abstracts of the 5th International Conference on Nuclear and Radiochemistry, Pontresina, September 3-8, (2000), p. 233.
12. Fricke, B.: Struct. Bond. **21**, (1975) 89.
13. Pyykkö, P.: Chem. Rev. **88**, (1988) 563.
14. Grant I.P.: In: The Effects of Relativity in Atoms, Molecules and the Solid State, Eds. Wilson, S., Grant, I.P. and Gyorffy, B.L., Plenum, New York, (1991), 17-43.

15. Pershina, V.: Chem. Rev. **96**, (1996) 1977.
16. Pershina, V. and Fricke, B.: "Electronic Structure and Chemistry of the Heaviest Elements". In: *Heavy Elements and Related New Phenomena,* Eds. Greiner, W., Gupta, R.K., World Scientific, Singapore, (1999) 194-162.
17. Schwerdtfeger, P. and Seth, M. : "Relativistic Effects on the Superheavy Elements".In: *Encyclopedia on Calculational Chemistry*, Wiley, New York, (1998), Vol. 4, 2480-2499.
18. Kohn, W., Becke, A.D., Parr, R.G.: J. Phys. Chem. **100**, (1996) 12974.
19. Grant, I. P.: Adv. At. Mol. Phys. **32**, (1994) 169.
20. Seaborg, G.T.: Metallurgical Laboratory Memorandum MUC-GTS-858, July 17, 1944; Seaborg, G.T.: Chem. Eng. News **23**, (1945) 2190.
21. Mann, J.B.: J. Chem. Phys. **51**, (1969) 841; Mann, J.B. and Waber, J.T.: ibid **53**, (1970) 2397.
22. Fricke, B., Greiner, W., Waber, J.T.: Theoret. Chim. Acta **21**, (1971) 235.
23. Desclaux, J.P.: At. Data Nucl. Data Tables, **12**, (1973) 311.
24. Desclaux, J.P. and Fricke, B.: J. Phys. **41**, (1980) 943.
25. Glebov, V.A., Kasztura, L., Nefedov, V.S., Zhuikov, B.L.: Radiochim. Acta **46**, (1989) 117.
26. Johnson, E., Fricke, B., Keller, Jr., O.L., Nestor, Jr., C.W., Ticker, T.C.: J. Chem. Phys. **93**, (1990) 8041.
27. Fricke, B., Johnson, E., Rivera, G. M.: Radiochim. Acta **62**, (1993) 17.
28. Johnson, E., Pershina, V., Fricke, B.: J. Phys. Chem. **103**, (1999) 8458.
29. Johnson, E., Fricke, B., Jacob, T., Dong, C.Z., Fritzsche, S., Pershina, V.: J. Phys. Chem. **116**, (2002) 1862.
30. Pyper, N.C. and Grant, I.P.: Proc. R. Soc. Lond. A **376**, (1981) 483.
31. Kaldor, U. and Eliav, E.: Adv. Quantum. Chem. **31**, (1998) 313.
32. Kaldor, U. and Eliav, E.: "Energies and Other Properties of Heavy Atoms and Molecules" In: *Quantum Systems in Chemistry and Physics*, Ed. Hernandes-Laguna, v. II, (2000) 185.
33. Eliav, E., Landau, A., Ishikawa, Y., Kaldor, U.: J. Phys. B. **35**, (2002) 1693.
34. Schwarz, W.H.E., van Wezenbeek, E.M., Baerends, E.J., Snijders, J.G. J. Phys. B: At. Mol. Opt. Phys. **22**, (1989) 1515; Baerends, E.J., Schwarz, W.H.E., Schwerdtfeger, P., Snijders, J.G.: J. Phys. B: At. Mol. Opt. Phys. **23**, (1990) 3225.
35. Eliav, E., Shmulyian, S., Kaldor, U., Ishikawa, Y.: J. Chem. Phys. **109**, (1998) 3954.
36. Lindgren, I.: Int. J. Quant. Chem. **57**, (1996) 683.
37. Pyykkö, P., Tokman, M., Labzowsky, L.N.: Phys. Rev. A, **57**, (1998) R689.
38. Pepper, M. and Bursten, B.E.: Chem. Rev. **91**, (1991) 719.
39. Mittleman, M.: Phys. Rev. A **4**, (1971) 89.
40. Sucher, J.: Phys. Rev. A **22**, (1980) 348.
41. Grant, I. P.: J. Phys. B **19**, (1986) 3187.
42. Desclaux, J. P.: Comp. Phys. Comm. **9**, (1975) 31.
43. Parpia, F. A., Froese-Fisher, S., Grant, I. P.: CPC **94**, (1996) 249.
44. Ishikawa, I. and Kaldor, U. In.: *Computational Chemistry, Reviews of Current Trends*, Ed. Leszynski, J., World Scientific, Singapore, (1996) Vol. 1.
45. Visscher, L., Visser, O., Aerts, H., Merenga, H., Nieuwpoort, W.C.: Comput. Phys. Comm. **81**, (1994) 120.
46. Collins, C.L., Dyall, K.G., Schaeffer, H. F., III, J. Chem. Phys. **102**, (1995) 2024; Dyall, K.G.: Chem. Phys. Lett. **224**, (1994) 186.
47. Visscher, L., Dyall, K.G., Lee, T.J.: Int. J. Quant. Chem. Quant. Chem. Symp. **29**, (1995) 411; Visscher, L., Lee, T.J., Dyall, K.G.: J. Chem. Phys. **105**, (1996) 8769.
48. Wood, C.P. and Pyper, N.C.: Chem. Phys. Lett. **84**, (1981) 614.
49. Saue, T., Faegri, K., Gropen, O.: Chem. Phys. Lett. **263**, (1996) 360.
50. Seth, M., Schwerdtfeger, P., Dolg, M., Faegri, K., Hess, B.A., Kaldor, U.: Chem. Phys. Lett. **250**, (1996) 461.

51. Faegri, K. and Saue, T.: J. Chem. Phys. **115**, (2001) 2456.
52. Faegri, K. and Saue, T.: to be published.
53. Ermler, W.C., Ross, R.B., Christiansen, P.A.: Adv. Quant. Chem. **19**, (1988) 139; Christiansen, P.A., Lee, Y.S., Pitzer, K.S.: J. Chem. Phys. **71**, (1979) 4445.
54. Nash, C.S., Bursten, B.E., Ermler, W.C.: J. Chem. Phys. **106**, (1997) 5133.
55. Mosyagin, N.S., Titov, A.V., Latajka, Z.: Int. J. Quant. Chem. **63**, (1997) 1107; Titov, A.V., Mosyagin, N.S.: Int. J. Quant. Chem. **71**, (1999) 359.
56. Mosyagin, N.S., Titov, A.V., Eliav, E., Kaldor, U.: to be published.
57. Dolg, M., Wedig, U., Stoll, H., Preuss, H.: J. Chem. Phys. **86**, (1987) 866; Andrae, D., Häussermann, U., Dolg, M., Stoll, H., Preuss, H.: Theor. Chem. Acta **77**, (1990) 123.
58. Schwerdtfeger, P., Dolg, M., Schwera, W.H.E., Bowmaker, G.A., Boyd, P.W.D.: J. Chem. Phys. **91**, (1989) 1762.
59. Dolg, M., Stoll, H., Preuss, H., Pitzer, R.M.: J. Phys. Chem. **97**, (1993) 5852.
60. Schwerdtfeger, P., Seth, M.: private communication.
61. Kahn, L.R., Hay, P.J., Cowan, R.D.: J. Chem. Phys. **68**, (1978) 2386.
62. Hay, P.J.: J. Chem. Phys. **79**, (1983) 5469; Schreckenbach, G., Hay, P.J., Martin, R.L.: Inorg. Chem. **37**, (1998) 4442.
63. Lee, H.S., Han. Y.K., Kim, M.C., Bae, C., Lee, Y.S.: Chem. Phys. Lett. **293**, (1998) 97.
64. Schwerdtfeger, P., Brown, J.R., Laerdahl, J.K., Stoll, H.: J. Chem. Phys. **113**, (2000) 7110.
65. Rosen, A.: Adv. Quant. Chem. **29**, (1997) 1.
66. Engel, E., Keller, S., Dreizler, R.M.: Phys. Rev. A **53**, (1996) 1367.
67. Rosen, A., Ellis, D.E.: J. Chem. Phys. **62**, (1975) 3039; Rosen, A.: Int. J. Quant. Chem. **13**, (1978) 509.
68. Bastug, T., Heinemann, D., Sepp, W.-D., Kolb, D., Fricke, B.: Chem. Phys. Lett. **211**, (1993) 119.
69. Varga, S., Engel, E., Sepp, W.-D., Fricke, B.: Phys. Rev. A **59**, (1999) 4288.
70. Varga, S., Fricke, B., Hirata, M., Bastug, T., Pershina, V., Fritzsche, S.: J. Phys. Chem. **104**, (2000) 6495.
71. Liu, W., Hong, G., Dai, D., Li, L., Dolg, M.: Theor. Chem. Acc. **96**, (1997) 75.
72. Liu, W. and van Wüllen, Ch.: J. Chem. Phys. **110**, (1999) 3730.
73. Liu, W., van Wüllen, Ch., Han, Y. K., Choi, Y. J., Lee, Y.S.: Adv. Quant. Chem. **39**, (2001) 325.
74. ADF 2.3, Theoretical Chemistry, Vrije Universiteit Amsterdam; te Velde, G. and Baerends, E.J.: J. Comput. Phys. **99**, (1992) 94.
75. Ziegler, T., Tschinke, V., Baerends, E. J., Snijders, J. G., Ravenek, W.: J. Phys. Chem. **93**, (1989) 3050.
76. van Lenthe, E., Baerends, E.J., Snijders, J.G.: J Chem. Phys. **101**, (1994) 9783.
77. Faas, S., van Lenthe, J. H., Hennum, A. C., Snijders, J. G.: J. Chem. Phys. **113**, (2000) 4052.
78. van Wüllen, Ch.: J. Chem. Phys. **109**, (1998) 392.
79. Nasluzov, V.A. and Rösch, N.: Chem. Phys. **210**, (1996) 413.
80. Häberlen, O.D., Chung, S.-C., Stener, M., Rösch, N.: J. Chem. Phys. 106, (1997) 5189.
81. Ziegler, T., Snijders, J.G., Baerends, E.J.: Chem. Phys. Lett. **75**, (1980) 1.
82. Case, D.A., and Yang, C.Y.: J. Chem. Phys. **72**, (1980) 4334.
83. Huber, K.P. and Herzberg, G.: *Constants of Diatomic Molecules*, Van Nostrand Reinhold, New York, 1979.
84. Hess, B.A.: Phys. Rev. A **33**, (1986) 3742.
85. Mann, J.B. Waber, J.T.: At. Data **5**, (1973) 201; Brewer, L.: J. Opt. Soc. Am. **61**, (1971) 1102.
86. Eliav, E., Kaldor, U., Ishikawa, Y.: Phys. Rev. A **52**, (1995) 291.
87. Keller, O.L.: Radiochim. Acta **37**, (1984) 169.
88. Eliav, E., Kaldor, U., Ishikawa, Y.: Phys. Rev. Lett. **74**, (1995) 1079.

89. Eliav, E., Kaldor, U., Schwerdtfeger, P., Hess, B.A., Ishikawa, Y.: Phys. Rev. Lett. **73**, (1994) 3203.
90. Eliav, E., Kaldor, U., Ishikawa, Y.: Phys. Rev. A **52**, (1995) 2765.
91. Eliav, E., Kaldor, U., Ishikawa, Y., Seth, M., Pyykkö, P.: Phys. Rev. A **53**, (1996) 3926.
92. Landau, A., Eliav, E., Ishikawa, Y., Kaldor, U.: J. Chem. Phys. **114**, (2001) 2977.
93. Eliav, E., Kaldor, U., Ishikawa, Y.: Mol. Phys. **94**, (1998) 181.
94. Landau, A., Eliav, E., Ishikawa, Y., Kaldor, U.: J. Chem. Phys. **115**, (2001) 2389.
95. Umemoto, K., Saito, S.: J. Phys. Soc. Jpn. **65**, (1996) 3175.
96. Seth, M., Faegri, K., Schwerdtfeger, P.: Angew. Chem. Int. Ed. Engl. **37**, (1998) 2493.
97. Balasubramanian, K.: Chem. Phys. Lett. **341**, (2001) 601; ibid. **351**, (2002) 161.
98. Nash, C. S., Bursten, B. E.: J. Phys. Chem. A **103**, (1999) 402.
99. Pershina, V., Johnson, E., Fricke, B.: J. Phys. Chem. A **103**, (1999) 8463.
100. Seth, M., Schwerdtfeger, P., Faegri, K.: J. Chem. Phys. **111**, (1999) 6422.
101. Eliav, E., Kaldor, U., Ishikawa, Y., M., Pyykkö, P.: Phys. Rev. Lett. **77**, (1996) 5350.
102. Lim, I., Pernpointer, M., Seth, M., Schwerdtfeger, P.: Phys. Rev. A. **60**, (1999) 2822.
103. Shannon, R.D.: Acta Crystallogr. , Sect. A **32**, (1976) 751.
104. Bilewicz, A.: Radiochim. Acta **88**, (2000) 833.
105. Seth, M., Dolg, M., Fulde, P., Schwerdtfeger, P.: J. Am. Chem. Soc. **117**, (1995) 6597.
106. Fricke, B.: J. Chem. Phys. **84**, (1986) 862.
107. Seth, M.: *The Chemistry of Superheavy Elements*, Thesis, University of Auckland, 1998.
108. Seth, M., Schwerdtfeger, P., Dolg, M.: J. Chem. Phys. **106**, (1997) 3623.
109. Pyykkö, P., Desclaux, J.P.: Chem. Phys. Lett. **50**, (1977) 503; Pyykkö, P., Desclaux, J.P.: Nature **226**, (1977) 336; Pyykkö, P., Desclaux, J.P.: Chem. Phys. **34**, (1978) 261.
110. Pershina, V.: "Electronic Structure and Chemistry of the Transactinides". In: Proceedings of "The Robert A. Welch Foundation Conference on Chemical Research XXXIV. Fifty Years with Transuranium Elements", Houston, Texas, 22-23 October 1990, pp. 167-194.
111. Varga, S., Fricke, B., Hirata, M., Bastug, T., Pershina, V., Fritzsche, S.: J. Phys. Chem. A **104**, (2000) 6495.
112. Pershina, V., Bastug, T., Fricke, B.: J. Alloy Comp. **271-273**, (1998) 283.
113. Pershina, V. and Bastug, T.: J. Chem. Phys, **113**, (2000) 1441.
114. Pershina, V., Bastug, T., Fricke, B., Varga, S.: J. Chem. Phys. **115**, (2001) 1.
115. Han, Y.-K., Son, S.-K., Choi, Y.J., Lee, Y.S.: J. Phys. Chem. **103**, (1999) 9109.
116. Malli, G. L. and Styszynski, J.: J. Chem. Phys. **109**, 4448 (1998).
117. Pershina, V. and Fricke, B.: J. Chem. Phys. **99**, (1993) 9720.
118. Girichev, G.V., Petrov, V.M., Giricheva, N.I., Utkin, A.N., Petrova, V.N.: Zh. Strukt. Khim. **22**, (1981) 65.
119. Brown, D.D. In: Comprehensive Inorganic Chemistry, Ed. Baylar, J.C., Pergamon Press, Oxford, (1973) Vol. 3, 553-622; Fergusson, M. ibid, p. 760; Aylett, B.J., ibid, p. 187; Rochow, E.G., ibid, Vol. 4, p. 1.
120. Lijima, K., Shibata, S.: Bull. Chem. Soc. Jpn. **48**, (1975) 666.
121. Jampolski, V.I.: Ph.D. Thesis, Moscow State University, 1973.
122. Mulliken, R. S.: J. Chem. Phys. **23**, (1955) 1833.
123. Pershina, V. and Fricke, B.: J. Phys. Chem. **100**, (1996) 8748, ibid. **99**, (1995) 144.
124. Schädel, M., Brüchle, W., Dressler, R., Eichler, B., Gäggeler, H. W., Günther, R., Gregorich, K. E., Hoffman, D. C., Hübener, S., Jost, D. T., Kratz, J. V., Paulus, W., Schumann, D., Timokhin, S., Trautmann, N., Türler, A., Wirth, G., Yakushev, A.: Nature (Letters) **388** (1997) 55.
125. Türler, A., Brüchle, W., Dressler, R., Eichler, B., Eichler, R., Gäggeler, H. W., Gärtner, M., Glatz, J.-P., Gregorich, K. E., Hübener, S., Jost, D. T., Lebedev, V. Ya., Pershina, V., Schädel, M., Taut, S., Timokhin, N., Trautmann, N., Vahle, A., Yakushev, A. B.: Angew. Chem. Int. Ed. **38**, (1999) 2212.
126. Amble, E., Miller, S.L., Schawlaw, A.L., Townes, C.H.: J. Chem. Phys. **20**, (1952) 192.

127. Krebs, B., Hasse, K.D. Acta Crystallogr., Sect. B:
Struct. Crystallogr. Cryst. Chem. B **32**, (1976) 1334.
128. CRC Handbook of Chemistry and Physics, 79th ed., Ed. Lide, D.R.,
CRC Boca Raton, (1998-1999), p. 10.
129. Burroughs, P., Evans, S., Hammnett, A., Orchard, A.F., Richardson, N.V.:
J. Chem. Soc., Faraday Trans. 2 **70**, (1974) 1895.
130. Pershina, V., Sepp, W.-D., Fricke, B., Kolb, D., Schädel, M., Ionova, G. V.:
J. Chem. Phys. **97**, (1992) 1116.
131. Pershina, V., Sepp. W.-D., Bastug, T., Fricke, B., Ionova, G. V.:
J. Chem. Phys. **97**, (1992) 1123.
132. Kadkhodayan, B., Türler, A., Gregorich, K. E., Baisden, P. A., Czerwinski, K. R., Eichler, B., Gäggeler, H. W., Hamilton, T. M., Jost, T. M., Kacher, C. D., Kovacs, A., Kreek, S. A., Lane, M. R., Mohar, M. F., Neu, M. P., Stoyer, N. J., Sylwester, E. R., Lee, D. M., Nurmia, M. J., Seaborg, G. T. , Hoffman, D. C.: Radiochim. Acta **72**, (1996) 169.
133. Türler, A., Eichler, B., Jost, D. T., Piguet, D., Gäggeler, H. W., Gregorich, K. E., Kadkhodayan, B., Kreek, S. A., Lee, D. M., Mohar, M., Sylwester, E., Hoffman, D. C., Hübener, S.: Radiochim. Acta **73**, (1996) 55.
134. Adamson, A.W.: *Physical Chemistry of Surfaces*, Wiley, New York, 1976.
135. Düllman, Ch. E., Eichler, R., Türler, A., Eichler, B. PSI Annual Report 2000, p. 5.
136. Rosen, A., Fricke, B., Morovic, T., Ellis, D.E.:
J. Phys. C-4, Suppl. 4, **40**, (1979) C-4/218.
137. Seth, M., Cooke, F., Schwerdtfeger, P., Heully, J.-L., Pelissier, M.:
J. Chem. Phys. **109**, (1998) 3935.
138. Eichler, B., Rossbach, H. and Eichler, B.: Radiochim. Acta **33**, (1983) 121.
139. Pitzer, K.S.: J. Chem. Phys. **63**, (1975) 1032.
140. Ionova, G. V., Pershina, V., Zuraeva, I. T., Suraeva, N. I.:
Sov. Radiochem. **37**, (1995) 282.
141. Eichler, R., Schädel, M.: J. Phys. Chem., in print.
142. Pershina, V., Bastug, T., Jacob, T., Fricke, B., Varga, S.: Chem. Phys. Lett., in print.
143. Rossbach, H. and Eichler, B.: ZfK –527, Akademie der Wissenschaft der DDR,
ISSN 0138-2950, Juni 1984.
144. Jacob, T., Geschke, D., Fritzsche, S., Sepp, W.-D., Fricke, B., Anton, J., Varga, S.:
Surf. Sci. **486,** (2001) 194.
145. Pershina, V., unpublished
146. Han, Y.-K., Bae, C., Son, S.-K., Lee, Y.S.: J. Chem. Phys. **112**, (2000) 2684; Choi, Y. J., Han, Y.K., Lee, Y.S.: J. Chem. Phys. **115**, (2001) 3448.
147. Han, Y.K., Bae, C., Lee, Y.S.: J. Chem. Phys. **110**, (1999) 8986.
148. Nash, C.: Ph. D. Thesis, The Ohio State University, 1996.
149. Wood, X. and Pyper, N.C.: Chem. Phys. Lett. **84**, (1981) 614.
150. Grant, I.P. and Pyper, N.C.: Nature **265**, (1977) 715.
151. Nash, C.S. and Bursten, B.E.: J. Phys. Chem. A. **103**, (1999) 632.
152. Visscher, L., Dyall., K.G.: J. Chem. Phys. **104**, (1996) 9040.
153. Malli, G.L, In: Relativistic and Electron Correlation Effects in Molecules and Solids, NATO ASI Series, Plenum, New York, Vol. 318, (1994).
154. Pitzer, K.S.: J. Chem. Soc., Chem. Commun. (1975) 760.
155. Han, Y.K. and Lee, Y.S.: J. Phys. Chem. A **103**, (1999) 1104.
156. Seaborg, G. T.: J. Chem. Soc., Dalton Trans. (1996) 3899.
157. Pershina, V. and Fricke, B.: J. Phys. Chem. **98**, (1994) 6468.
158. Ionova, G. V., Pershina, V., Johnson, E., Fricke, B., Schädel, M.:
J. Phys. Chem. **96**, (1992) 11096.
159. Bratsch, S.G.: J. Phys. Chem. Ref. Data **18**, (1989) 1.
160. Bratsch, S. G., Lagowski, J. J.: J. Phys. Chem. 90, (1986) 307.
161. Johnson, E., Fricke, B.: J. Phys. Chem. 95, (1991) 7082.

162. Ahrland, S., Liljenzin, J. O., Rydberg, J. In: Comprehensive Inorganic Chemistry, ed. Bailar, J., Pergamon Press, Oxford, 1973, Vol. 5, pp. 519-542.
163. Baes, Jr., C.F. and Mesmer, R.E. *The Hydrolysis of Cations*, John Wiley, New York, 1976.
164. Pershina, V.: Radiochim. Acta **80**, (1998) 65.
165. Pershina, V.: Radiochim. Acta **80**, (1998) 75.
166. Pershina, V. and Bastug, T.: Radiochim. Acta **84**, (1999) 79.
167. Pershina, V. and Kratz, J.V.: Inorg. Chem. **40**, (2001) 776.
168. Pershina, V., Trubert, D., Le Naour, C., Kratz, J. V. Radiochim. Acta, in print.
169. Kratz, V. and Pershina, V.: "Experimental and Theoretical Study of the Chemistry of the Heaviest Elements". In: Relativistic Effects in Heavy-Element Chemistry and Physics, Ed. Hess, B.A., John Wiley and Sons, Ltd. (2002), in print.
170. Kassiakoff, A. and Harker, D.: J. Am. Chem. Soc. **60**, (1938) 2047.
171. Czerwinski, K. R.:" Studies of Fundamental Properties of Rutherfordium (Element 104) Using Organic Complexing Agents", Doctoral Thesis, LBL Berkeley, 1992.
172. Strub, E., Kratz, J.V., Kronenberg, A., Nähler, A., Thörle, P., Zauner, S., Brüchle, W., Jäger, E., Schädel, M., Schausten. B., Schimpf, E., Zongwei, Li, Kirbach, U., Schumann, D., Jost, D., Türler, A., Asai, M., Nagame, Y., Sakara, M., Tsukada, K., Gäggeler, H.W., Glanz, J.P.: Radiochim. Acta **88**, (2000) 265.
173. Trubert, D., Le Naour, C., Hussonois, M., Brillard, L., Montroy Gutman, F., Le Du, J. F., Constantinescu, O., Barci, V., Weiss, B., Gasparro, J., Ardisson, G.: In: Abstracts of the 1st Intern. Conf. on Chemistry and Physics of the Transactinides, Seeheim, September 26-30, 1999.
174. Haba, H., Tsukada, K., Asai, M., Goto, S., Toyoshima, A., Nishinaka, I., Akiyama, K., Hirata, M., Ichikawa, S., Nagame, Y., Shoji, Y., Shigekawa, M., Koike, T., Iwasaki, M., Shinohara, A., Kaneko, T., Maruyama, T., Ono, S., Kudo, H., Oura, Y., Sueki, K., Nakahara, H., Sakama, M., Yokoyama, A., Kratz, J.V., Schädel, M., Brüchle, W.: Proceedings of the 2nd ASR2001 Conference, Tokai, JAERY, November 2001.
175. Günther, R., Paulus, W., Kratz, J. V., Seibert, A., Thörle, P., Zauner, S., Brüchle, W., Jäger, E., Pershina, V., Schädel, M., Schausten, B., Schumann, D., Eichler, B., Gäggeler, H. W., Jost, D. T., Türler, A.: Radiochim. Acta **80**, (1998) 121.
176. Kratz, J.V., Zimmermann, H.P., Scherer, U.W., Schädel, M., Brüchle, W., Gregorich, K.E., Gannett, C.M., Hall, H.L., Henderson, R.A., Lee, D.M., Leyba, J.D., Nurmia, M., Hoffman, D.C., Gäggeler, H.W., Jost, D., Baltensperger, U., Ya Nai-Qi, Türler, A., Lienert, Ch.: Radiochim. Acta **48**, (1989) 121.
177. Paulus, W., Kratz, J.V., Strub, E., Zauner, S., Brüchle, W., Pershina, V., Schädel, M., Schausten, B., Adams, J.L., Gregorich, K.E., Hoffman, D.C., Lane, M.R., Laue, C., Lee, D.M., McGrath, C.A., Shaughnessy, D.K., Strellis, D.A., Sylwester, E.R.: Radiochim. Acta **84**, (1999) 69.
178. Schädel, M., Brüchle, W., Jäger, E., Schausten, B., Wirth, G., Paulus, W., Günther, R., Eberhardt, K., Kratz, J.V., Seibert, A., Strub, E., Thörle, P., Trautmann, N., Waldek, W., Zauner, S., Schumann, D., Kirbach, U., Kubica, B., Misiak, R., Nagame, Y., Gregorich, K.E.: Radiochim. Acta **83**, (1998) 163.
179. Pershina, V., to be published.
180. Kronenberg, A.: "Entwicklung einer online-Chromatographie für Element 106 (Seaborgium)", Doctoral Thesis, University of Mainz, 2001.
181. Caletka, R. and Krivan, V.: J. Radioanal. Nucl. Chem. **142**, (1990) 239.
182. *Solvent Extraction Reviews*, Ed. Markus, Y., Marcel Dekker, New York, (1971) Vol. 5.
183. Aylett, B. J.: In: Comprehensive Inorganic Chemistry, Ed. Baylar, J.C., Pergamon Press, Oxford, (1973) Vol. 3, p. 187.
184. Abel, E. W.: In: Comprehensive Inorganic Chemistry, Ed. Baylar, J.C., Pergamon Press, Oxford, (1973) Vol. 2, p. 105.

Index

Chapter 3

Fundamental Aspects of Single Atom Chemistry

D. Trubert and C. Le Naour
Institut de Physique Nucléaire, Radiochimie, F-91406 Orsay Cedex, France

1. Introduction

The discovery of a new transactinide or superheavy element (SHE) is based on the identification of the atomic and mass number of the produced isotope. However, the knowledge of Z is not sufficient to predict its chemical behavior: the expected occurrence of relativistic effects (see V. Pershina in Chapter 2) may lead to deviations in the regularity of the periodic table. Thus, the chemical behavior of transactinide elements cannot be extrapolated straightforward from the properties of their lighter chemical homologues. Experiments, to assess the chemical properties and therefore the place of the element in the periodic table, are essential. SHE have short half-lives and low production rates, therefore, chemical experiments have to be conducted with a few atoms or even a single atom. Such extreme conditions represent a huge experimental challenge for radiochemists [1,2,3].

The question is whether the concepts of thermodynamics and kinetics used in classical chemistry are relevant at tracer scale, ultra-traces and for a single atom ?

The tracer scale covers usually element concentrations ranging from 10^{-10} to 10^{-16} M [2]. At such low concentrations, the number of entities present in

M. Schädel (ed.), The Chemistry of Superheavy Elements, 95-116.

1 cm^3 lies between 10^{10} and 10^4. This number is large enough to apply the mass action law without any restriction, obviously as far as the relevant reaction is kinetically allowed. Consequently, the Stirling approximation (Equation 1) used for the calculation of the free enthalpy of the microcomponent in the gaseous phase and therefore in the derivation of the thermodynamic activity in a condensed phase, is perfectly valid.

$$\ln(N!) = N \ln(N) - N \quad (1)$$

Thus, because the mass action law applies, the behavior of the microcomponent is expected to be the same at normal concentration and at the tracer scale. In ultra diluted systems, the thermodynamical behavior of an element does not depend on concentrations. However, some discrepancies can occur in particular cases, for example as a result of the different degrees of consumption of ligands leading to unusual complexation reactions or in unexpected redox processes [4]. Tracer scale chemistry is also characterized by a kinetics hindrance for reactions between two microcomponents in a given system: this trend excludes polymerization reactions or disproportionation for ultra trace chemistry.

The aim of the present chapter is to describe what happens when the concentration of the microcomponent decreases to some tens of atoms and finally to the ultimate limit of dilution: a single atom. Only aqueous chemistry will be evoked, involving true homogeneous solution without pseudo-colloids [2]. In the following considerations, only reactions of a microcomponent with macrocomponent will be treated, as they constituted the general case for SHE chemistry; micro-microcomponent reaction will be evoked in Section 3. Experimental limits of the system (container) are always finite. Since edge effects are difficult to quantify, they will not be taken into account in the fundamental (thermodynamics and kinetics) concepts.

This chapter is divided in four parts. The first one is devoted to thermodynamic considerations, the second deals with kinetic aspects and one-atom-at-a-time chemistry. Then, experimental approaches and finally effects of the media and their influence on aqueous chemistry will be discussed.

2. Thermodynamics at Extreme Dilution and for a Single Atom

When the number of atoms decreases to a few tens of atoms and a fortiori to one atom, the fundamental microscopic and macroscopic concepts of the thermodynamics, valid at tracer level, cannot be applied directly:

i) Macroscopically, the basic notion of phases disappears since the repartition of the entities E_i becomes less homogeneous, when the number of

atoms N_i decreases. The definition of the partition coefficient must also be revised since at the extreme limit of a single atom, this latter can only be in one phase at a time.
ii) From the microscopic point of view, the number of entities N_{Ei} cannot be considered as a continuous variable. Therefore, the Stirling equation can not be used to calculate the factorial factors included in the partition function $Z(T, V, N_{Ei})$ of the entities E_i. Moreover, derivation of the partition function $Z(T, V, N_{Ei})$ with respect to N_{Ei}, can not be performed in order to determine the free enthalpy of the entities E_i. This means that the equilibrium between E_i and the other entities of the system cannot be described as usually by using the chemical potentials of E_i. Therefore, to access thermodynamic values, it is necessary to substitute the concept of a time-independent statistical ensemble by the concept of a time dependant statistical behavior of elementary entities.

Moreover, with only a few atoms the probability of an encounter between two microcomponents is infinitely low. Therefore in the following discussion, reactions involving the formation of polynuclear compounds are excluded; only the reactions where the stoichiometric coefficient of the microcomponent (m) is equal to unity will be examined.
The reaction can be written in the general form:

$$\sum_{i=0}^{i} E_{i(m)} + \sum_{i=0,j=0}^{i,j} \nu_{iMj} E_{iMj} = 0 \qquad (2)$$

Where ν_i are the stoichiometric coefficients (positive for adducts and negative for products) and M_j denotes the macrocomponent. The corresponding mass action law is obviously:

$$\prod a_i^{\nu_i} = K \quad \text{and} \ \Delta G_0 = -RT\, Ln(K) \qquad (3)$$

Where K is the equilibrium constant and a_i is the thermodynamical activity of each entity (macro and microcomponent) involved in the equilibrium. Equation 3 can also be expressed in terms of concentrations (corrected with activity coefficients).

All the following considerations refer to pure thermodynamics (phenomenologic, classic and statistic) and do not take into account kinetic aspects of the equilibrium. These aspects will be described in Section 3. It should be noted that statistical mechanics offer a powerful tool for mathematical treatment of both thermodynamics and kinetics. However, the application of statistical mechanics in kinetics will not be demonstrated in this chapter. In the following, it is assumed that the reactional volume is small enough and that the time devoted to the reaction is larger than the time required for establishment of equilibrium.

Before dealing with the ultimate case of single atom chemistry, one has to establish first, the thermodynamical relations with some tens or hundreds of atoms and then extrapolate these concepts to a single atom. Only the equations necessary for the basic understanding of the demonstration are presented in this paragraph. Details of the mathematical treatment can be found in Appendix 1.

In the case of a small number of entities $N_{E(M)}$, the notion of a mean concentration $C_{E(M)}$ must be introduced. According to statistical thermodynamics, in a system, S(T,V,N), containing N entities included in only one type of molecule, of phase volume V and temperature T, the thermodynamical values of the system S are related to the mean values of the number of entities $\overline{N_{E(M)}}$ [2,5]:

$$\overline{C_{E(M)}} = \frac{\overline{N_{E(M)}}}{V} \text{ with } \overline{N_{E(M)}} = \sum P(N)\, N_{E(M)} \quad (4)$$

Where P(N) is the canonical probability of realization of the states of the closed system which constitutes this phase. The probability P(N) is defined from all the partition functions of the included entities:

$$P\ (N) = (Z_0/Z)\, Z(T,V,N) \quad (5)$$

Where Z_0 is the partition function of all species $E_{i(mj)}$ and $E_{i(Mj)}$ whose numbers are far above hundreds and the Z(T,V,N) is partition function for species whose number is less than 100. Z is the partition function of all species, and thus acts as a normalization factor.
In order to compare the expression of the mass action law for microcomponent and macrocomponent (in the former case activities can be replaced by concentrations), let ρ be the ratio of these two quantities:

$$\prod c_i^{v_i} = \overline{K} \text{ and } \prod a_i^{v_i} = K \text{ and } \rho = \frac{\overline{K}}{K} \quad (6)$$

This expression of P(N) (Equation 5) allows to calculate for every case the values of $\overline{N_{E(M)}}$ and ρ from the known values of $\overline{N_{E(M)}}$ and K.

All these considerations assume that the element is stable or has a half-life large enough to allow the occupation of all the quantum states corresponding to the different degrees of freedom of each entity (m). This condition is fulfilled since generally the residing times are very short. It is also assumed that a dynamic equilibrium is achieved, i.e. the entities Ei(mj) have time to occupy the quantum states of the system S.

From the general equations of P and Z, we will focus on the case of a single microcomponent involved in a reaction of stoichiometry 1-1, which is the only case encountered in transactinide chemistry. The Equation 2 reduces to:

$$E_1 + E_2 = 0 \tag{7}$$

Let i be the extent of the reaction, N_1^0, N_2^0 the initial numbers of E_1 and E_2 in presence when i = 0. When the extent of the reaction is i $N_{E2} = N_2^0 + i$, E_2 species considered as the product of the reaction and $N_{E1} = N_1^0 - i$, E_1 species that coexist together with other species. The canonical probability for realizing of the micro-states of the system, according to the general expression P(N), turns to P(i); see Appendix 1.

$$P(i) = \frac{Z_0}{Z}\left(\frac{Z_1 V}{\Gamma_1}\right)^{N_1^0 - i}\left(\frac{Z_2 V}{\Gamma_2}\right)^{N_2^0 + i}\frac{1}{(N_1^0 - i)!\,(N_2^0 + i)!} \tag{8}$$

Where $\Gamma_{E(m)}$ is a coefficient characteristic of E(m) for the system S which depends only on the macrocomponents ($\Gamma_{E(m)}$ is equal to unity for an ideal gaseous phase) [5]:

The parameter χ , characteristic of the system, can be defined as :

$$\chi = \left(\frac{Z_2}{Z_1}\frac{\Gamma_1}{\Gamma_2}\right) \tag{9}$$

After rearrangement and normalization (Appendix 1), Equation 9 becomes:

$$\frac{\overline{N_2}}{\overline{N_1}} = \chi \tag{10}$$

When N_2 and N_1 increase, the situation of the classical equilibrium is reached for which the most probable canonical state is that of the highest P(i). The successive values of P(i) and P(i)-1 come closer and closer and the limiting value of the ratio given by Equation 10 is unity.

$$\frac{P(i)}{P(i)-1} = \chi\frac{(N_1^0 - i)+1}{i} \tag{11}$$

Then:

$$\chi = \frac{i}{N-i} = \frac{N_2}{N_1} \tag{12}$$

It becomes:

$$\rho = \frac{\overline{N_2}/\overline{N_1}}{N_2/N_1} = \frac{\chi}{\chi} = 1 \tag{13}$$

<u>Thus, in the case of a reaction of stoichiometry 1-1, and only in this case, the mass action law holds, regardless of the number of present species, if the mean concentrations are used</u>. Down to the case of a single atom that can only appear <u>in one species at a time</u>, the probability becomes:

$$P_0 = \frac{1}{1+\chi} \tag{14}$$

While the probability of the single atom into another species is

$$P_1 = \frac{\chi}{1+\chi} \tag{15}$$

Based on the above, for a single atom and in the case of a reaction of stoichiometry 1-1, the mass action law can be used with the mean concentrations proportional to P_0 and P_1.

The case of a microcomponent with a reaction of stoichiometry ν_1–ν_2 (different from 1-1) is much more complicated. The ratio ρ is a function of N_1^0 and K, which means that deviation from the law of mass action can occur. Thus, there is no general solution to describe the variations of ρ with N_1^0 and each particular case needs to seek its own solution.

3. Kinetic Aspects

The kinetics at extreme dilution between two microcomponents is strongly hindered. It will be discussed whether the notion of half-time τ associated with such a reaction is valid (Equation 16) in ultra trace chemistry. Assuming a general reaction of two entities containing a microcomponent:

$$E_1 + E_2 \leftarrow\!\!\!-\!\!\!-\!\!\!\rightarrow E_3 + E_4 \qquad k_1\ k_2 \tag{16}$$

k_1 and k_2 being the forward and the backward kinetic constant of the equilibrium, respectively. For sake of simplification, stoichiometric coefficients will be taken as unity. As a first approximation, a macroscopic model of activated complex is based on the free energy of activation of the system and especially on the collision frequency between reactants [6,7].

The half-time τ of the reaction is determined by the concentrations of the species E_1 and E_2 and the rate constant k_1 and k_2. The general case with $k_1 \neq k_2$ and $[E_1] \neq [E_2]$ leads to a rather complicated expression of τ. A hypothesis must be introduced to simplify the equation: at time zero the concentrations of E_3 and E_4 are nil, the concentration of $[E_1]^0=[E_2]^0=[E]^0$ and $k_1 >> k_2$. With these assumptions, the half-time of the reaction is simply [2,4]:

$$\tau = \frac{1}{k_1[E]^0} \tag{17}$$

It can be seen immediately that a decrease of the concentration of the species E_1 and E_2 results in an increase of τ in the same proportions. R. Guillaumont and J.P. Adloff [4] have shown, for 100 atoms of plutonium, assuming a very fast reaction rate ($k = 10^{11}$ M s^{-1}), the half-time τ is equal to 50 μs and 1.6 y in a volume of 1 μl and 1 cm^3 respectively. Obviously, a decrease in the rate constant would yield much higher reaction times (exponential). Diffusion phenomena have negligible effects since for a typical coefficient value of 10^{-5} cm^2 s^{-1}, 10^9 diffusive displacements occur per second, which is far greater than the exchange frequency required to establish an equilibrium.

Summarizing, it can be seen that the reaction rate of a given number of atoms of a micro component decreases drastically with the spatial expansion of the system in which they are contained (i.e. the number of atoms per volume unit). Therefore, at critical concentrations, micro-microcomponent kinetics prevails largely over thermodynamics in a reaction.

The aspects of kinetics of a reaction between a microcomponent with a macrocomponent, are much simpler. As there is no consumption of the macrocomponent, kinetics appears of the "pseudo" first order with respect to the microcomponent [2,8]. The rate of the reaction of species E containing the microcomponent with macrocomponent M_1, M_2,... takes the form:

$$\frac{d[E]}{dt} = -k\,[E]\,[M_1]^a\,[M_2]^b\,.... \tag{18}$$

Since all terms on the right side of Equation 18, including macrocomponent (M_1, M_2,...) concentrations, remain constant under the given conditions, Equation 18 becomes:

$$\frac{d[E]}{dt} = -k'\,[E] \tag{19}$$

Where k' becomes a "conditional" rate constant of the "pseudo" first order [8]. Hence, the only kinetics data that can be reached at extreme dilution are those related to reactions between the microcomponent and a macrocomponent.

R.J. Borg and G.J. Dienes [9] have attempted to show the validity of single atom chemistry from a kinetics point of view. Their approach was based on activation energy (activated complex), combined with the frequency of exchange between entities. This approach has led to the conclusion that the equilibrium cannot be reached if the activation energy is greater than about 85 kJ. However, the concepts developed in Borg's work are only valid for large number of atoms. Energies of activated complexes can only be used for sufficiently high quantities since this concept involves energies estimated from statistical values. Applying a Maxwell-Boltzmann distribution requires that an individual atom or species can acquire energies levels quite different from its most probable or mean values [7]. This fundamental concept forms the basis of the fluctuation theory. However, classical or formal thermodynamics and kinetics never take fluctuations into account. Without presenting the mathematical treatment [7], the relative mean deflection of a property of a system (ζ) can be estimated according to [6,7]:

$$\zeta \cong \frac{1}{\sqrt{N}} \tag{20}$$

N is the number of atoms or species. Therefore for a single atom, the fluctuation has the same order of magnitude as the property itself. The interested reader can find a brief development of this concept in Appendix 2.

For the development of concepts of kinetics and of thermodynamics, considered in this chapter, the single atom was assumed to be stable. In fact, the probability for a unique atom to exist in specific "specie" can only be determined in a chemical experiment if the atom is radioactive. Therefore, if the atom is stable or has a large half-life, any radiochemical techniques can be used for single atom chemistry.

The concept of "one-atom-at-a-time chemistry" is different from that of a single atom chemistry. The main difference is the time. An "observable" chemical reaction of a given radio nuclide can be defined with the following temporal quantities:
$T_{1/2}$: half-life of the radionuclide
t: time necessary for reactants to compete
τ: half-time of equilibrium or steady state chemical reaction
Δt: time for collecting data
A reaction will be observable if: $T_{1/2} > t > \tau > \Delta t$
At the experimental scale, τ and t are macroscopically related.
Problems associated with the disintegration of a single radioactive atom will not be developed here However, an understanding of the decay appears rather important in one-atom-at-a-time chemistry [2,10].

4. Conclusive Remarks on Fundamental Aspects of Single Atom Chemistry

Since the SHE chemistry is correlated with one-atom-at-a-time chemistry, one may ask if it is meaningful to carry out experiments with a single atom. From a theoretical point of view, as it was demonstrated above, for a 1:1 stoichiometry reaction the mass action law and the kinetics laws are valid. However, as there is no macrocomponent consumption, such reaction appears as of pseudo first order. Note that reactions with a 1:1 stoichiometry include all reactions between the microcomponent and a single macrocomponent. This concept can also be extended to stepwise reactions such as successive formation of metal complexes (hydrolysis, halide complexation); for example:

$$M(H_2O)_x^{n+} + OH^- \leftrightarrows MOH(H_2O)_{x-1}^{(n-1)+}$$
$$\bullet\bullet\bullet\bullet\bullet \qquad (21)$$
$$MOH(H_2O)_{x-y}^{(n-y)+} + OH^- \leftrightarrows MOH(H_2O)_{x-y-1}^{(n-y-1)+}$$

For higher stoichiometry, these concepts are not longer valid, both for thermodynamics and for kinetics, and each case becomes a particular case seeking its own solution.

All the previous theoretical considerations have been established assuming an "ideal system" without any boundary conditions. It should be pointed out however that in practice, all the studied systems, especially in SHE chemistry, have finite dimensions (time and volume). As only ideal system were considered, edge effects, pseudo-colloid formation, sorption phenomena, redox processes with impurities or surfaces, medium effects have not been taken into account. All these effects, representing the most important part from the deviation to ideality, cannot be predicted with formal thermodynamics and/or kinetics. Thus, radiochemists who intend to perform experiments at the scale of one atom must be aware that the presence of any solid phase (walls of capillary tubes, vessels, etc.) can perturb the experimental system. It is important to check that these edge effects are negligible at tracer level before performing experiments at the scale of the atom [11]. The following section describes experimental techniques used in SHE chemistry.

5. Experimental Approaches

Each study of the chemical property of a transactinide element requires the development of radiochemical technique that allows collecting data relevant to the scale of one atom. Experimental approaches are therefore severely limited.

In principle, chemical information on a system can only be obtained with methods that do not alter the species present in solution. However, in order to get this information, an external perturbation must be applied to the system and its response must be analyzed. In the case of radioactive tracer, where the radioactivity measurement is the only way to detect the element (but it does not allow the identification of the form of the species), two types of external perturbation can be applied: (i) by contacting the system with a second phase and subsequently observing the distribution of the radio-nuclides between the two phases (static or dynamic partition) or (ii) by applying an electrical potential or a chemical gradient (transport methods). So far, transport methods have not yet been used in one-atom-at-a-time chemistry since they are not only difficult to carry out but also extremely time consuming.

The aim of a partition experiment, regardless of the concentration involved, is the determination of the partition coefficient (D) of an element between two phases. For this purpose, measurements of average concentrations of the entities present in each phase have to be performed. Although partition methods are widely used in the field of radiochemistry, chemical properties can only be determined through the variations of D with characteristic parameters of the system (ionic strength, ligand concentrations, acidity, temperature, etc). In the case of a few atoms, it has been shown that the law of mass action involves probabilities for the atom to form a given species in a given phase. Obviously for a single atom, the distribution coefficient is defined in terms of a probability for the element to be in one phase or in the other. This means that measurements must be conducted with the whole phase since the system contains only one entity.

Prior to experiments with SHE, systematic studies with homologues at the tracer scale have to be carried out to select the experimental conditions (solvent extraction, ion exchanger, aqueous media, ...). On line experiments with short lived isotopes of homologues are also necessary to improve the setup, e.g. to evaluate the eventual impact of edge effects (sorption, ...). Furthermore, on line experiments involving SHE may preferably be performed with homologues since it is the only way to ensure strictly identical experimental conditions for all present elements.

5.1 STATIC PARTITION

In a static partition, the atom (necessarily radioactive in the present context) is distributed between two immiscible phases (liquid and/or solid). Since this procedure is sequential, the accuracy in the measurement of the average concentrations increases with the number of trials. For a given system, under given conditions, the determination of one partition coefficient (D) requires numerous repetitive experiments, even for the more simple case involving only one chemical species in each phase. The experimental conditions must always ensure that at the end of the experiment, the atom has reached permanent partition equilibrium between the two phases. Moreover, the short half-life of the nucleus does not bring any perturbation since there is only one alternative: either the measurement indicates in which phase the atom is or the atom has disintegrated before the measurement and no information is obtained.

Among the static partition techniques, solvent extraction was the most used for heavy elements studies. For instance R.J. Silva et al. [12] using a mixture of TTA / MIBK have demonstrated the 3+ and 2+ oxidation state for Lr and No respectively. In a single experiment, a sample of about 10 atoms was available, of which only a tenth remained for detection at the end of the experiment. A statistically reliable result was obtained by performing about 200 identical partition experiments. Conversely, results of K.R. Czerwinski et al. [13] regarding extraction of Rf and homologues in HCl media with TBP, have to be considered with caution since the experimental conditions have not been maintained identical i.e.: i) Hf was not included in the batch experiments, ii) Hf was studied on line with experimental procedures different from the batch ones (volume, contact times, stirring mode), iii) in the Rf study only the organic phase was analyzed, which is in contradiction with the basic principle mentioned above. The variety of experimental conditions does not ensure the absence of edge effect and as the aqueous phase was not measured in this Rf experiments, it did not allow determining a reliable D value. The low D value of Hf obtained in Czerwinski's work was not confirmed by C. D. Kacher et al. [14]. The previously observed low extraction yield was correlated with sorption of Hf.

If solvent extraction and ion exchange batch experiments are commonly considered as multi-step procedures, allowing therefore the collection of information on the chemical properties of a microcomponent, the same conclusion concerning "coprecipitation reactions" cannot be drawn systematically. The term "coprecipitation" has no precise meaning since it involves many phenomena as syncrystallization (homogeneous incorporation of a microcomponent in the lattice of an isomorphous macro-compound), adsorption (chemisorption, electrostatic adsorption) and surface precipitation [2, 15]. When dealing with a few atoms, syncrystallization is strongly hindered. In case of sorption, even if a multi-step mechanism (frequent and rapid sorption – desorption steps) is involved, this technique depends strongly on to many experimental parameters to allow reproducible results. These factors are related to the properties of the solid (surface charge, grain size) and of the electrolyte solution (ion nature, proton concentration,...): especially, the isoelectric point of the solid [16], that determines the sorption properties, is strongly correlated with the pH. In view of the numerous experimental conditions to reproduce and to parameterize, a distribution coefficient can hardly be formulated [2]. F.J. Reischmann et al. [17,18] have studied the influence of $^{210, 218}$Po concentration (10^8 to 40 atoms) on "coprecipitation " with tellurium and arsenic sulphide. Although no evidence for a concentration-dependent behavior of Po was observed, no chemical information can be derived from these experiments.

However, at tracer level, coprecipitation – even if the mechanism remains unknown -- is a convenient method for the separation or concentration of different species. The coprecipitation of carrier free Tc and Re with

Ph_4AsReO_4 in HNO_3/HF media was used successfully as an additional separation step of these group 7 elements from Nb and Ta [19]. At the same time, the precipitate was used to prepare an alpha source.

5.2 DYNAMIC PARTITION

A dynamic partition can be considered as a succession of static partitions. In other words such partition is characterized by a large number of successive equilibria, governed as a first approximation, by the same constant K. In this type of experiments, the displacement of the species, controlled by K, is directly related to the probability of the presence of the atom in the mobile phase. In a single experiment, a significant result can be obtained since it is in its own purely statistical. Techniques - in which dynamic partitions are performed - such as extraction (reverse phase) chromatography, elution chromatography and thermochromatography (out of the scope of this chapter – see Chapter 7), are particularly suitable for one-atom-at-a-time chemistry. In solution chromatography, the atom can be observed in a well-defined elution fraction, implying that the position of the elution band can be determined with a single atom.

The first significant experiments on the behavior of SHE in aqueous solutions were performed in order to collect comparative data between SHE species and actinides or/and homologues. For instance, using extraction chromatography, E.K. Hulet et al., were able to demonstrate, based on the detection of 6 events, the similarity of chloro-complexes of Hf and Rf, and their difference from the trivalent actinides Cm and Fm [20]. With elution chromatography, using the HPLC setup of ARCA II, M. Schädel et al. [21] unambiguously showed the stability of the 5+ oxidation state of dubnium similar to Nb, Ta and Pa, and unlike group-4 elements and trivalent actinides. These two illustrative works concern not only the SHE at the atom scale, but also the behavior of homologues in batch and on line experiments. From their results, no concentration-dependant behavior was observed, validating therefore the method used and the conclusions drawn on the chemical behavior of SHE. To go further in this topic, extensive research was conducted using the same chemical techniques in order to get quantitative data. Distribution coefficients (K_d) of Rf [22, 23] in HF/HNO3 media and HF/HCl [24] were determined from ion exchange experiments. The validity of these experiments has been checked previously with systematic studies on homologues off line and simultaneously on line with Rf, ensuring identical experimental conditions.

The feasibility of K_d determination in the context of one-atom-at-a-time chemistry is very promising since the collection of K_d values will allow to establish reliable variations of the chemical properties (complexation, hydrolysis) of elements within a group, to compare with theoretical predictions, and to determine perhaps thermodynamic constants. Moreover, other information could be derived from chromatography experiments. The mathematical treatment of elution curves can be carried out with various models, especially Glückauf's, which offers the advantages of using simple equations and takes into account the possible dissymmetry of elution bands [25,26]. The parameters included in Glückauf's equations allow the determination of the distribution coefficient (K_d), and possibly of the diffusion coefficient of the species in the mobile phase through the knowledge of the effective height of theoretical plate (EHTP) [27]. This latter parameter is characteristic of the efficiency of the column. A systematic study of the width of the elution band as a function of elution rate is required to determine the variations of EHTP. Unfortunately, these types of experiments are very cumbersome and have never been carried out with heavy elements. However, they could bring information about the diffusion constant and the related Stoke's radius of the species in solution [26,27].

6. Medium Effects

The media used in aqueous chemistry have a huge influence on the reaction thermodynamics and must be taken into account. These effects were often ignored in a large number of papers in the past. The purpose of this paragraph is to remind the reader that the effects of media govern the reaction constants not only in pure aqueous solutions but also in partition experiments [28,29,30]. Temperature effects will not be treated here, but they are of significant importance [29].

The effects of the media are imposed by macrocomponents and by supporting electrolytes if present. All thermodynamical laws are valid and must be applied without restrictions. The microcomponent does not act in the calculations of the ionic strength (I_m). Coulomb forces between charged ions mainly impose deviation of experimental constants from zero ionic strength under well-defined conditions. Such deviations have been described in the literature by numerous models, which will not be reiterated here [29,31]. For high concentrations, neutral species and multiple interactions between species must also be taken into account.

Since it is independent of the nature of the supporting electrolyte, an extrapolation to zero ionic strength remains the only universal thermodynamical value for a given equilibrium. However, such extrapolation requires the knowledge of numerous parameters, sometimes difficult to determine and/or to estimate. Recently a model was proposed by V. Neck et al. [32] to estimate successive hydrolysis constants using the inter-ligand electrostatic repulsion term. This model gave accurate results for actinides (Th, U, Np and Pu) and can be extrapolated to other complexation reactions as long as data are available on chemical homologues.

In case of low charged species, and approximately below 3 mol kg^{-1} the Specific Ion interaction Theory (SIT) [29] can be applied for the calculations of activity coefficients. Data available on interaction coefficients are scarce. But, paradoxically for actinide ions such data are relatively well known. However, in certain cases, they can be estimated from the model developed by L. Ciaviatta [33,34].

The knowledge of both thermodynamic constants at zero ionic strength and of the specific interaction coefficients will allow the speciation diagram of the element in the considered medium to be established. At higher electrolyte concentrations, more sophisticated theories (Pitzer, MSA, ... [29,31]) have been developed. However, they involve a larger number of characteristic parameters, which unfortunately are unknown for the majority of chemical elements.

Except for some recent work, most data on SHE chemistry were determined in concentrated media with sometimes undefined ionic strength. A comparison of hydrolysis or complexation constants of homologues, available in the literature at various ionic strengths, must be considered with caution. Most of these constants were determined with weighable amount of elements. Even in one-atom-at-a-time experiments involving the same aqueous media, a direct comparison of results issued from different chemical procedures is delicate. In their work, R. Günther et al. [35] have re-investigated the behavior of Rf and homologues in the system HCl/TBP using solvent extraction and extraction chromatography. In batch experiments they obtained similar results as those of C.D. Kacher et al. [14]. Their on line experiments with Rf, Zr and Hf are reliable since no edge effects have been proved to occur. Part of their discussion is devoted to a comparison with the work of K.R. Czerwinski et al. [13]. However, even if the methodology used by the latter would be adequate (which is not the case – see above), data collected in liquid-liquid solvent extraction and extraction chromatography cannot be directly compared, essentially because of medium effects. A system involving liquid-liquid partition can be perfectly defined in thermodynamical terms. From Sergievskii's model indeed, the activity coefficient of a species in the organic phase depends only on the water activity a_{H2O} of the bi-phasic system [28]. Therefore, if the water activity of the system is kept constant, the influence of the aqueous phase on the activity coefficients in the organic phase is also constant. Nevertheless, maintaining the water activity constant (or the ionic strength) does not neutralize all causes of deviations from an ideal solution. In extraction chromatography (or reversed phase chromatography) and in ion exchange chromatography, the water activity cannot rigorously be defined and maintained constant [28]. However, at low ionic strength, the variations of the water activity in both phases can be neglected. This leads to meaningful thermodynamic data such as the recently determined K_d value of Rf [22,23,24].

7. Conclusion

In the case of a single atom, the mass action law can be applied with mean concentrations proportional to the probabilities of finding an atom in one species or another. However, this derivation of the mass action law can only be achieved for reactions with a 1-1 stoichiometry. Reactions between a microcomponent and macrocomponent are always allowed kinetically, the rate of the reaction being of the "pseudo" first order. But, at extreme dilution, the kinetics of reactions involving two microcomponents is strongly hindered.

Experimentally, limitations are mainly imposed by the one-atom-at-a-time concept since the time devoted to the collection of data may be important. The collection of experimental data must also include effects of the media and the temperature (if used). Prior to experiments to be carried out at the level of a single atom, the absence of edge effects must be checked carefully at the tracer scale.

Appendix 1

The mean number of entity $\overline{N_{E(M)}}$ and the mean partition function $\overline{Z}$ of a system S is given by:

$$\overline{Z} = \frac{Z(T,V,1)^{\overline{N_{E(M)}}}}{\overline{N_{E(M)}}^{\overline{N_{E(M)}}}} \exp\left(\overline{N_{E(M)}}\right) \tag{22}$$

Z(T,V,1) being the molecular partition function. Obviously, mean values are not only integers. For large values of $\overline{N_{E(M)}}$, Equation 22 becomes:

$$\frac{\exp\left(N_{E(M)}\right)}{N_{E(M)}^{N_{E(M)}}} = \frac{1}{N_{E(M)}\,!} \tag{23}$$

Equation 23 is a form of the Stirling equation, only valid if $N_{E(M)}$ has an integer value. In this case, which corresponds to usual concentrations, the partition function of S becomes that of a canonical ensemble without mean values:

$$Z = \frac{Z(T,V,1)^{N_{E(M)}}}{N_{E(M)}!} \tag{24}$$

Back to the mean values of the number of entities, when $\overline{N_{E(M)}}$ has a small value, only the mean values of the concentrations are relevant to describe the system, then:

$$\overline{C_{E(M)}} = \frac{\overline{N_{E(M)}}}{V} \text{ with } \overline{N_{E(M)}} = \sum P(N)\, N_{E(M)} \tag{25}$$

Where P(N) is the canonical probability of realization of the states of the closed system which constitutes this phase, P(N) is defined from all the partition functions of the included entities.

Comparing the expression of the mass action law for microcomponent and macrocomponent (in the former case activities can be replaced by concentrations), the ratio of these two quantities is defined as ρ:

$$\prod c_i^{v_i} = K \text{ and } \prod a_i^{v_i} = K \text{ and } \rho = \frac{\overline{K}}{K} \tag{26}$$

To establish the expressions of K and $\overline{K}$, one must know the expression of the canonical probability P(N) in order to determine the mean values $\overline{N_{E(M)}}$.

In a well defined closed system S(T,V,N), the presence of some species E(m) of a microcomponent in a defined phase does not perturb the system. Then, the properties of the system are only fixed by the amount of the macrocomponent and then the energy levels of a single entity E(m) are given by:

$$\varepsilon_n = \varepsilon_n^* + \varepsilon_n^\infty \tag{27}$$

Where ε_n^* and ε_n^∞ are the energy levels in an ideal system and an additional term only dependent of the whole properties of the phase, respectively. The molecular partition function becomes:

$$Z(T,V,1) = \sum_{n=0}^{\infty} \exp\left(\frac{-\varepsilon_n}{k_b T}\right) = \sum_{n=0}^{\infty} \exp\left(\frac{-\varepsilon_n^*}{k_b T}\right) + \sum_{n=0}^{\infty} \exp\left(\frac{-\varepsilon_n^\infty}{k_b T}\right) \tag{28}$$

Where k_b is the Boltzmann constant.

Introducing the partition function, $Z_{E(m)}(T,V,1)$, of a single species E(m) in an ideal system and $\Gamma_{E(m)}$ a coefficient characteristic of E(m) for the system S which depends only on the macrocomponents ($\Gamma_{E(m)}$ is equal to unity for an ideal gaseous phase) [5]:

$$Z_{E(m)}(T,V,1) = \sum_{n=0}^{\infty} \exp\left(\frac{-\varepsilon_n^*}{k_b T}\right) \text{ and } \Gamma_{E(m)} = \sum_{n=0}^{\infty} \exp\left(\frac{-\varepsilon_n^\infty}{k_b T}\right) \tag{29}$$

And

$$Z(T,V,1) = \frac{Z_{E(m)}(T)\ V}{\Gamma_{E(m)}} \quad (30)$$

Therefore, the partition function for the $N_{E(m)}$ species E(m) is:

$$Z(T,V,N_{E(m)}) = \left[\frac{Z_{E(m)}(T, N_{E(m)})\ V}{\Gamma_{E(m)}}\right]^{N_{E(m)}} \frac{1}{N_{E(m)}!} \quad (31)$$

For several i species for j elements:

$$Z(T,V,N) = \prod_{i,j} \left[\frac{Z_{E(i)}(m_j)(T)\ V}{\Gamma_{E(i)}(m_j)}\right]^{N_{E(i)}(m_j)} \frac{1}{N_{E(i)}(m_j)!} \quad (32)$$

This expression of the law of the realization of the microscopic states can be separated in two parts: the partition function Z_0 of all species $E_{i(mj)}$ and $E_{i(Mj)}$ of which numbers are far above hundreds and the partition function Z(T,V,N) for species whose number is less than 100. The probability P(N) is then given by:

$$P\ (N) = (Z_0/Z)\ Z(T,V,N) \quad (33)$$

Z is the partition function of all the species, and acts as a normalization factor.

For the sake of simplification of Equation 8, let pose $N_2^0 = 0$ for i = 0, Equation 8 becomes:

$$P(i) = \frac{Z_0}{Z}\left(\frac{Z_1 V}{\Gamma_1}\right)^{N_1^0 - i}\left(\frac{Z_2 V}{\Gamma_2}\right)^{i} \frac{1}{(N_1^0 - i)!i!} \quad (34)$$

and therefore:

$$P(i) = \frac{Z_0}{Z}\left(\frac{Z_1 V}{\Gamma_1}\right)^{N_1^0}\left(\frac{Z_2}{Z_1}\frac{\Gamma_1}{\Gamma_2}\right)^{i} \frac{1}{(N_1^0 - i)!i!} \quad (35)$$

Setting:

$$K^* = \frac{Z_0}{Z}\left(\frac{Z_1 V}{\Gamma_1}\right)^{N_1^0}, \ \chi = \left(\frac{Z_2}{Z_1}\frac{\Gamma_1}{\Gamma_2}\right) \text{ and } \theta = \frac{\chi^i}{(N_1^0 - i)!i!} \quad (36)$$

χ being a parameter characteristic of the system. Equation 35 turns to:

$$P(i) = K^*\ \theta_i \quad (37)$$

The K* value is obtained by normalizing P(i) from $\sum_{i=0}^{N_1^0} P(i) = 1$, then it follows:

$$K^* = \left(\sum_{i=0}^{N_1^0} \theta_i \right)^{-1} = \frac{1}{S} \qquad (38)$$

The average values of $\overline{N_1}$ and $\overline{N_2}$ are given by:

$$S\left(T, V, \overline{N_1}\right) = \sum_{i=0}^{N_1^0} (N_1^0 - i)\theta_i \quad \text{and} \quad S\left(T, V, \overline{N_2}\right) = \sum_{i=0}^{N_1^0} i\, \theta_i \qquad (39)$$

Appendix 2

The transition state model considers a chemical reaction as an equilibrium between the ground state of the reactants and their state on the top of the potential barrier of reaction (activated complex) [6,7].

$$A + B \leftrightarrows AB^* \leftrightarrows C \qquad (40)$$

The final transformation of the activated complex AB* in product (C) is governed by oscillations of low frequencies ω [6].

Based on the molecular collision model, that describes successfully experimental data for a large number of bimolecular reactions, the rate of the reaction can be calculated as the number of collisions of molecules having energy higher than the required value E* [7]:

$$d[A]/dt = [A][B] \left((r_A + r_B)^2 \left(8\pi k_b T / \mu_{AB}\right)^{1/2} \right) e^{-\frac{E^*}{RT}} \qquad (41)$$

Where μ_{AB} is the reduced mass and r_A, r_B the radii of the molecules A and B, respectively. Therefore, as it can be seen in Equation 41, the reaction rate is strongly dependent of the height on the potential barrier E*, that can be estimated according to:

$$E^* \cong \mathbf{N}\, \hbar\, S_N \omega \qquad (42)$$

S_N represents the number of closely spaced energy levels and **N** the Avogadro number. Within this model, a chemical reaction can be described as a stochastic process of molecules crossing a potential barrier through many quantum excitations [7]. The interested reader can find details of these theories and further considerations in references [6,7].

Moreover, fluctuations represent one of the most fundamental concepts of statistical mechanics and/or thermodynamics (not included in traditional thermodynamics). When the atom number is large enough (at least some hundred of atoms), the properties of the systems can be calculated in the most probable state. As quoted in Section 3 of the present chapter, the relative mean deflection of a property of a system (ζ) can be estimated according to [6,7]:

$$\zeta \cong \frac{1}{\sqrt{N}} \tag{43}$$

Where N is the number of atoms or species. Thus for a single atom, the fluctuations have the same order of magnitude than the property itself and formal statistical mechanics equation cannot be applied.

References

1. Hoffman, D.C.: Radiochim. Acta **61**, (1993) 123.
2. Adloff, J.P., Guillaumont, R., *Fundamentals of Radiochemistry,* CRC Press, Boca Raton, 1993.
3. Kratz, J.V.: "Chemical properties of the Transactinide Elements". In: *Heavy Elements and Related New Phenomena,* Eds. Greiner, W., Gupta, R.K., World Scientific, Singapore, (1999) 129-193.
4. Guillaumont, R., Adloff, J.P.: Radiochim. Acta **58/59**, (1992) 53.
5. Guillaumont, R., Adloff, J.P., Peneloux, A., Delamoye, P.: Radiochim. Acta **54**, (1991) 1.
6. Liboff, R.L., *Introduction to the theory of kinetic equations,* John Wiley & sons, New York, 1969.
7. Koudriavtsev, A.B., Jameson, R.F. and Linert, W., *The law of Mass Action,* Springer-Verlag, Berlin, 2001.
8. Guillaumont, R., Adloff, J.P., Peneloux, A.: Radiochim. Acta **46**, (1989) 169.
9. Borg, R.J., Dienes, G.J.: J. Inorg. Nucl. Chem. **43**, (1981) 1129.
10. Schmidt, K.H., Sahm, C.C., Pielenz, K., Clerc, H.G.: Z. Phys. A **316**, (1984) 19.
11. Guillaumont, R., Bouissieres, G.: Bull. Soc. Chim. France **12**, (1972) 4555.
12. Silva, R., Sikkeland, T., Nurmia, M., Ghiorso A.: Inorg. Nucl. Chem. Letters **6**, (1970) 733.
13. Czerwinski, K.R., Kacher, C.D., Gregorich, K.E., Hamilton, T.M., Hannink, N.J., Kadkhodayan, B.A., Kreek, S.A., Lee, D.M., Nurmia, M.J., Türler, A., Seaborg, G.T., Hoffman, D.C.: Radiochim. Acta **64**, (1994) 29.
14. Kacher, C.D., Gregorich, K.E., Lee, D.M., Watanabe, Y., Kadkhodayan, B.A., Wierczinski, B., Lane, M.R., Sylwester, E.R., Keeney, D.A., Hendricks, M., Stoyer, N.J., Yang, J., Hsu, M., Hoffman, D.C., Bilewicz, A.: Radiochim. Acta **75**, (1996) 127.
15. Benes, P. and Majer, V., *Trace chemistry of aqueous solutions,* Elsevier, Amsterdam, 1980.
16. Stumm, W.: *Chemistry of the Solid-Water Interface,* Wiley Interscience Publication, New York, 1992.
17. Reischmann, F.J., Trautmann, N., Herrmann, G.: Radiochim. Acta **36**, (1984) 139.
18. Reischmann, F. J., Rumler, B., Trautmann, N. Herrmann, G.: Radiochim. Acta **39**, (1986) 185.
19. Schumann, D., Novgorodov, A.F., Misiak, R., Wunderlich, G.: Radiochim. Acta **87**, (1999) 7.
20. Hulet, E.K., Lougheed, R.W., Wild, J.F., Nitschke, J.M., Ghiorso, A.: J. Inorg. Nucl. Chem. **42**, (1980) 79.
21. Schädel, M., Brüchle, W., Schimpf, E., Zimmermann, H.P., Gober, M.K., Kratz, J. V., Trautmann, N., Gäggeler, H., Jost, D., Kovacs, J., Scherer, U.W., Weber, A., Gregorich, K.E., Türler, A., Czerwinski, K.R., Hannink, N.J., Kadkhodayan, B.A., Lee, D.M., Nurmia, M.J., Hoffman, D.C.: Radiochim. Acta **57**, (1992) 85.
22. Strub, E., Kratz, J.V., Kronenberg, A., Nähler, A., Thörle, P., Zauner, S., Brüchle, W., Jäger, E., Schädel, M., Schausten, B., Schimpf, E., Li Zongwei, Kirbach, U., Schumann, D., Jost, D.T., Türler, A., Assai, M., Nagame, Y., Sakama, M., Tsukada, K., Gäggeler, H.W., Glatz, J.P.: Radiochim. Acta **88**, (2000) 265
23. Pfrepper, G., Pfrepper, R., Krauss, D., Yakushev, A.B., Timokhin, S.N., Zvara I.: Radiochim. Acta **80**, (1998) 7
24. Trubert, D., Le Naour, C., Hussonnois, M., Brillard, L., Monroy Guzman, F., Le Du, J.F., Constantinescu, M., Barci, V., Weiss, B., Gasparo, J., Ardisson G. : Extended abstracts. 1 St International Conference on the Chemistry and Physics of the Transactinide Elements, Seeheim, September 26-30 (1999). Contribution PW-20
25. Glückauf, E.: Trans. Faraday Soc. **51**, (1955) 34.

26. Glückauf, E.: "*Ion exchange and its application*", Society of Chemical Industry, London, 1955.
27. Helferich, F.: "*Ion exchange*", Mc Graw-Hill, New York, 1962.
28. Marcus, Y. and Kertes, A.S., *Ion exchange and solvent extraction of metal complexes* , Wiley-interscience, London, 1970.
29. Grenthe, I., Puigdomenech, I. (Editors), *Modeling in aquatic chemistry,* OECD/NEA, Paris (France), 1997.
30. Duplessis, J., Guillaumont, R.: Radiochem. Radioanal. Letters **31**, (1977) 283.
31. Barthel, J.M.G., Krienke, H., Kunz, W.: *Physical chemistry of electrolyte solutions modern Aspects,* Springer-Verlag, Steinkopff Darmstadt, 1998.
32. Neck, V., Kim, J.I.: Radiochim. Acta **89**, (2001) 1.
33. Ciaviata, L.: Ann. Chim (Roma) **70**, (1980) 551.
34. Ciaviata, L.: Ann. Chim (Roma) **80**, (1990) 255.
35. Günther, R., Paulus, W., Kratz, J.V., Seibert, A., Thörle, P., Zauner, S., Brüchle, W., Jäger, E., Pershina, V., Schädel, M., Schausten, B., Schumann, D., Eichler, B., Gäggeler, H.W., Jost, D.T., Türler, A.: Radiochim. Acta **80**, (1998) 121.

Index

Chapter 4

Experimental Techniques

A. Türler [a], K.E. Gregorich [b]

[a] *Institut für Radiochemie, Technische Universität München, D-85748 Garching, Germany*
[b] *Nuclear Science Division, Lawrence Berkeley National Laboratory, Berkeley, CA 94720*

1. Introduction

The study of the chemical properties of the heaviest known elements in the Periodic Table is an extremely challenging task and requires the development of unique experimental methods, but also the persistence to continuously improve all the techniques and components involved. The difficulties are numerous. First, elements at the upper end of the Periodic Table can only be artificially synthesized "one-atom-at-a-time" at heavy ion accelerators, requiring highest possible sensitivity. Second, due to the relatively short half-lives of all known transactinide nuclides, very rapid and at the same time selective and efficient separation procedures have to be developed. Finally, sophisticated detection systems are needed which allow the efficient detection of the nuclear decay of the separated species and therefore offer unequivocal proof that the observed decay signature originated indeed form a single atom of a transactinide element.

Fast chemical isolation procedures to study the chemical and physical properties of short-lived radioactive nuclides have a long tradition and were applied as early as 1900 by E. Rutherford [1] to determine the half-life of ^{220}Rn. A rapid development of fast chemical separation techniques [2,3,4,5,6,7, and Ref. 8 for an in-depth review] occurred with the discovery

M. Schädel (ed.), The Chemistry of Superheavy Elements, 117-157.

of nuclear fission [9]. Indeed, the discovery of new elements up to Z=101 was accomplished by chemical means [10]. Only from there on physical methods prevailed. Nevertheless, rapid gas phase chemistry played an important role in the claim to discovery of Rf and Db [11]. As of today, the fastest chemical separation systems allow access to the study of α-decaying nuclides within less than 1 s as demonstrated by the investigation of ^{224}Pa with a half-life of 0.85 s [12]. Reviews on rapid chemical methods for the identification and study of short-lived nuclides from heavy element synthesis, can be found in [13,14,15,16,17,18,19].

A chemistry experiment with a transactinide element can be divided in four basic steps:

1.) Synthesis of the transactinide element.
2.) Rapid transport of the synthesized nuclide to the chemical apparatus.
3.) Fast chemical isolation of the desired nuclide and preparation of a sample suitable for nuclear spectroscopy.
4.) Detection of the nuclide via its characteristic nuclear decay properties.

Already the synthesis of heavy and superheavy elements is from the technological point of view very demanding. In order to gain access to the longer-lived isotopes of transactinide elements, exotic, highly radioactive target nuclides such as ^{244}Pu, ^{243}Am, ^{248}Cm, ^{249}Bk, ^{249}Cf or ^{254}Es are bombarded with intense heavy ion beams such as ^{18}O, ^{22}Ne, and ^{26}Mg. On the one hand as intense beams as possible are to be used; on the other hand the destruction of the very valuable and highly radioactive targets has to be avoided. Experimental arrangements that have successfully been used to irradiate exotic targets and to rapidly transport the synthesized transactinide nuclei to a chemistry set-up are discussed in Section 2.

The choice of the chemical separation system has to be based on a number of prerequisites that have to be fulfilled simultaneously to reach the required sensitivity:

• Speed: Due to the short half-lives of even the longest-lived currently known transactinide nuclides, see Chapter 1, the time required between the production of a nuclide and the start of the measurement is one of the main factors determining the overall yield. In contrast to kinematic separators, chemical instrumentation currently allows the investigation of transactinide nuclides with half-lives of a few seconds or longer. The importance of the speed of separation may diminish in future experiments with the reported longer-lived isotopes of superheavy elements 108, 110, and 112.

• Selectivity: Due to the low production cross sections, ranging from the level of a few nanobarns for the production of nuclides of elements Rf and Db, down to a level of a few picobarns or even femtobarns for the production of elements with $Z > 108$, the selectivity of the chemical procedure for the specific element must be very high. Two groups of elements are of major concern as contaminants: Due to the fact that many Po isotopes have similar half-lives and/or α-decay energies as transactinide elements, the separation from these nuclides must be particularly good. Some short-lived Po isotopes are observed as daughter nuclides of precursors i.e. Pb, Bi, and Rn, also some At isotopes are of concern. These elements are formed in multinucleon transfer reactions with Pb impurities in the target material and/or the target assembly but also with the target material itself. Here a chemical purification of the actinide target materials and a careful selection of the materials used in the target assembly can already reduce the production of unwanted nuclides by orders of magnitude. Nevertheless, in order to chemically investigate the recently discovered superheavy elements, a kinematic pre-separation might be required to efficiently suppress e.g. Rn transfer products. A second group of elements, which interfere with the detection of transactinide nuclei, are heavy actinides that decay by spontaneous fission (SF). These are inevitably produced with comparably large cross sections in multinucleon transfer reactions. The separation from heavy actinides must be particularly good if SF is the only registered decay mode and if no other information, such as the half-life of the nuclide can be derived from the measurement.

• Single atom chemistry: Due to the very low production rates, transactinide nuclei must be chemically processed on a "one-atom-at-a-time" scale. Thus, the classical derivation of the law of mass action is no longer valid, see Chapter 3. R. Guillaumont et al. [20] have derived an expression equivalent of the law of mass action in which concentrations are replaced by probabilities of finding the species in a given state and a given phase. The consequence for single atom chemistry is that the studied atom must be subjected to a repetitive partition experiment to ensure a statistically significant behavior. Here, chromatography experiments are preferred.

• Repetition: Since the moment in time at which a single transactinide atom is synthesized can currently not be determined and chemical procedures often work discontinuously, the chemical separation has to be repeated with a high repetition rate. Thus, thousands of experiments have to be performed. This inevitably led to the construction of highly automated chemistry set-ups. Due to the fact, that the studied transactinide elements as well as the interfering contaminants are radioactive and decay with a certain half-life, also continuously operating chromatography systems were developed.

- Detection: The unambiguous detection of the separated atom is the most essential part of the whole experiment. Even though some techniques, such as atomic force microscopy (AFM), have reached the sensitivity to manipulate single atoms or molecules (out of many), the detection of the characteristic nuclear decay signature of transactinide nuclei remains currently the only possibility to unambiguously detect the presence of a single atom of a transactinide element after chemical separation. Thus, final samples must be suitable for high resolution α-particle and SF-spectroscopy (coincident detection of SF fragments). Most transactinide nuclei show characteristic decay chains that involve the emission of α-particles or the SF of daughter nuclei. The detection of such correlated decay chains requires the event-by-event recording of the data. In experiments with physical separators, the use of position sensitive detectors further enhanced the discrimination against randomly correlated events.

- Speciation: Due to the fact that transactinide nuclei are detected after chemical separation via their nuclear decay, the speciation cannot be determined. Currently, the speciation in all transactinide chemistry experiments has to be inferred by carefully studying the behavior of lighter, homologue elements. The chemical system must be chosen in such a manner that a certain chemical state is probable and stabilized by the chemical environment.

Basically, four different approaches which involve the direct detection of the nuclear decay of the isolated nuclides have been successful in studying the chemical properties of transactinide elements. Two of the systems work in the liquid phase, as discussed in Section 3, whereas the other two are designed to investigate volatile transactinide compounds in the gas phase, as discussed in Section 4. Chemical information can also be obtained by studying the distribution of long-lived α-decay grand daughters after completion of the on-line experiment. Also these systems will be included in the current review.

2. Targets, Recoil Techniques and Gas-jets

2.1 ACTINIDE TARGETS

To date, the highest production rates of isotopes of transactinide elements have been achieved in compound nucleus reactions between light-heavy ion beams (^{18}O, ^{19}F, ^{22}Ne, ^{26}Mg, and ^{48}Ca) with actinide targets (^{238}U, ^{244}Pu, ^{248}Cm, ^{249}Bk, $^{249,250}Cf$, and ^{254}Es). Production rates are proportional to both the target thickness and the beam intensity. For nuclear reactions involving the lighter projectiles, target thickness is limiting the compound nucleus recoil ranges (<1 mg/cm^2), because the separation techniques used require the compound nucleus products to recoil out of the target. With the higher-Z projectiles, the recoil ranges are larger, but because of the higher rate of projectile energy loss in the target material and the narrow projectile energy range effective for heavy element production, the effective target thicknesses are once again limited to ~1 mg/cm^2. The high levels of radioactivity associated with the milligram amounts of these actinide isotopes presents unique challenges for safe handling and irradiation of these targets. The targets must be physically strong to maintain integrity through handling and irradiation. Passage of the beams through the targets produces large amounts of heat, which must be dissipated. In addition, the target material must be chemically stable in the highly ionizing environment created by passage of the beam.

2.1.1 *Target production techniques*

Two methods have been used in the fabrication of actinide targets for heavy element studies. They produce targets of similar quality.

Evaporation of volatile actinide compounds. Targets for heavy element chemistry experiments have been produced by the vacuum evaporation of volatile actinide compounds onto stable metallic backing foils. A review of many techniques has been published by A.H.F Muggleton [21], although he concentrated on high-temperature vacuum evaporation. For vacuum evaporation, a relatively volatile compound of the target element is heated to near its vaporization point, and the evaporated molecules form a uniform deposit on the nearby cooled target backing foil. Heating can be achieved by passing electric current through a refractory metal boat or using resistance heating on various crucibles. For production of targets from metals and more refractory materials, heating can be achieved by electron bombardment. These techniques can produce very pure and uniform targets. The main disadvantage is the relatively low efficiency for collecting the evaporated target material on the target backing foil. This can be a serious disadvantage with the use of extremely rare and radioactive heavy actinide target materials.

Molecular plating. Electrodeposition or molecular plating has been the method of choice for production of the small-area actinide targets for use in production of heavy element isotopes for chemical studies. Nitrate or chloride compounds of the actinide element are suspended in a small volume of organic solution (usually isopropanol or acetone), and a high voltage is applied between the solution and the target backing foil. A typical plating cell is shown in Figure 1. Targets up to 1 mg/cm^2 have been produced by plating successive 0.1 mg/cm^2 layers, and converting the deposit to the oxide form by heating to ~500°C before plaiting the next layer.

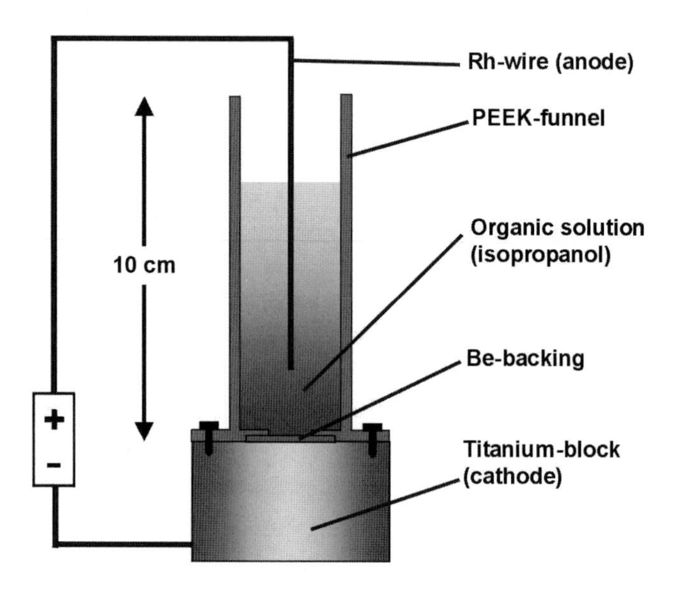

Fig. 1. Schematic of cell used [29] for molecular plating of actinide targets.

Such techniques have been described by several authors [22,23,24,25]. The molecular plating technique can easily be accomplished inside a glove box for containment of the radioactive target material. Sub-milligram quantities of actinide elements have been used and the deposition yield can approach 100%. The small volumes used facilitate recovery of actinide materials. Extreme care must be taken to produce targets free of impurities.

Electrospray. With the electrospray, see the review by A.H.F. Muggleton [21] and references therein, an actinide compound is dissolved in a non-conducting organic solvent. An extremely fine glass capillary is drawn and an electrode is inserted into the capillary. The capillary is filled with the solvent containing the actinide compound, and the capillary tip is placed within a few centimeters of the target backing foil. Once a high voltage is applied between the electrode and the metal target backing foil, a fine spray is emitted from the capillary which immediately dries on the backing foil.

While large area and thick targets have been produced by this technique, with actinide target materials, the spraying of aerosol-sized particles of an extremely radioactive solution poses challenging safety problems. This technique has essentially been supplanted by the molecular plating process.

Ink-jet techniques. Piezoelectric pulsed drop jet devices have been used for target production [26]. These devices are similar to those used in ink-jet printers.

2.1.2 *Target cooling*

The projectile beam loses energy upon passing through the target backing and the target, resulting in deposition of heat in the target. The heat generated must be removed to prevent damage to the target. To allow the highest beam intensities, highly efficient target cooling is necessary.

Double-window systems and forced gas cooling. The double-window target system has been developed [27] for use with targets on relatively thick backing foils. In the double window system, a vacuum isolation foil is placed just upstream of the target backing foil (with the target material on the downstream side of the backing foil). Cooling gas at pressures near 1 bar is forced at high velocity through a narrow gap between the vacuum isolation foil and the target backing foil, cooling both foils.

Since both the vacuum isolation foil and the target backing foil must hold a pressure difference of greater than 1 bar, relatively thick metal foils, such as 2.5 mg/cm^2 Be or 1.8 mg/cm^2 HAVAR, have been used. These thick foils are especially attractive when considering the mechanical stability of extremely radioactive actinide targets. Variations on the double-window target system have traditionally been used for heavy element production with actinide target materials.

Because of the mechanical stresses associated with the pressure differences across the foils, target areas have been limited to < 1 cm^2. Heating of the target by passage of the beam is inversely proportional to the cross sectional area of the beam. Clearly, an increase in the target size (or area) would allow the use of higher beam intensities.

A. Yakushev [28] has overcome the small target area limitation by placing a supporting grid over a large-area ^{238}U target and using an electrostatic beam wobbler to spread the beam out over a much larger area. While some fraction of the beam is lost on this grid, much larger beam intensities, and therefore higher production rates were achieved.

J.M. Nitschke has shown that when using a pulsed beam (with beam pulse widths on the order of 10 ms), even though the instantaneous beam intensity on the target is higher, similar average beam intensities can be used without damage to the target [27]. Thus, as illustrated in Figure 2, it should be possible to use electrostatic deflection plates together with pulsed high-voltage supplies to deliver the beam alternately among an array of several targets. In Figure 2, a high intensity beam is distributed among a square array of nine targets on a 5-ms timescale. In this way, an effectively larger area target could be used without the beam loss and heating associated with a target grid. Each of the targets would be irradiated at approximately an 11% duty factor.

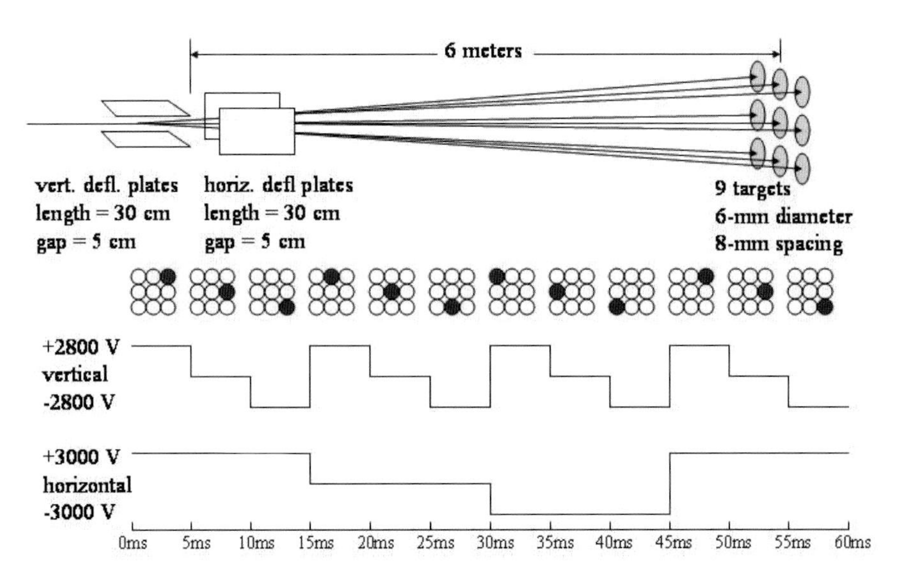

Fig. 2. Pulsed voltage on deflection plates to distribute beam among nine targets on a 5-ms timescale.

2.1.3 *Rotating targets*

Another method for effectively spreading the beam heating over a larger target area is the use of rotating targets. A rotating ^{248}Cm target has recently been used in heavy element chemistry experiments at the Gesellschaft für Schwerionenforschung (GSI) [29]. This rotating target has allowed the use of modern high-intensity accelerated beams for heavy element chemistry experiments [30]. A picture of this rotating ^{248}Cm target after irradiation with up to 1 pμA of a ^{26}Mg beam (a factor of four larger than possible with a fixed target) is presented in Figure 3.

Fig. 3. The GSI rotating ^{248}Cm target wheel (photo taken from Ref. 29).

2.2 RECOIL TECHNIQUES

In the early days of accelerator-based radiochemical separations, thick targets were irradiated, allowing the compound nucleus products to stop in the target material. At the end of an irradiation, the long-lived radionuclides could be chemically separated from the target. Obvious limitations arise with the use of highly radioactive heavy actinide targets in the search for short-lived heavy element isotopes.

2.2.1 *Recoil catcher foils*

When a compound nucleus is formed in the reaction of a projectile beam with a target nucleus, the compound nucleus is formed with the momentum of the beam particle. If the target layer is thin enough, this recoil momentum is sufficient to eject the compound nucleus product from the target. These recoiling reaction products can be stopped in a metal foil placed directly downstream of the target. By using this recoil catcher method [10], chemical separation of the heaviest elements is facilitated because only a small fraction of the radioactive target material is transferred to the recoil catcher foil, and the recoil catcher foil material can be chosen to facilitate its dissolution and chemical removal. Perhaps the first use of the recoil catcher technique for heavy element studies was with the chemical separations used in the 1955 discovery of Mendelevium [31]. With the recoil catcher foil technique, the time required for removal of the foil, and dissolution and separation of the foil material results in chemical separation times longer than a few minutes. Since the half-lives of transactinide isotopes are on the order of one minute or less, this technique has not been used for transactinide chemistry experiments.

2.2.2 *Aerosol gas-jets*

To achieve faster chemical separation times, the aerosol gas-jet transport technique has been used to deliver transactinide isotopes form the target chamber to various chemical separation devices. Transport times on the order of one second have been achieved. The principles behind the aerosol gas-jet transport technique have been presented by H. Wollnik [32]. Products of nuclear reactions are allowed to recoil out of a thin target, and are stopped in a gas at a pressure usually above 1 bar. The gas is usually helium because of the relatively long recoil ranges. The He gas is seeded with aerosol particles. After stopping in the gas, the reaction products become attached to the aerosol particles. The gas and activity-laden aerosol particles are sucked through a capillary tube to a remote site, by applying vacuum to the downstream end of the capillary. At the downstream end of the capillary, the He goes through a supersonic expansion into the vacuum, exiting in a broad cone. The aerosol particles, having much lower random thermal velocities are ejected from the tip of the capillary in a narrow cone, and can be collected on a foil or filter. The collected aerosol particles, containing the nuclear reaction products, can be made rapidly available for chemical separation.

Many aerosol materials have been used, and the aerosol material can be specifically chosen to minimize interference with the chemical separation being conducted. Widely used were KCl aerosols which can easily be generated by sublimation of KCl from a porcelain boat within a tube furnace. By choosing a temperature between 650°C and 670°C, specially tailored aerosols with a mean mobility diameter of about 100 nm and number concentrations of few times 10^6 particles/cm^3 could be generated. The same technique could be applied to produce MoO_3 aerosols. Carbon aerosol particles of similar dimensions were generated by spark discharge between two carbon electrodes.

The large transport efficiencies through the capillary for the aerosol gas-jet technique can be explained in terms of the laminar flow profile of the gas inside the capillary [32]. According to Bernoulli's law, the gas pressure at the center of the capillary, where the velocity is highest, is lower than near the capillary walls. Thus, when the sub µm-sized aerosol particles drift away from the center of the capillary, they are subject to a restoring force toward the center of the capillary. Transport efficiencies of over 50% have routinely been achieved for transport capillary lengths over 20 meters.

2.2.3 *Kinematic separators and gas-jets*

Atoms of the transactinide elements are produced at extremely low rates: atoms per minute for Rf and Db, down to atoms per day for elements 106-108. They are produced among much larger amounts of "background" activities which hinder the detection and identification of decay of the transactinide atoms of interest. For these reasons, there is a recognized need for a physical pre-separation of the transactinide element atoms before chemical separation. Thus, it has been proposed that a kinematic separator could be coupled to a transactinide chemistry system with an aerosol gas-jet device, and a workshop has been held to discuss the merits and design of such a device [33].

Coupling of the Berkeley Gas-Filled Separator (BGS) with a transactinide chemical separation system has been accomplished at the Lawrence Berkeley National Laboratory in some proof-of-principle experiments [34]. For these experiments, 4.0-s ^{257}Rf was produced in the ^{208}Pb(^{50}Ti,n)^{257}Rf reaction, and separated from all other reaction products with the Berkeley Gas-filled Separator [35]. In the focal plane area of the BGS, the ^{257}Rf recoils passed through a 6-μm MYLAR foil and were stopped in a volume of He at a pressure of 2 bar in the Recoil Transfer Chamber (RTC) [36]. The He gas in the RTC was seeded with KCl aerosol particles, and the ^{257}Rf atoms became attached to the aerosols, and were transported through a 20-m capillary to the SISAK chemical separation system, where the chemical separation and detection were performed. A schematic of the experiment is presented in Figure 4. These proof-of-principle experiments showed the practicality of a physical pre-separation of transactinide nuclides before chemical separation.

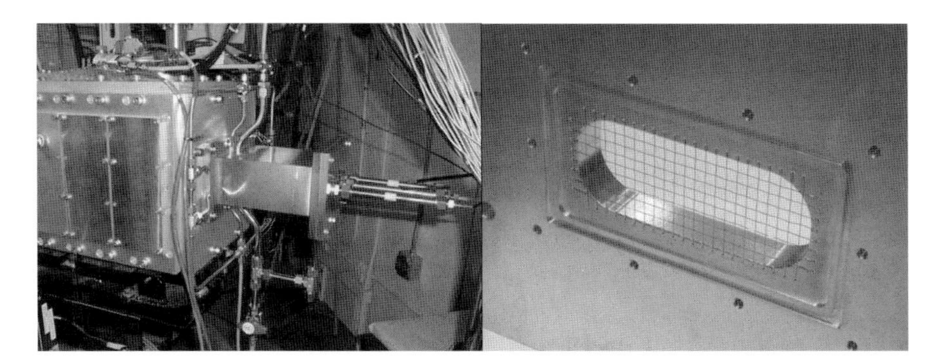

Fig. 4. The Recoil Transfer Chamber installed at the BGS focal plane (left), and the support grid for the MYLAR foil (right).

3. Techniques and Instruments for Liquid-Phase Chemistry

Liquid-phase chemical separation techniques have been used for over 100 years, thus their utility for separation and isolation of the chemical elements has been demonstrated. Adaptations of these well-understood separation techniques have been applied to the transactinide elements. These adaptations have been developed to overcome the single-atom and short half-life limitations inherent in the study of transactinide element chemical properties.

3.1 MANUAL LIQUID-PHASE CHEMICAL SEPARATIONS

The early chemical separations with the heaviest elements were performed manually. Today, most of the research has turned to automated chemical separation techniques.

3.1.1 *Manual liquid-liquid extractions*

Manually performed liquid-liquid extractions have been used for the study of chemical properties of Rf [37,38,39,40] and Db [41] in Berkeley. The micro-scale liquid-liquid extraction technique used in these studies was based on techniques developed for the study of Lr chemical properties [42,43]. To minimize the separation and sample preparation time, phase volumes were kept to ~20 μl. For these experiments, several specially developed techniques and apparatus were necessary. These include: a) a special collection turntable for easy collection of the gas-jet samples which also signaled the data acquisition computer at the start and end of collection for each sample; b) small syringes with hand-made transfer pipettes for rapid pipetting of volumes as small as 15 μl; c) the ultrasonic mixer, which was made by modification of a commercially available home ultrasonic humidifier; d) a specially modified centrifuge which reached full speed and returned to a full stop in a few seconds and also signaled the data acquisition computer to record the chemical separation time; e) techniques for rapidly and reliably drying the organic phase at the center of an Al disc or glass plate; and f) a set of α-particle detection chambers which could be quickly loaded, evacuated, and which signaled the acquisition computer as to the start and stop of the measurement for each sample. The time from end of collection of aerosols to the beginning of counting of the transactinide chemical fractions was as short as 50 seconds, and the collection-separation-counting cycle could be repeated every 60-90 seconds.

For the Rf chemistry, 78-second ^{261}Rf was produced by the ^{248}Cm(^{18}O,5n)^{261}Rf reaction. Identification of ^{261}Rf was made by measuring the 8.29-MeV α particles from the decay of ^{261}Rf and/or the 8.22-8.32 MeV α particles for the decay of the 26-second ^{257}No daughter. The detection was made more certain by detecting correlated pairs of α decays from both the ^{261}Rf parent and the ^{257}No daughter. Detection rates for the ^{261}Rf were as high as 5 events per hour of experiments. For the Db chemistry, 34-second ^{262}Db was produced by the ^{249}Bk(^{18}O,5n)^{262}Db reaction. Identification of ^{262}Db was made by measuring the 8.45-MeV α particles from the decay of ^{262}Db and/or the 8.59-8.65 MeV α particles for the decay of the 3.9-second ^{258}Lr daughter. The detection was made more certain by detecting correlated pairs of α decays from both the ^{262}Db parent and the ^{258}Lr daughter. Detection rates for the ^{262}Db were approximately one event per hour of experiments.

Because of interference from the radioactive decay of other nuclides (which are typically formed with much higher yields), extraction systems with relatively high decontamination factors from actinides, Bi, and Po must be chosen, and the transactinide activity can only be measured in the selectively extracting organic phase. For this reason, measurement of distribution coefficients is somewhat difficult. By comparing the Rf or Db detection rate under a certain set of chemical conditions to the rate observed under chemical control conditions known to give near 100% yield, distribution coefficients between about 0.2 and 5 can be determined. If the control experiments are performed nearly concurrently, many systematic errors, such as gas-jet efficiency and experimenter technique, are cancelled out. Additionally, extraction systems which come to equilibrium in the 5-10 second phase contact time must be chosen.

3.1.2 *Manual column chromatography*

The first liquid-phase transactinide chemical separations were manually performed Rf cation exchange separations [42] using α-hydroxyisobutyrate (α-HIB) as eluent, performed by R.J. Silva et al. in 1970. The newly discovered 78-second ^{261}Rf was produced in the ^{248}Cm(^{18}O,5n)^{261}Rf reaction, the recoils were stopped on NH_4Cl-coated Pt foils which were transported to the chemical separation area with a rabbit system. The ^{261}Rf and other products from the nuclear reaction, along with the NH_4Cl were collected from the Pt disk in a small volume of α-HIB, and run through a small cation exchange resin column. Under these conditions, all cations with charge states of 4+ or higher were complexed with the α-HIB and eluted from the column. These experiments showed that Rf had a charge state of 4+ (or higher) and that its chemical properties are distinctly different from those of the actinides.

Although manually performed column chromatography separations were used [43] for studies with 3-minute ^{260}Lr, further use of column chromatography for studying the chemical properties of the transactinides awaited the development of the automated techniques described below.

3.1.3 *Future techniques for manual liquid-phase separations*

Manually performed liquid-phase chemical separations will continue to be used, especially for investigations of Rf and Db, with their relatively large production rates. New devices and techniques will be developed to minimize the separation and sample preparation time and to minimize the repetitive labor required. One idea being explored is the use of extraction chromatography packings inside small syringes. By picking up the gas-jet sample in a small volume of an appropriate aqueous phase, and pulling it through the chemically selective packing inside the syringe body, and then forcing it back through the packing on to the sample preparation disk, a very fast and simple equivalent of a column separation could be performed.

With the advent of the use of kinematic pre-separators, as described in Section 2.2.2 above, the requirements of the chemical separation have been relaxed. It is no longer necessary to have the highest separation factors from interfering Bi, Po, and actinide radioactivities, so simpler separations which are more specific to the transactinide element being studied can be used. These relaxed separation requirements will allow development of simpler chemical separation techniques, and may lead to a new interest in manually performed chemical separations.

3.2 AUTOMATED LIQUID-PHASE CHEMICAL SEPARATIONS

As chemical investigations progress from Z=104-105 (with detection rates of atoms per hour), through Z=106-108 (with detection rates of atoms per week), and on to even heavier elements (with expected detection rates of only a few atoms per month), manually performed chemical separations become impractical. With the automated liquid-phase chemical separation systems that have been developed to date, faster chemical separation and sample preparation times have been achieved. In addition, the precision and reproducibility of the chemical separations has been improved over that obtainable via manually performed separations.

3.2.1 *Automated Rapid Chemistry Apparatus (ARCA)*

Extensive studies of chemical properties of transactinide elements, Rf, Db, and Sg have been performed with the Automated Rapid Chemistry Apparatus (ARCA), see summaries in Ref. [15, 44, 45]. To improve the speed and reduce cross contamination, the ARCA II was built, featuring two magazines of 20 miniaturized ion exchange columns [46]. With the large number of columns, cross contamination between samples can be prevented by using each column only once. By miniaturizing the columns, the elution volumes, and therefore, the time needed to dry the final sample to produce a source for α spectroscopy, is much reduced. A photograph and a schematic of the ARCA microprocessor-controlled column chromatography system are presented in Figure 5.

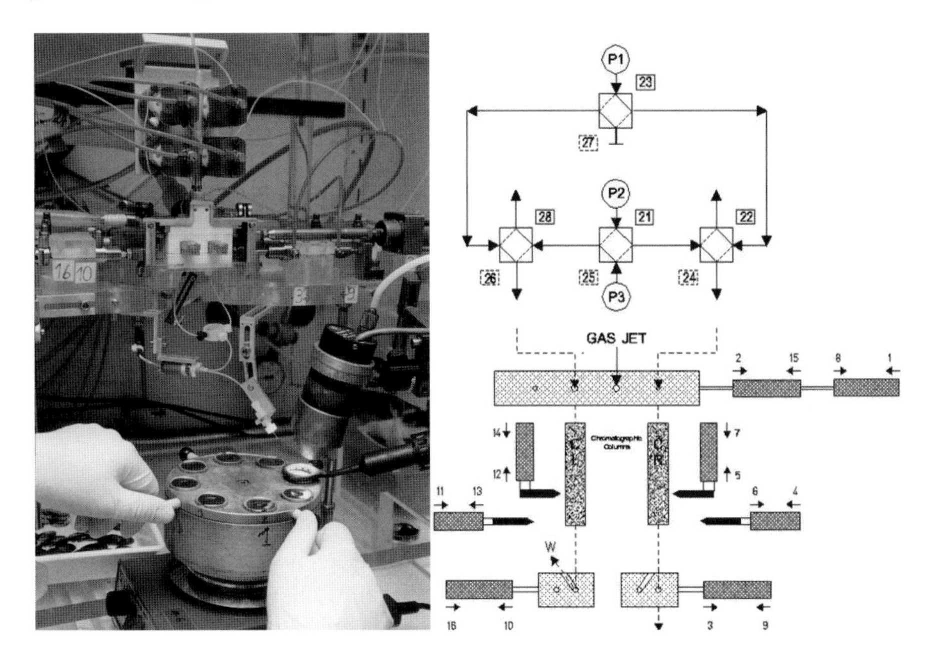

Fig. 5. Photograph (left) and schematic (right) of the ARCA

With ARCA II, activity-bearing aerosols are collected from a gas-jet in a small spot on a slider (seen at the center of the schematic). At the end of a suitable collection time, the slider is moved to position the collection site above one of the miniature ion exchange columns. A suitable aqueous solution is used to dissolve the aerosols and load the activities onto the ion exchange material in the column below. Selective elutions of transactinide elements are carried out by directing appropriate solutions through the column. A slider below the ion exchange column is moved at the appropriate time to collect the chemical fraction of interest on a hot Ta disk. A sample suitable for α-particle pulse–height analysis is prepared by rapid evaporation of the chemical fraction on the Ta disk, which is heated from below by a hot-plate, and from above by a flow of hot He gas and a high-intensity lamp. The final samples are then manually placed in a detector chamber for α-particle pulse-height spectroscopy. The two magazines of chromatography columns (CL and CR in the schematic) can be moved independently. During a chemical separation on CL, the right column, CR, is prepared by flowing an appropriate solution through it. After the separation on CL, the magazine is moved forward, placing a new column in the CL position, the next separation is performed on CR while CL is being prepared for the subsequent separation. In this way, up to 40 separations can be carried out, at time intervals of less than one minute, with each separation performed on a freshly prepared, unused column.

Transactinde chemical separations with ARCA II have been performed with Rf [47,48], Db [49,50,51,52,53], and Sg [54,55]. More details and results are discussed in Chapter 5.

3.2.2 *Automated Ion-exchange Separation Apparatus coupled with the Detection System for Alpha Spectroscopy (AIDA)*

Building upon the design of ARCA II an automated column separation apparatus, AIDA [56,57,58] has recently been built. While the apparatus for collection of aerosols, and performing multiple chemical separations on magazines of miniaturized ion exchange columns is very similar to that in ARCA II, AIDA has automated the tasks of sample preparation and placing the samples in the detector chambers. Using robotic technology, the selected fractions are dried on metal planchettes, and are then placed in vacuum chambers containing large-area PIPS α-particle detectors.

This robotic sample preparation and counting technology, together with mechanical improvements in the chemical separation system, has resulted in an automated column chromatography system that can run almost autonomously, whereas several people were required to operate the ARCA II system for a transactinide chemistry experiment.

Transactinide chemistry experiments with Rf have been carried out with the AIDA apparatus [59].

3.2.3 *Continuous on-line chromatography – identifying the heavy element daughters*

One of the difficulties in studying the chemical properties of the transactinide elements is presented by their relatively short half-lives. The time and labor required for performing batch-wise experiments and preparing samples suitable for α-particle pulse-height spectroscopy presents a daunting task. Since the transactinide isotopes typically decay by α-particle emission to relatively long-lived actinides, a technique has been developed to perform continuous chemical separations on the transactinide element isotopes, and detect their presence by observing the α-particle decay of the long-lived daughters. Since the daughter activities are typically produced via multinucleon transfer reactions at rates much greater than the transactinide element production rates, special techniques must be used. The three-column technique has been developed and applied to the study of the solution chemistry of Rf by Z. Szeglowski et al. [60].

In the three column technique, as used by D. Trubert et al. for the study of chemical properties of Db [61], 34-s ^{262}Db was produced in the ^{248}Cm(^{19}F,5n)^{262}Db reaction and transported to the chemical separation apparatus with an aerosol gas-jet. The activities delivered by the gas-jet were continuously dissolved in an appropriate aqueous solution and passed through a series of three ion exchange columns. The first column was used to separate all of the directly-produced daughter activities, allowing the Db atoms to pass through to the second column, where they are quantitatively retained. As the Db atoms decay to Lr, Md, and Fm (via α- and EC-decay), they are desorbed from the second column and pass to the third column. The third column quantitatively retains the longer-lived daughter activities. Since all directly-produced Lr, Md, and Fm activities were retained on the first column, this third column should contain only Lr, Md, and Fm atoms that are the decay descendents of Db atoms which were retained on the second column. At the end of a suitable production and chemical separation cycle, the daughter activities are chemically separated from this third column, and assayed for the α-decay of the ^{254}Fm great-granddaughter of ^{262}Db.

The separation technique used for the Db study was adapted from the slightly more complicated technique developed by G. Pfrepper et al. [62,63]. In this more general version, once again, the first column is used to remove all of the daughter activities initially present. The chemical conditions are chosen so that the retention time for the nuclide of interest is on the order of its half-life on the second column, and very short on the third column. If the parent nuclide decays while on the second column, the daughter is strongly adsorbed on the third. On the other hand, if the parent atom survives passage through the second column (and the third), the daughter atom will be found in the effluent from the third column. In this way, the retention time on the second column (and thus the distribution coefficient) of the nuclide of interest is can be measured by comparing the relative amounts of the long-lived daughter activity on the third column and in the effluent of the third column. Three-column systems have been developed in hopes of performing chemical studies with other transactinides [64,65]; see also Chapter 5.

3.2.4 *SISAK*

The SISAK (Short-lived Isotopes Studied by the AKUFVE-technique, where AKUFVE is s Swedish acronym for an arrangement of continuous investigations of distribution ratios in liquid extraction) system performs continuous liquid-liquid extractions using small-volume separator centrifuges [66]. Activities are delivered to the apparatus with an aerosol gas-jet. The gas-jet is mixed with the aqueous solution to dissolve the activity-bearing aerosols, and the carrier gas is removed in a degasser centrifuge. The aqueous solution is then mixed with an organic solution and the two liquid phases are separated in a separator centrifuge. A scintillation cocktail is then mixed with the organic solution, and this is passed through a detector system to perform liquid scintillation alpha pulse-height spectroscopy on the flowing solution.

This modular separation and detection system allows the use of well-understood liquid-liquid extraction separations on timescales of a few seconds, with detection efficiencies near 100%. This extremely fast chemical separation and detection system has been used with a sub-second α-active nuclide [67,68]. However, for the transactinide elements, which are produced in much lower yields with larger amounts of interfering β-activities, detection of the α-decay of the transactinide isotopes failed. As described in Section 2.2.3, pre-separation with the Berkeley Gas-filled Separator before transport to and separation with SISAK allowed the chemical separation and detection of 4-s ^{257}Rf [34]. A schematic of the BGS-RTC-SISAK apparatus is presented in Figure 6. These proof-of-principle experiments have paved the way for detailed liquid-liquid extraction experiments on short-lived transactinide element isotopes.

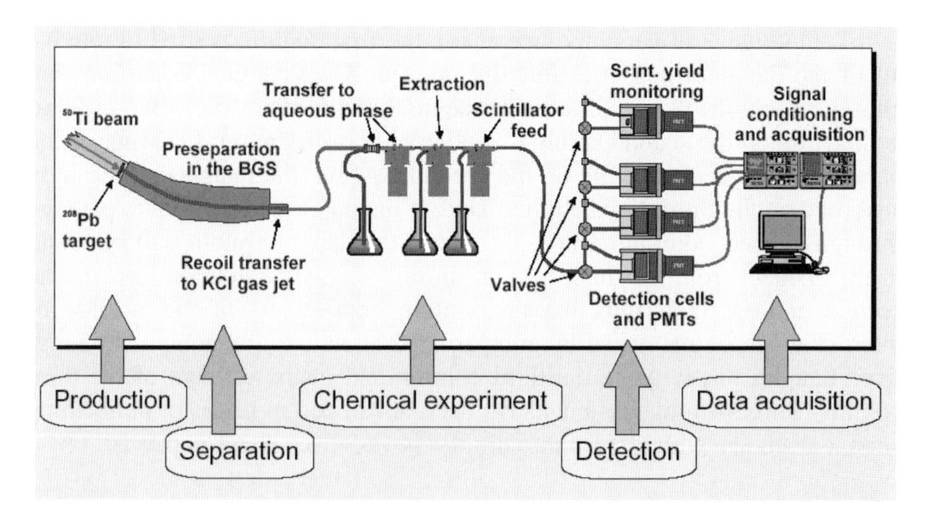

Fig. 6. Schematic of the SISAK liquid-liquid extraction system using the Berkeley Gas-filled Separator as a pre-separator.

3.2.5 *Future techniques for automated liquid-phase chemical separations*
Experiments have been carried out to measure the chemical properties of elements through Hs, element 108. Recent experiments have resulted in claims of isotopes of elements 112 and 114 with half-lives of at least a few seconds [69,70,71,72]: long enough for future chemical studies. For isotopes of these superheavy elements, production rates are as low as a few atoms per month, making any chemical separations with these elements especially difficult. New highly efficient aqueous-phase chemical separation systems will have to be developed. Since these separations will be operating for months at a time, they must be designed to run autonomously, presenting unique new challenges to the aqueous-phase chemists.

4. Techniques and Instruments for Gas-Phase Chemistry

Despite the fact that only few inorganic compounds of the transition elements exist, that are appreciably volatile below an experimentally still easily manageable temperature of about 1000°C, gas-phase chemical separations played and still play an important role in chemical investigations of transactinide elements. A number of prerequisites that need to be fulfilled simultaneously to accomplish a successful chemical experiment with a transactinide element are almost ideally met by gas chromatography of volatile inorganic compounds. Since the synthesis of transactinide nuclei usually implies a thermalization of the reaction products in a gas volume, a recoil chamber can be connected with a capillary directly to a gas chromatographic system. Gas phase separation procedures are fast, efficient and can be performed continuously, which is highly desirable in order to achieve high overall yields. Finally, nearly weightless samples can be prepared on thin foils, which allow α-spectroscopy and SF-spectroscopy of the separated products with good energy resolution and in high, nearly 4π, detection geometry.

Early on, gas-phase chemical separations played an important role in the investigation of the chemical properties of transactinide elements. The technique was pioneered by I. Zvara and co-workers at the Dubna laboratory and involved first chemical studies of volatile Rf, Db and Sg halides and/or oxyhalides [73,74,75]; see Chapter 7 for a detailed discussion. The experimental set-ups and the techniques involved are presented in Section 4.2. A new technique, named OLGA (On-line Gas chromatography Apparatus), which allowed the α-spectrometric measurement of final products, developed by H.W. Gäggeler and co-workers was then successful in studying volatile transactinide compounds from Rf up to Bh [76,77,78,79], see Chapter 7 for a detailed discussion. In all these experiments, the isolated transactinide nuclides were unambiguously identified by registering their characteristic nuclear decay, see Section 4.3. Only recently, the technique of synthesizing volatile species *in-situ* in the recoil chamber combined with a new cryo thermochromatography detector [80] was successful in the first chemical identification of Hs (element 108) [30] and appears to be a valid concept for first chemical investigations of element 112 in the elemental state, see Section 4.4.

4.1 THERMOCHRMOMATOGRAPHY AND ISOTHERMAL CHROMATOGRAPHY

For the experimental investigation of volatile transactinide compounds two different types of chromatographic separations have been developed, thermochromatography and isothermal chromatography. Sometimes also combinations of the two have been applied. The basic principles of thermochromatography and isothermal chromatography are explained in Figure 7.

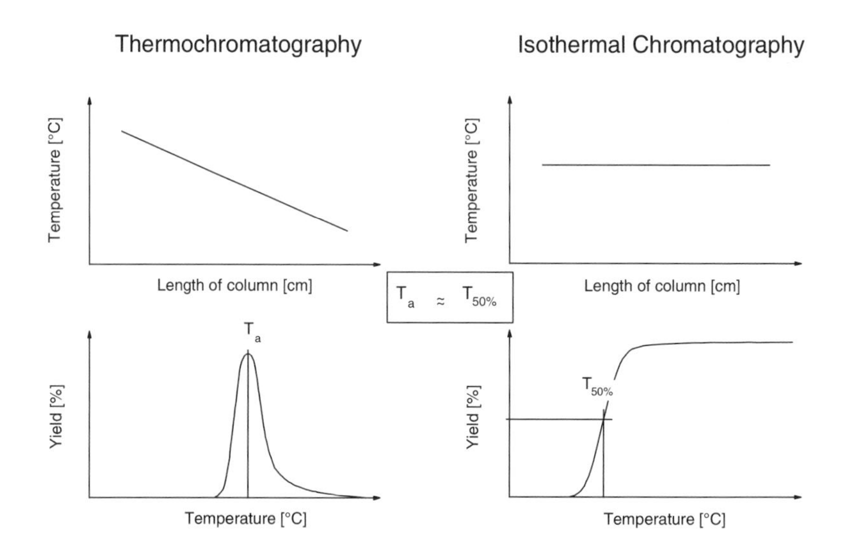

Fig. 7. Upper panel: temperature profiles employed in thermochromatography and isothermal chromatography; lower panel: deposition peak and integral chromatogram resulting from thermochromatography and isothermal chromatography, respectively.

4.1.1 *Thermochromatography*

In thermochromatography, see e.g. Reference 14 for a review, a carrier gas is flowing through a chromatography column, to which a negative longitudinal temperature gradient has been applied. Open or filled columns can be employed. Species, that are volatile at the starting point, are transported downstream of the column by the carrier gas flow. Due to the decreasing temperature in the column, the time the species spend in the adsorbed state increases exponentially. Different species form distinct deposition peaks, depending on their adsorption enthalpy (ΔH_a^0) on the column surface and are thus separated from each other. A characteristic quantity is the deposition temperature (T_a), which depends on various experimental parameters, see Chapter 6. The mixture of species to be separated can be injected continuously into the column [81,82,83,84], or the experiment can be performed discontinuously by inserting the mixture of species through the hot end of the chromatography column, and removing the column through the cold end after completion of the separation. The two variants (continuous or discontinuous) result in slightly different peak shapes. The chromatographic resolution is somewhat worse for the continuous variant. Thermochromatographic separations are the method of choice to investigate species containing long-lived nuclides that decay either by γ-emission, EC- or β^+ decay, or by the emission of highly energetic β^- particles [85,86,87,88,89,90,91,92]. Thus, the emitted radiation can easily be detected by scanning the length of the column with a detector. The detection of nuclides decaying by α-particle emission or SF decay is more complicated. By inserting SF track detectors into the column, SF decays of short- and long-lived nuclides can be registered throughout the duration of the experiment. After completion of the experiment, the track detectors are removed and etched to reveal the latent SF tracks. Columns made from fused silica have also been used as SF track detectors [75]. However, the temperature range for which SF track detectors can be applied is limited, due to the annealing of tracks with time. It should also be noted, that in thermochromatography all information about the half-life of the deposited nuclide is lost, which is a serious disadvantage, since SF is a non-specific decay mode of many actinide and transactinide nuclides. However, thermochromatography experiments with transactinides decaying by SF have an unsurpassed sensitivity (provided that the chromatographic separation from actinides is sufficient), since all species are eventually adsorbed in the column and the decay of each nuclide is registered. Thus, the position of each decay in the column contributes chemical information about ΔH_a^0 of the investigated species.

4.1.2 *Isothermal chromatography*

In isothermal chromatography, a carrier gas is flowing through a chromatography column of constant, isothermal temperature. Open or filled columns can be employed. Depending on the temperature and on ΔH_a^0 of the species on the column surface, the species travel slower through the length of the column than the carrier gas. This retention time can be determined either by injecting a short pulse of the species into the carrier gas and measuring the time at which it emerges through the exit of the column [93,94], or by continuously introducing a short-lived nuclide into the column and detecting the fraction of nuclides that have decayed at the exit of the column [95,96,97,98,99]. A characteristic quantity is the temperature at which half of the introduced nuclides are detected at the exit ($T_{50\%}$). In this case, the retention time in the column is equal to the half-life of the introduced nuclide. The half-life of the nuclide is thus used as an internal clock of the system. The $T_{50\%}$ temperature depends on various experimental parameters (see Chapter 6). It can be shown, that for similar gas flow rates and column dimensions $T_a \approx T_{50\%}$. By varying the isothermal temperature, an integral chromatogram is obtained. The yield of the species at the exit of the column changes within a short interval of isothermal temperatures from zero to maximum yield. A variant of isothermal chromatography using long-lived radionuclides is temperature programmed chromatography. The yield of different species at the exit is measured as a function of the continuously, but isothermally, increasing temperature [93,100,101,102,103].

On-line isothermal chromatography is ideally suited to rapidly and continuously separate short-lived radionuclides in the form of volatile species from less volatile ones. Since volatile species rapidly emerge at the exit of the column, they can be condensed and assayed with nuclear spectroscopic methods. Less volatile species are retained much longer and the radionuclides eventually decay inside the column.

A disadvantage of isothermal chromatography concerns the determination of ΔH_a^0 on the column surface of transactinide nuclei. In order to determine the $T_{50\%}$ temperature, a measurement sufficiently above and below this temperature is required. Since for transactinide elements this temperature is a priori unknown, several measurements at different isothermal temperatures must be performed, which means that long measurements are required below the $T_{50\%}$ temperature that demonstrate that the transactinide compound is retained long enough that most of the nuclei decayed in the column. Such an approach is very beam time consuming. Furthermore, it must be demonstrated that the experiment was performing as expected and the non-observation of transactinide nuclei was not due to a malfunctioning of the apparatus.

4.2 INSTRUMENTATION FOR EARLY GAS-PHASE CHEMISTRY EXPERIMENTS WITH TRANSACTINIDE ELEMENTS

A schematic drawing of the chemical apparatus constructed for the first chemical isolation of element 104 in Dubna is shown in Figure 8 [104].

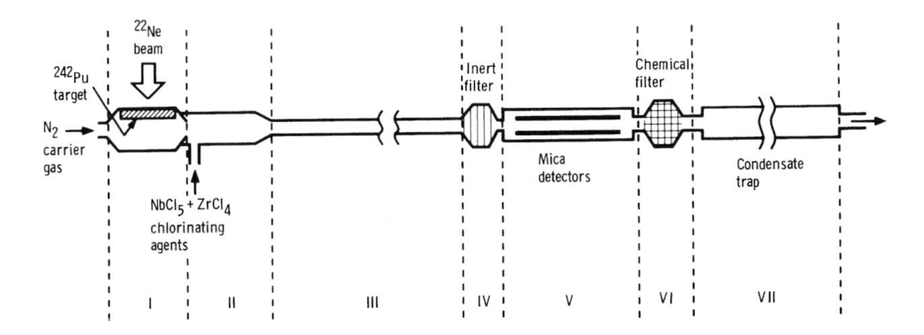

Fig. 8. Schematic of the first gas chromatography apparatus used to chemically isolate element 104 in the form of volatile chlorides (Figure from Ref. 11, adapted from Ref. 105).

A diagram showing different sections of the apparatus is displayed in Figure 9.

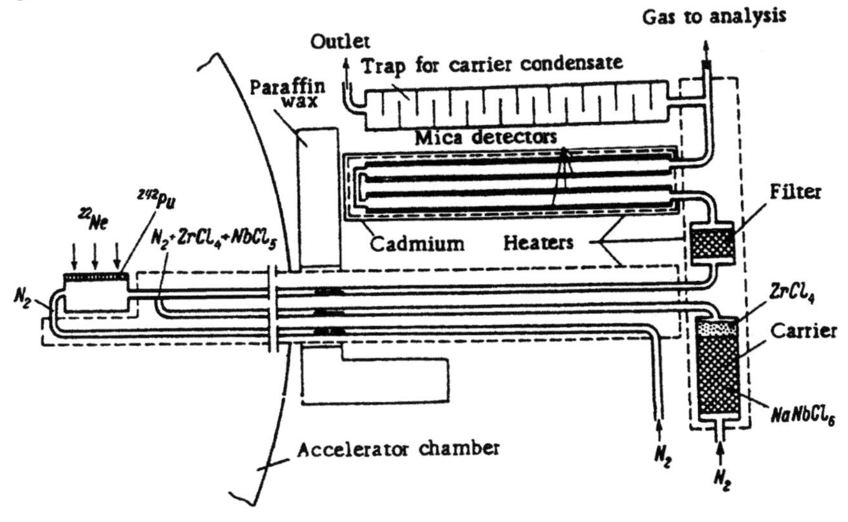

Fig. 9. Diagram of the experimental set-up to isolate short-lived, volatile Rf compounds at the internal beam of the U-300 cyclotron in Dubna [104].

In section I, a ^{242}Pu target was bombarded at the inner beam of the U-300 cyclotron at Dubna with ^{22}Ne ions. The ^{242}Pu with a thickness of about 800 μg/cm^2 was deposited as oxide on aluminum foil of 6-9 μm thickness. The target was held between two plates made of an aluminum alloy, into which a number of closely spaced holes of 1.5 mm diameter were drilled. Both sides of the target were flushed with nitrogen at 250°C or 300°C. In the outer housing of the target chamber, a number of closely spaced holes of 1.5 mm were drilled, which matched the holes of the target holder. An Al foil of 9 μm thickness served as vacuum window. The whole target block was heated with a heater. The space behind the target was limited with an Al foil and was about 11 mm deep and had a volume of about 2 cm^3.

Reaction products recoiling from the target were thermalized in a rapidly flowing stream of N_2 (18 l/min) and were transferred to section II, where the chlorination of the reaction products took place. The transfer efficiency was measured to be > 75% and the transfer time was only about 10^{-2} s. In the reaction chamber of section II, vapors of $NbCl_5$ and $ZrCl_4$ were continuously added as chlorinating agents and as carriers [104]. It was found that carriers with a vapor pressure similar to that of the investigated compound yielded optimum transfer efficiencies [81]. The addition of carriers was essential since in this manner the most reactive adsorption sites could be passivated.

Volatile reaction products were flushed into the chromatographic section (section III), which was consisted of a 4 m long column with an inner diameter of 3.5 mm. This section consisted of an outer steel column into which tubular inserts of various materials (Teflon™, glass) could be inserted [104]. Section IV contained a filter. This filter had a stainless steel jacket into which, as a rule, crushed column material from section III was filled [104]. This filter was intended to trap large aggregate particles [105]. Aerosols were apparently formed by the interaction of the chloride vapors ($NbCl_5$) with oxygen present in the nitrogen carrier gas. Volatile products passing the filter in section IV now entered the detector in section V. This detector consisted of a narrow channel of mica plates, which recorded SF fragments of the SF decay of a Rf nuclide. The mica plates were removed after completion of the experiment, etched and analyzed for latent fission tracks. The filter in section VI served for the chemisorption of Hf nuclides in ancillary work with long-lived nuclides. In section VII the chloride carriers were condensed.

The whole apparatus was built to chemically identify an isotope of Rf decaying by SF with a half-life of 0.3 seconds, that had previously been synthesized and identified by a team of physicists at Dubna. In a number of experiments, I. Zvara and co-workers identified multiple SF tracks in the mica detectors when they used glass surfaces and temperatures of 300°C [104]. They had shown in preparatory experiments with Hf, that indeed the transfer of Hf through the apparatus occurred within less than 0.3 s, and thus, that the experimental set-up was suited to study the short-lived Rf isotope [81]. A number of possible sources of SF tracks in the mica detectors other than the SF decay of an Rf isotope were discussed and ruled out. Further experiments with a slightly modified apparatus [106] were conducted immediately after the experiments described here. A total of 63 SF events were attributed to the decay of an Rf nuclide.

Similar thermochromatography set-ups, always relying on the registration of SF tracks, were employed by I. Zvara and co-workers to chemically identify the next heavier transactinide elements Db [13,74,107], and Sg [75,108,109,110], whereas experiments to chemically identify Bh [111] yielded negative results. However, due to the fact that in all these experiments the separated nuclides were identified by the non-characteristic SF decay, and no further information such as the half-life of the investigated nuclide could be measured, most of the experiments fell short of fully convincing the scientific community, that indeed a transactinide element was chemically isolated [11,112,113]. Nevertheless, the ideas conceived at Dubna and the techniques invented to chemically study short-lived single atoms, paved the way to gas chromatography experiments that allowed the unambiguous identification of the separated transactinide nuclei.

4.3 ON-LINE GAS-PHASE CHEMISTRY WITH DIRECT IDENTIFICATION OF TRANSACTINIDE NUCLEI

One of the most successful approaches to the study of volatile transactinide compounds is the so-called OLGA (On-Line Gas chromatography Apparatus) technique. Contrary to the technique in Dubna, reaction products are rapidly transported through a thin capillary to the chromatography setup with the aid of an aerosol gas-jet transport system, see Subsection 2.2.2. With typical He flow rates of 1 to 2 l/min and inner diameters of the capillaries of 1.5 to 2 mm, transport times of less than 10 s were easily achieved. This way, the chromatography system and also the detection equipment could be set up in an accessible, fully equipped chemistry laboratory close to the shielded irradiation vault. The aerosols carrying the reaction products are collected on quartz wool inside a reaction oven. Reactive gases are introduced to form volatile species, which are transported downstream by the carrier gas flow to an adjoining isothermal section of the column, where the chromatographic separation takes place.

A first version of OLGA (I) was developed and built by H.W. Gäggeler and co-workers for the search of volatile superheavy elements, and tested with 25 s ^{211m}Po [114]. Volatile elements were separated in a stream of He and hydrogen gas at 1000°C from non volatile actinides and other elements. At the exit of the column, the separated nuclei were condensed on thin metal foils mounted on a rotating wheel (ROtating Wheel Multidetector Apparatus, ROMA [115,116]) and periodically moved in front of solid state detectors, where α-particles and SF events were registered in an event-by-event mode. A first attempt to identify the nuclide ^{261}Rf after chemical isolation as volatile $RfCl_4$ is described by R.D. von Dincklage and co-workers [96]. While the gas chemical isolation of short-lived α-particle emitting Hf nuclides was successful, the experiment with Rf failed, since the employed surface barrier detectors were destroyed by the prolonged exposure to the chlorinating agent.

The first successful gas chemical studies of volatile halides of Rf [76] involving the unambiguous detection of time correlated nuclear decay chains $^{261}Rf \xrightarrow{\alpha} {}^{257}No \xrightarrow{\alpha} {}^{253}Fm$ was accomplished with OLGA(II) [98]. Instead of condensing the separated molecules on metal foils after passing the chromatography column, they were attached to new aerosol particles and transported through a thin capillary to the detection system. This so-called reclustering process was very effective and allowed to collect the aerosols on thin (≈ 40 μg/cm^2) polypropylene foils in the counting system (ROMA or MG, Merry Go round [117]). Thus, samples could be assayed from both sides in a 4π geometry, which doubled the counting efficiency. A crucial detail was the use of then newly available PIPS (passivated ion-implanted planar silicon) detectors (instead of surface barrier detectors) which were resistant to the harsh chemical environment. At the same time the Paul Scherrer Institute (PSI) tape system was developed [98], which allowed to significantly reducing the background of long-lived SF activities that accumulated on the wheel systems. However, only a 2π counting geometry could be realized.

An improved version of OLGA(II) was built at Berkeley and was named Heavy Element Volatility Instrument (HEVI) [99]. With both instruments the time needed for separation and transport to detection was about 20 s, the time consuming process being the reclustering. In order to improve the chromatographic resolution and increase the speed of separation, OLGA(III) was developed [17]. Using a commercial gas chromatography oven and a 2 m long quartz column which ended in a much smaller, redesigned recluster unit, the overall separation time could be reduced by one order of magnitude, while the chromatographic resolution was much better. A schematic of OLGA(III) connected to either a rotating wheel or a tape detection system is shown in Figure 10.

OLGA(III) has very successfully been applied to study volatile halides and/or oxyhalides of Rf [118], Db [119], Sg [78,120] and Bh [121], see Chapter 7 for a detailed discussion. In all these experiments the separated transactinide nuclides were unambiguously identified via their nuclear decay properties. An improved version of OLGA(I), named HIgh-Temperature on-line Gas chromatography ApparatuS (HITGAS), has been developed at Forschungszentrum Rossendorf and successfully applied to study oxide hydroxides of group 6 elements including Sg [122,123].

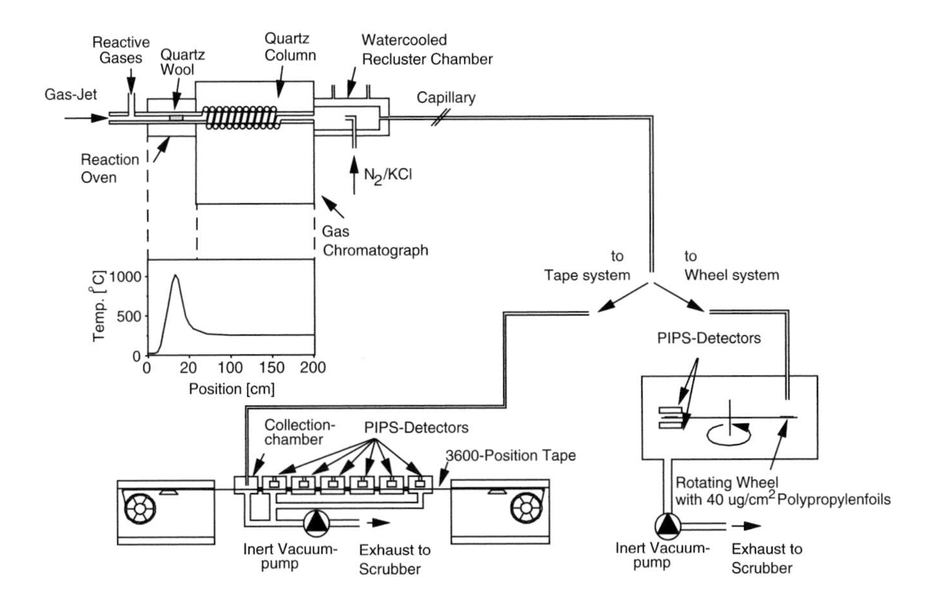

Fig. 10. Schematic of OLGA(III) in combination with the tape detection system or the MG or ROMA wheel detection system. See text for a more detailed description.

In order to further reduce the background of unwanted α-decaying nuclides, the so-called parent-daughter recoil counting modus was implemented at the rotating wheel systems. Since the investigated transactinide nuclei decay with a characteristic decay sequence involving the α decay and/or SF decay of daughter nuclei, the significance of the observed decay sequence can be enhanced by observing the daughter decays in a nearly background free counting regime. This can be accomplished in the following manner, see example with ^{267}Bh in Figure 11. In the parent mode, a ^{267}Bh atom is deposited on the top of a thin foil together with a sample of the aerosol transport material. The wheel is double-stepped at preset time intervals to position the collected samples successively between pairs of α-particle detectors. When the ^{267}Bh α-decay is detected in the bottom of a detector pair, it is assumed that the ^{263}Db daughter has recoiled into the face of the top detector. The wheel is single-stepped to remove the sources from between the detector pairs, and a search for the ^{263}Db and ^{259}Lr daughters is made for a second preset time interval, before single-stepping the wheel again to resume the search for decays of ^{267}Bh.

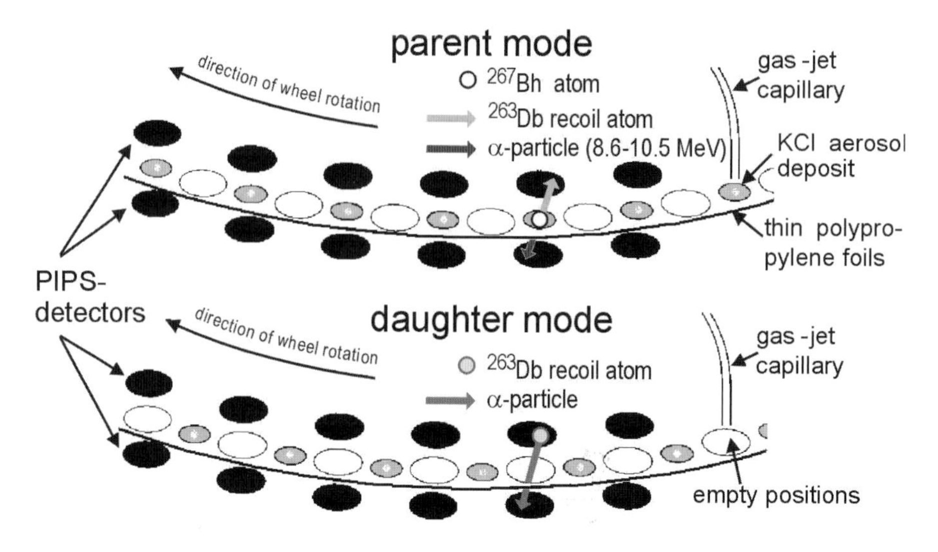

Fig. 11. Parent-daughter mode for rotating wheel systems. See the text for detailed description of operation.

4.4 IN-SITU VOLATILIZATION AND ON-LINE DETECTION

Even though the OLGA technique has very successfully been applied in gas phase chemical studies of elements Rf through Bh, the overall efficiency (including the detection of a two member decay chain) was only about 4%, too low to continue studies at the picobarn cross section level. Obviously, the large number of steps involved, each having a yield below 100%, lead to a poor overall efficiency. These include thermalization of the recoil products in a gas, attachment of products in ionic or atomic form to the surface of aerosol particles, transport of these particles through capillaries, hetero-chemical reactions of the attached species with reactive gases to form volatile compounds, gas adsorption chromatography of the compounds with the surface of the chromatography column, re-attachment of ejected molecules to new particles followed by a transport to the counting device and, finally, deposition of the particles on thin foils via impaction.

Therefore, it would be advantageous to perform the chemical synthesis of the volatile molecule in-situ in the recoil chamber. This approach has already been used in the very first chemical studies of transactinides, see Section 4.3. Chlorinating and brominating agents were added to a carrier gas in order to form volatile halides of the 6d elements. However, in these early experiments only tracks of SF events were revealed after completion of the experiment. Obviously, such a technique has the disadvantage of not yielding any on-line information during an ongoing experiment. Moreover, most isotopes of transactinide elements decay primarily by α-particle emission.

For studies of element 108 (Hs) and element 112 the new device named In-situ Volatilization and On-line detection (IVO) was developed [124]. By adding O_2 to the He carrier gas, volatile tetroxides of group-8 elements were formed in-situ in the recoil chamber. A quartz column containing a quartz wool plug heated to 600°C was mounted as close as possible to the recoil chamber. The hot quartz wool served as an aerosol filter and provided a surface to complete the oxidation reaction. For future studies with element 112 pure He or even a reducing He/H_2 mixture can be employed.

In order to efficiently detect the nuclear decay of isolated nuclides as well as to obtain chemical information about the volatility of the investigated compounds a completely new technique was devised. This new development is cryo thermochromatography of very volatile species on positive implanted N-type silicon (PIN) diode surfaces that allow α-particle and SF spectrometry [80]. The carrier gas containing volatile atoms (i.e. At, Rn, Hg) or molecules (i.e. OsO_4, HsO_4) is flowing through a narrow channel formed by a series of planar silicon diodes. Along this channel a longitudinal negative temperature gradient is established. Due to the close proximity of the silicon diodes facing each other, the probability to register a complete decay chain consisting of a series of α decays is rather high. A first Cryo Thermochromatographic Separator (CTS) was constructed by U. Kirbach and coworkers [80] at the Lawrence Berkeley National Laboratory. A schematic of the experimental set-up used in the first successful chemical identification of Hs as volatile tetroxide is shown in Figure 12.

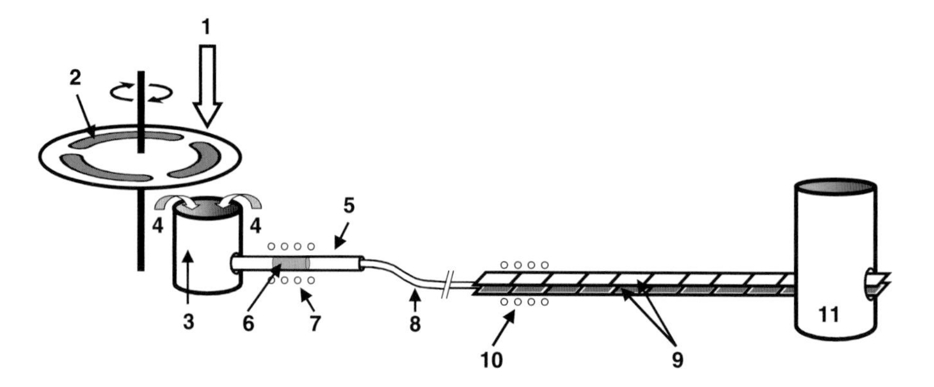

Fig. 12. The ^{26}Mg-beam (1) passed through the rotating vacuum window and ^{248}Cm-target (2) assembly. In the fusion reaction ^{269,270}Hs nuclei were formed which recoiled out of the target into a gas volume (3) and were flushed with a He/O_2 mixture (4) to a quartz column (5) containing a quartz wool plug (6) heated to 600 °C by an oven (7). There, Hs was converted to HsO_4 which is volatile at room temperature and transported with the gas flow through a perfluoroalkoxy (PFA) capillary (8) to the COLD detector array registering the nuclear decay (α and spontaneous fission) of the Hs nuclides. The array consisted of 36 detectors arranged in 12 pairs (9), each detector pair consisted of 3 PIN diode sandwiches. Always 3 individual PIN diodes (top and bottom) were electrically coupled. A thermostat (10) kept the entrance of the array at 20 °C; the exit was cooled to -170 °C by means of liquid nitrogen (11). Depending on the volatility of HsO_4, the molecules adsorbed at a characteristic temperature. Figure reproduced from [30].

In the actual Hs experiment [30] an improved version namely the Cryo On-Line Detector (COLD) was used. This detector was constructed at PSI and consisted of 2 times 36 PIN diodes. A schematic of the COLD is shown in Figure 13.

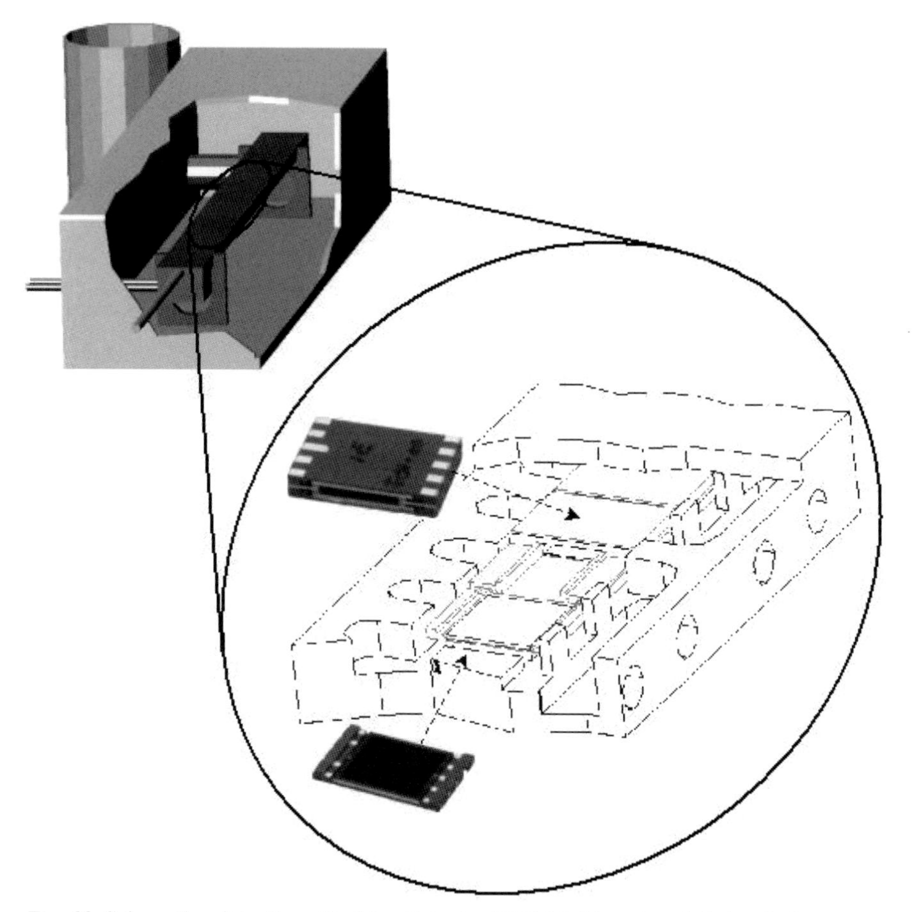

Fig. 13. Schematic of the Cryo On-Line Detector (COLD). For a detailed description see text. Figure from [125].

The COLD featured a steeper temperature gradient and reached a lower end temperature of -170 °C. Two PIN-diodes mounted on ceramic supports were glued together facing each other. Two T-shaped spacers made from silicon confined the gas flow to the active surface of the diodes. The gap between the PIN diodes in the COLD was reduced to 1.5 mm, increasing thus the detection efficiency. The PIN diode sandwiches were enclosed in a Teflon™ coated copper bar, which was placed in a stainless steel housing that was purged with dry N_2. The copper bar was heated at the entrance side with a thermostat to +20 °C and cooled at the exit with a liquid N_2 cold finger. The geometrical efficiency for detecting a single α-particle emitted by a species adsorbed inside the detector array was 77%. The detectors of the COLD array were calibrated on-line with α-decaying ^{219}Rn and its daughters ^{215}Po and ^{211}Bi using a ^{227}Ac source.

References

1. Rutherford, E.: "Radioactive Substance emitted from Thorium compounds", Philosophical Magazine, January 1900, ser. 5, xlix, (1900) 1.
2. Herrmann, G., Trautmann, N.: Annu. Rev. Nucl. Part. Sci. **32**, (1982) 117.
3. Meyer, R.A., J. Radioanal. Nucl. Chem. **142**, (1990) 135.
4. Skarnemark, G., Alstad, J., Kaffrell, N., Trautmann, N.,: J. Radioanal. Nucl. Chem. **142**, (1990) 145.
5. Rengan, K., J. Radioanal. Nucl. Chem. **142**, (1990) 173.
6. Baker, J.D., Maikrantz, D.H., Gehrke, R.J., Greenwood, R.C.: J. Radioanal. Nucl. Chem. **142**, (1990) 159.
7. Gäggeler, H.W., J. Radioanal. Nucl. Chem. **183**, (1994) 261.
8. Rengan, K., Meyer, R.A.: *Ultrafast Chemical Separations*, Nuclear Science Series, NAS-NS-3118, National Academy Press, Washington (1993).
9. Hahn, O., Strassmann, F.: Naturwissenschaften **27,** (1939) 11.
10. Seaborg, G.T., Loveland, W.D., *The Elements Beyond Uranium*, Wiley & Sons, New York, 1990.
11. Hyde, E.K., Hoffman, D.C., Keller Jr., O.L.: Radiochim. Acta **42**, (1987) 57.
12. Wierczinski, B., Gregorich, K.E., Kadkhodayan, B., Lee, D.M., Beauvais, L.G., Hendricks, M.B., Kacher, C.D., Lane, M.R., Keeney-Shaughnessy, D.A., Stoyer, N.J., Strellis, D.A., Sylwester, E.R., Wilk, P.A., Hoffman, D.C., Malmbeck, R., Skarnemark, G., Alstad, J., Omtvedt, J.P., Eberhardt, K., Mendel, M., Nähler, A., Trautmann, N.: J. Radioanal. Nucl. Chem. **247**, (2001) 57.
13. Zvara, I., Eichler, B., Belov, V.Z., Zvarova, T.S., Korotkin, Yu.S., Shalaevsky, M.R., Shchegolev, V.A., Hussonois, M.: Radiokhimiya **16**, (1974) 720.
14. Zvara, I.: Isotopenpraxis **26**, (1990) 251.
15. Schädel, M.: Radiochim. Acta **70/71**, (1995) 207.
16. Trautmann, N.: Radiochim. Acta **70/71**, (1995) 237.
17. Türler, A.: Radiochim. Acta **72**, (1995) 7.
18. Gäggeler, H.W.: "Fast Chemical Separation Procedures for Transactinides". In: Proceedings of "The Robert A. Welch Foundation 41st Conference on Chemical Research. The Transactinide Elements", Houston, Texas, 27-28 October 1997, pp. 43-63.
19. Hoffman, D.C., Lee, D.M., J. Chem. Ed. **76**, (1999) 331.
20. Guillaumont, R., Adloff, J.P., Peneloux, A.: Radiochim. Acta **46**, (1989) 169.
21. Muggleton, A.H.F., J. Phys. E **12**,781 (1979).
22. Evans, J.E., Lougheed, R.W., Coops, M.S., Hoff, R.W., Hulet, E.K.: Nucl. Instrum. Meth. **102**, (1972) 389.
23. Aumann, D.C., Müllen G.: Nucl. Instrum. Meth. **115**, (1973) 75.
24. Müllen, G., Aumann, C.: Nucl. Instrum, Meth. **128**, (1975) 425.
25. Trautmann, N., Folger, H.: Nucl. Instrum. Meth. **A282**, (1989) 102.
26. Coppieters, L., Van Campenhout, J., Triffaux, J., Van Audenhove, J.: Nucl. Instrum. Meth. **224**, (1984) 41.
27. Nitschke, J.M.: Nucl. Instrum. Meth. **138**, (1976) 393.
28. Yakushev, A: http://159.93.28.88/www-nekho/chem_112.html, Yakushev, A.: "First Chemistry Experiments with Element 112". In "Workshop on Recoil Separator for Superheavy Element Chemistry", GSI, Darmstadt, Germany, March 20-21, 2002, http://www.gsi.de/chemsep/images/Element%20112.pdf

29. Eberhardt, K.: "Preparation of Lanthanide and Actinide Targets for the new GSI Rotaing Wheel Target Assembly". In: "Workshop on Recoil Separator for Superheavy Element Chemistry", GSI, Darmstadt, Germany, March 20-21, 2002, http://www.gsi.de/chemsep/images/Eberhardt%20V2.pdf
30. Düllmann, C.E., Brüchle, W., Dressler, R., Eberhardt, K., Eichler, B., Eichler, R., Gäggeler, H.W., Ginter, T.N., Glaus, F., Gregorich, K.E., Hoffman, D.C., Jäger, E., Jost, D.T., Kirbach, U.W., Lee, D.M., Nitsche, H., Patin, J.B., Pershina, V., Piguet, D., Qin, Z., Schädel, M., Schausten, B., Schimpf, E., Schott, H.J., Soverna, S., Sudowe, R., Thörle, P., Timokhin, S.N., Trautmann, N., Türler, A., Vahle, A., Wirth, G., Yakushev, A.B., Zielinski, P.M.: Nature **418**, (2002) 859.
31. Ghiorso, A., Harvey, B., Choppin, G.R., Thompson, S.G., Seaborg, G.T.: Phys. Rev. **98**, (1955) 1518.
32. Wollnik, H: Nucl. Instrum. Meth. **139**, (1976) 311.
33. Proceedings of the "Workshop on Recoil Separator for Superheavy Element Chemistry", GSI, Darmstadt, Germany, March 20-21, 2002, http://www.gsi.de/chemsep/
34. Omtvedt, J.P., Alstad, J., Brevik, H, Dyve, J.E., Eberhardt, K., Folden III, C.M., Ginter, K., Gregorich, K.E., Hult, E.A., Johansson, M., Kirbach, U.W., Lee, D.M., Mendel, M., Nähler, A., Ninov, V, Omtvedt, L.A., Patin, J.B., Skarnemark, G., Stavsetra, L., Sudowe, R., Wiehl, N., Wierczinski, B., Wilk, P.A., Zielinski, P.A., Kratz, J.V., Trautmann, N., Nitsche, H., Hoffman, D.C.: J. Nucl. Radioanalyt. Sci. **3**,121 (2002) 121.
35. Ninov, V., Gregorich, K.E.: "Berkeley Gas-filled Separator". In: *ENAM98*, Eds. Sherrill, B.M., Morrissey, D.J., Davids, C.N., AIP, Woodbury, (1999) 704.
36. Kirbach, U.W., Gregorich, K.E., Ninov, V., Lee, D.M., Patin, J.B., Shaughnessy, D.A., Strellis, D.A., Wilk, P.A., Hoffman, D.C., Nitsche, H.: Lawrence Berkeley National Laboratory Nuclear Science Division Annual Report 1999, http://www-nsd.lbl.gov/nsd/annual/ydc/nsd1999/lowenergy/nuclchem/pdf/AnnualReport99_RTC.pdf
37. Czerwinski, K.R., Gregorich, K.E., Hannink, N.J., Kacher, C.D., Kadkhodayan, B.A., Kreek, S.A., Lee, D.M., Nurmia, M.J., Türler, A., Seaborg, G.T., Hoffman, D.C.: Radiochim. Acta, **64**, (1994) 23.
38. Czerwinski, K.R., Kacher, C.D., Gregorich, K.E., Hamilton, T.M., Hannink, N.J., Kadkhodayan, B.A., Kreek, S.A., Lee, D.M., Nurmia, M.J., Türler, A., Seaborg, G.T., Hoffman, D.C.: Radiochim. Acta **64**, (1994) 29.
39. Kacher, C.D., Gregorich, K. E., Lee, D.M., Watanabe, Y., Kadkhodayan, B., Wierczinski, B., Lane, M.R., Sylwester, E.R., Keeney, D.A., Hendricks, M., Stoyer, N.J., Yang, J., Hsu, M., Hoffman, D.C., Bilewicz, A.: Radiochim. Acta **75**, (1996) 127.
40. Kacher, C.D., Gregorich, K.E., Lee, D.M., Watanabe, Y., Kadkhodayan, B., Wierczinski, B., Lane, M.R., Sylwester, E.R., Keeney, D.A., Hendricks, M., Hoffman, D.C., Bilewicz, A.: Radiochim. Acta **75**, (1996) 135.
41. Gregorich, K.E., Henderson, R. A., Lee, D.M., Nurmia, M. J., Chasteler, R. M., Hall, H.L., Bennett, D.A., Gannet, C.M., Chadwick, R.B., Leyba, J.D., Hoffman, D.C., Herrmann, G.: Radiochim. Acta **43**, (1988) 223.
42. Silva, R., Sikkeland, T., Nurmia, M., Ghiorso, A.: Inorg. Nucl. Chem. Letters **6**, (1970) 733.
43. Hoffman, D.C., Henderson, R.A., Gregorich, K.E., Bennett, D.A., Chasteler, R.M., Gannet, C.M., Hall, H.L., Lee, D.M., Nurmia, M.J., Cai, S., Agarwal, Y.K., Charlop, A.W., Chu, Y.Y., Silva, R.J.: J. Radioanal. Nucl. Chem. **124**, (1988) 135.
44. Schädel, M.: Radiochim. Acta **89**, (2001) 721.
45. Schädel, M.: J. Nucl. Radiochem. Sci. **3**, (2002) 113.

46. Schädel, M., Brüchle, W., Jäger, E., Schimpf, E., Kratz, J.V., Scherer, U.W., Zimmermann, H.P.: Radiochim. Acta **48**, (1999) 171.
47. Günther, R., Paulus, W., Kratz, J.V., Seibert, A., Thörle, P., Zauner, S., Brüchle, W., Jäger, E., Pershina, V., Schädel, M., Schausten, B., Schumann, D., Eichler, B., Gäggeler, H.W., Jost, D.T., Türler, A.: Radiochim. Acta **80**, (1998) 121.
48. Strub, E., Kratz, J.V., Kronenberg, A., Nähler, A., Thörle, P., Zauner, S., Brüchle, W., Jäger, E., Schädel, M., Schausten, B., Schimpf, E., Li, Z.W., Kirbach, U., Schumann, D., Jost, D., Türler, A., Asai, M., Nagame, Y., Sakama, M., Tsukada, K., Gäggeler, H.W., Glatz, J.P.: Radiochim. Acta **88**, (2000) 265.
49. Kratz, J.V., Gober, M.K., Zimmermann, H.P., Schädel, M., Brüchle, W., Schimpf, E., Gregorich, K.E., Türler, A., Hannink, N.J., Czerwinski, K.R., Kadkhodayan, B., Lee, D.M., Nurmia, M.J., Hoffman, D.C., Gäggeler, H., Jost, D., Kovacs, J., Scherer, U.W., Weber, A.: Phys. Rev. C**45**, (1992) 1064.
50. Schädel, M., Brüchle, W., Schimpf, E., Zimmermann, H.P., Gober, M.K., Kratz, J.V., Trautmann, N., Gäggeler, H., Jost, D., Kovacs, J., Scherer, U.W., Weber, A., Gregorich, K.E., Türler, A., Czerwinski, K.R., Hannink, N.J., Kadkhodayan, B., Lee, D.M., Nurmia, M.J., Hoffman, D.C.: Radiochim. Acta **57**, (1992) 85.
51. Gober, M.K., Kratz, J.V., Zimmermann, H.P., Schädel, M., Brüchle, W., Schimpf, E., Gregorich, K.E., Türler, A., Hannink, N.J., Czerwinski, K.R., Kadkhodayan, B., Lee, D.M., Nurmia, M.J., Hoffman, D.C., Gäggeler, H., Jost, D., Kovacs, J., Scherer, U.W., Weber, A.: Radiochim. Acta **57**, (1992) 77.
52. Zimmermann, H.P., Gober, M.K., Kratz, J.V., Schädel, M., Brüchle, W., Schimpf, E., Gregorich, K.E., Türler, A., Czerwinski, K.R., Hannink, N.J., Kadkhodayan, B., Lee, D.M., Nurmia, M.J., Hoffman, D.C., Gäggeler, H., Jost, D., Kovacs, J., Scherer, U.W., Weber, A.: Radiochim. Acta **60**, (1993) 11.
53. Paulus, W., Kratz, J.V., Strub, E., Zauner, S., Brüchle, W., Pershina, V., Schädel, M., Schausten, B., Adams, J.L., Gregorich, K.E., Hoffman, D.C., Lane, M.R., Lane, C., Lee, D.M., McGrath, C.A., Shaughnessy, D.K., Strellis, D.A., Sylwester, E.R.: Radiochim. Acta **84**, (1999) 69.
54. Schädel, M., Brüchle, W., Schausten, B., Schimpf, E., Jäger, E., Wirth, G., Günther, R., Kratz, J.V., Paulus, W., Seibert, A., Thörle, P., Trautmann, N., Zauner, S., Schumann, D., Andrassy, M., Misiak, R., Gregorich, K.E., Hoffman, D.C., Lee, D.M., Sylwester, E.R., Nagame, Y., Oura, Y.: Radiochim. Acta **77**, (1997) 149.
55. Schädel, M., Brüchle, W., Jäger, E., Schausten, B., Wirth, G., Paulus, W., Günther, R., Eberhardt, K., Kratz, J.V., Seibert, A., Strub, E., Thörle, P., Trautmann, N., Waldek, A., Zauner, S., Schumann, D., Kirbach, U., Kubica, B., Misiak, R., Nagame, Y., Gregorich, K.E.: Radiochim. Acta **83**, (1998) 163.
56. Haba, H., Tsukuda, K., Asai, M., Goto, S. Toyoshima, A., Nishinaka, I., Akiyama, K., Hirata, M., Ichikawa, S., Nagame, Y., Shoji, Y., Shigekawa, M., Koike, T., Iwasaki, M., Shinohara, A., Kaneko, T., Maruyama, T., Ono, S., Kudo, H., Oura, Y., Sueki, K., Nakahara, H., Sakama, M., Yokoyama, A., Kratz, J.V., Schädel, M., Brüchle, W.: Radiochim. Acta **89**, (2001) 733.
57. Kazuaki, T., Nishinaka, I, Asai, M., Goto, S., Sakama, M., Haba, H., Ichikawa, S., Nagame, Y., Schädel, M.: "Automated Liquid Chromatography Apparatus Coupled with an On-line Alpha-Particle Detection System". In: "2nd International Symp. on Advanced Sci. Res., Advances in Heavy Element Research – ASR-2001", Tokai, Ibaraki, Japan, Nov. 13-15, 2001 http://asr.tokai.jaeri.go.jp/asr2001/abstract/Tsukada.pdf

58. Nagame, Y., Asai, M., Haba, H., Kazuaki, T., Goto, S., Sakama, M., Nishinaka, I, Toyoshima, A., Akayama, K., Ichikawa, S.: J. Nucl. Radiochem. Sci. **3**, (2002) 129.
59. Haba, H., Tsukuda, K., Asai, M., Goto, S., Toyoshima, A., Nishinaka, I., Akiyama, K., Hirata, M., Ichikawa, S., Nagame, Y., Shoji, Y., Shigekawa, M., Koike, T., Iwasaki, M., Shinohara, A., Kaneko, T., Maruyama, T., Ono, S., Kudo, H., Oura, Y., Sueki, K., Nakahara, H., Sakama, M., Yokoyama, A., Kratz, J.V., Schädel, M., Brüchle, W.: J. Nucl. Radiochem. Sci. **3**, (2002) 143.
60. Szeglowski, Z., Bruchertseifer, H., Domanov, V.P., Gleisberg, B., Guseva, L.J., Hussonnois, M., Tikhomirova, G.S., Zvara, I., Oganessian, Yu.Ts.: Radiochim. Acta **51**, (1990) 71.
61. Trubert, D., Hussonnois,M., Brillard, L., Barci, V., Ardisson, G., Szeglowski, Z., Constantinescu, O.: Radiochim. Acta **69**, (1995) 149.
62. Pfrepper, G., Pfrepper, R., Yakushev, A.B., Timokhin, S.N., Zvara, I.: Radiochim. Acta **77**, (1997) 201.
63. Pfrepper, G., Pfrepper, R., Krauss, D., Yakushev, A.B., Timokhin, S.N., Zvara, I.: Radiochim. Acta **80**, (1998) 7.
64. Pfrepper, G., Pfrepper, R., Kronenberg, A., Kratz, J.V., Nähler, A., Brüchle, W., Schädel, M.: Radiochim. Acta **88**, (2000) 273.
65. Kronenberg, A., Kratz, J.V., Strub, E., Brüchle, W., Jäger, E., Schädel, M., Schausten, B.: GSI Scientific Report 1988, GSI 99-1, (1999) p. 171.
66. Omtvedt, J.P., Alstad, J., Eberhardt, K., Fure, K., Malmbeck, R., Mendel, M., Nähler, A., Skarnemark, G., Trautmann, N., Wiehl, N. Wierczinski, B.: J. Alloys Comp. **271**, (1998) 303.
67. Wilk, P.A., Gregorich, K.E., Hendricks, M.B., Lane, M.R., Lee, D.M., McGrath, C.A., Shaughnessy, D.A., Strellis, D.A., Sylwester, E.R., Hoffman, D.C.: Phys. Rev. C **56**, (1997) 1626.
68. Wilk, P.A., Gregorich, K.E., Hendricks, M.B., Lane, M.R., Lee, D.M., McGrath, C.A., Shaughnessy, D.A., Strellis, D.A., Sylwester, E.R., Hoffman, D.C.: Phys. Rev. C **58**, (1998) 1352.
69. Oganessian, Y.T., Yeremin, A.V., Gulbekian, G.G., Bogomolov, S.L., Chepigin, V.I., Gikal, B.N., Gorshkov, V.A., Itkis, M.G., Kabachenko, A.P., Kutner, V.B., Lavrentev, A.Y., Malyshev, O.N., Popeko, A.G., Rohac, J., Sagaidak, R.N., Hofmann, S., Münzenberg, G., Veselsky, M., Saro, S., Iwasa, N., Morita, K.: Eur. Phys. J. A **5**, (1999) 63.
70. Oganessian, Y.T., Yeremin, A.V., Popeko, A.G., Bogomolov, S.L., Buklanov, G.V., Chelnokov, M.L., Chepigin, V.I., Gikal, B.N., Gorshkov, V.A., Gulbekian, G.G., Itkis, M.G., Kabachenko, A.P., Lavrentev, A.Y., Malyshev, O.N., Rohac, J., Sagaidak, R.N., Hofmann, S., Saro, S., Giardina, G., Morita, K.: Nature **400**, (1999) 242.
71. Oganessian, Y.T., Utyonkov, V.K., Lobanov, Y.V., Abdullin, F.S., Polyakov, A.N., Shirokovsky, I.V., Tsyganov, Y.S., Gulbekian, G.G., Bogomolov, S.L., Gikal, B.N., Mezentsev, A.N., Iliev, S., Subbotin, V.G., Sukhov, A.M., Buklanov, G.V., Subotic, K., Itkis, M.G., Moody, K.J., Wild, J.F., Stoyer, N.J., Stoyer, M.A., Lougheed, R.W.: Phys. Rev. Lett. **83**, (1999) 3154.
72. Oganessian, Y.T., Utyonkov, V.K., Lobanov, Y.V., Abdullin, F.S., Polyakov, A.N., Shirokovsky, I.V., Tsyganov, Y.S., Gulbekian, G.G., Bogomolov, S.L., Gikal, B.N., Mezentsev, A.N., Iliev, S., Subbotin, V.G., Sukhov, A.M., Ivanov, O.V., Buklanov, G.V., Subotic, K., Itkis, M.G., Moody, K.J., Wild, J.F., Stoyer, N.J., Stoyer, M.A., Lougheed, R.W.: Phys. Rev. C **62**, (2000) 1604.

73. Zvara, I., Chuburkov, Yu.T., Tsaletka, R., Zvarova, T.S., Shalaevsky, M.R., Shilov B.V.: At. Energ. **21,** (1966) 83.
74. Zvara, I., Belov, V.Z., Korotkin, Yu.S., Shalayevsky, M.R., Shchegolev, V.A., Hussonnois, M., Zager, B.A.: "Experiments on Chemical Identification of Spontaneously Fissionable Isotope of Element 105". In: "Communications of the Joint Institute for Nuclear Research" P12-5120, Dubna (1970), 13 p.
75. Zvara, I., Yakushev, A.B., Timokhin, S.N., Honggui, Xu, Perelygin, V.P., Chuburkov, Yu.T.: Radiochim. Acta **81**, (1998) 179.
76. Türler, A., Gäggeler, H.W., Gregorich, K.E., Barth, H., Brüchle, W., Czerwinski, K.R., Gober, M.K., Hannink, N.J., Henderson, R.A., Hoffman, D.C., Jost, D.T, Kacher, C.D., Kadkhodayan, B., Kovacs, J., Kratz, J.V., Kreek, S.A., Lee, D.M., Leyba, J.D., Nurmia, M.J., Schädel, M., Scherer, U.W., Schimpf, E., Vermeulen, D., Weber, A., Zimmermann, H.P., Zvara, I.: J. Radioanal. Nucl. Chem. **160**, (1992) 327.
77. Gäggeler, H.W., Jost, D.T., Kovacs, J., Scherer, U.W., Weber, A., Vermeulen, D., Türler, A., Gregorich, K.E., Henderson, R.A., Czerwinski, K.R., Kadkhodayan, B., Lee, D.M., Nurmia, M., Hoffman, D.C., Kratz, J.V., Gober, M.K., Zimmermann, H.P., Schädel, M., Brüchle, W., Schimpf, E., Zvara, I.: Radiochim. Acta **57**, (1992) 93.
78. Schädel, M., Brüchle, W., Dressler, R., Eichler, B., Gäggeler, H.W., Günther, R., Gregorich, K.E., Hoffman, D.C., Hübener, S., Jost, D.T., Kratz, J.V., Paulus, W., Schumann, D., Timokhin, S., Trautmann, N., Türler, A., Wirth, G., Yakushev, A.: Nature **388**, (1997) 55.
79. Eichler, R., Brüchle, W., Dressler, R., Düllmann, Ch.E., Eichler, B., Gäggeler, H.W., Gregorich, K.E., Hoffman, D.C., Hübener, S., Jost, D.T., Kirbach, U.W., Laue, C.A., Lavanchy, V.M., Nitsche, H., Patin, J.B., Piguet, D., Schädel, M., Shaugnessy, D.A., Strellis, D.A., Taut, S., Tobler, L., Tsyganov, Y.S., Türler, A., Vahle, A., Wilk, P.A., Yakushev, A.B.: Nature **407**, (2000) 63.
80. Kirbach, U.W., Folden III, C.M., Ginter, T.N., Gregorich, K.E., Lee, D.M., Ninov, V., Omtvedt, J.P., Patin, J.B., Seward, N.K., Strellis, D.A., Sudowe, R., Türler, A., Wilk, P.A., Zielinski, P.M., Hoffman, D.C., Nitsche, H.: Nucl. Instr. Meth. A **484,** (2002) 587.
81. Zvara, I., Chuburkov, Yu.T., Zvarova, T.S., Tsaletka, R.: Radiokhimiya **11**, (1969) 154.
82. Trautmann, N., Zendel, M., Dittner, P.F., Stender, E., Ahrens, H., Silva, R.J.: Nucl. Instr. Meth. **147**, (1977) 371.
83. Hickmann, U., Greulich, N., Trautmann, N., Gäggeler, H.W., Gäggeler-Koch, H., Eichler, B., Herrmann, G.: Nucl. Instr. Meth. **174** (1980) 507.
84. Zvara, I., Keller Jr., O.L., Silva, R.J., Tarrant, J.R.: J. Chromatogr. **103**, (1975) 77.
85. Eichler, B., Domanov, V.P.: J. Radioanal. Chem. **28**, (1975) 143.
86. Helas, G., Hoffmann, P., Bächmann, K.: Radiochem. Radioanal. Lett. **30**, (1977) 371.
87. Eichler,B., Maltseva, N.S.: Sov. Radiochemistry **20**, (1978) 70.
88. Tsalas, S., Bächmann, K.: Anal. Chim. Acta **98**, (1978) 17.
89. Bayar, B., Novgorodov, A.F., Vocilka, I., Zaitseva, N.G.: Radiochem. Radioanal. Lett. **35**, (1978) 109.
90. Polyakov, E.V.: Sov. Radiochem. **32**, (1990) 548.
91. Bayar, B., Vocilka, I., Zaitseva, N.G., Novgorodov, A.F.: J. Inorg. Nucl. Chem. **40**, (1978) 1461.
92. U Zin, Kim, Timokhin, S.N., Zvara, I.: Isotopenpraxis **24**, (1988) 30.
93. Rudolph, J., Bächmann, K.: Microchim. Acta **I**, (1979) 477.
94. Rudolph, J., Bächmann, K.: J. Radioanal. Chem. **43**, (1978) 113.
95. Rudolph, J., Bächmann, K.: Radiochim. Acta **27**, (1980) 105.

96. von Dincklage, R.D., Schrewe, U.J., Schmidt-Ott, W.D., Fehse, H.F., Bächmann, K.: Nucl. Instr. Meth. **176**, (1980) 529.
97. Gäggeler, H., Dornhöfer, H., Schmidt-Ott, W.D., Greulich, N., Eichler, B.: Radiochim. Acta **38**, (1985) 103.
98. Gäggeler, H.W., Jost, D.T., Baltensperger, U., Weber, A., Kovacs, A., Vermeulen, D., Türler, A.: Nucl. Instr. Meth. **A309**, (1991) 201.
99. Kadkhodayan, B., Türler, A., Gregorich, K.E., Nurmia, M.J., Lee, D.M., Hoffman, D.C.: Nucl. Instr. Meth. **A317**, (1992) 254.
100. Eichler, B.: Radiochem. Radioanal. Lett. **22**, (1975) 147.
101. Rudolph, J., Bächmann, K., Steffen, A., Tsalas, S.: Microchim. Acta **I** (1978) 471.
102. Rudolph, J., Bächmann, K.: J. Chromatogr. **178**, (1979) 459.
103. Tsalas, S., Bächmann, K., Heinlein, G.: Radiochim. Acta **29**, (1981) 217.
104. Zvara, I., Chuburkov, Yu.T., Tsaletka, R., Shalaevskii, M.R.: Radiokhimiya **11**, (1969) 163.
105. Chuburkov, Yu.T., Zvara, I., Shilov, B.V.: Radiokhimiya **11**, (1969) 174.
106. Zvara, I., Chuburkov, Yu.T., Belov, V.Z., Buklanov, G.V., Zakhvataev, B.B., Zvarova, T.S., Maslov, O.D., Caletka, R., Shalaevsky, M.R.: J. Inorg. Nucl. Chem. **32**, (1970) 1885.
107. Zvara, I., Belov, V.Z., Domanov, V.P., Shalaevskii, M.R.: Radiokhimiya **18**, (1976) 371.
108. Timokhin, S.N., Yakushev, A.B., Perelygin, V.P., Zvara, I.: "Chemical Identification of Element 106 by the Thermochromatographic Method". In: Proceedings of the "International School Seminar on Heavy Ion Physics", Dubna, 10-15 May 1993, pp. 204-206.
109. Timokhin, S.N., Yakushev, A.B., Honggui, XU, Perelygin, V.P., Zvara, I.: J. Radioanal. Nucl. Chem., Lett. **212**, (1996) 31.
110. Yakushev, A.B., Timokhin, S.N., Vedeneev, M.V., Honggui, XU, Zvara, I.: J. Radioanal. Nucl. Chem. **205**, (1996) 63.
111. Zvara, I., Domanov, V.P., Hübener, S., Shalaevskii, M.R., Timokhin, S.N., Zhuikov, B.L., Eichler, B., Buklanov, G.V.: Radiokhimiya **26**, (1984) 76.
112. Ghiorso, A., Nurmia, M., Eskola, K., Eskola, P.: Inorg. Nucl. Chem. Letters **7**, (1971) 1117.
113. Kratz, J.V.: "Chemistry of Seaborgiuim and Prospects for Chemical Studies of Bohrium and Hassium", In: Proceedings of "The Robert A. Welch Foundation 41st Conference on Chemical Research, The Transactinide Elements", Houston, Texas, 27-28 October 1997, pp. 65.
114. Gäggeler, H., Dornhöfer, H., Schmidt-Ott, W.D., Greulich, N., Eichler, B.: Radiochim. Acta **38**, (1985) 103.
115. Sümmerer, K., Brügger, M., Brüchle, W., Gäggeler, H., Jäger, E., Schädel, M., Schardt, D., Schimpf, E.: "ROMA - A Rotating Wheel Multidetector Apparatus used in Experiments with ^{254}Es as a Target", In: GSI Annual Report 1983, 84-1, Darmstadt (1984) 246.
116. Schädel, M., Jäger, E., Schimpf, E., Brüchle, W.: Radiochim. Acta **68**, (1995) 1.
117. Hoffman, D.C., Lee, D.M., Ghiorso, A., Nurmia, M.J., Aleklett, K., Leino, M.: Phys. Rev. **C24**, (1981) 495.
118. Türler, A., Buklanov, G.V., Eichler, B., Gäggeler, H.W., Grantz, M., Hübener, S., Jost, D.T., Lebedev V.Ya., Piguet, D., Timokhin, S.N., Yakushev, A.B., Zvara, I.: J. Alloys Comp. **271-273,** (1998) 287.

119. Türler, A. Eichler, B. Jost, D.T., Piguet, D., Gäggeler, H.W., Gregorich, K.E., Kadkhodayan, B., Kreek, S.A., Lee, D.M., Mohar, M., Sylwester, E., Hoffman, D.C., Hübener, S.: Radiochim. Acta **73**, (1996) 55.
120. Türler, A. Brüchle, W., Dressler, R., Eichler, B., Eichler, R., Gäggeler, H.W., Gärtner, M., Gregorich, K.E., Hübener, S., Jost, D.T., Lebedev, V.Y., Pershina, V.G., Schädel, M., Taut, S., Timokhin, S.N., Trautmann, N., Vahle, A., Yakushev, A.B.: Angew. Chem. Int. Ed. **38**, (1999) 2212.
121. Eichler, R., Brüchle, W., Dressler, R., Düllmann, Ch. E., Eichler, B., Gäggeler, H.W., Gregorich, K.E., Hoffman, D.C., Hübener, S., Jost, D.T., Kirbach, U.W., Laue, C.A., Lavanchy, V.M., Nitsche, H., Patin, J.B., Piguet, D., Schädel, M., Shaughnessy, D.A., Strellis, D.A. Taut, S., Tobler, L., Tsyganov, Y.S., Türler, A., Vahle, A., Wilk, P.A., Yakushev, A.B.: Nature **407**, 63 (2000).
122. Vahle, A., Hübener, S., Dressler, R., Grantz, M.: Nucl. Instr. Meth. A **481**, (2002) 637.
123. Hübener, S., Taut, S., Vahle, A., Dressler, R., Eichler, B., Gäggeler, H.W., Jost, D.T., Piguet, D., Türler, A., Brüchle, W., Jäger, E., Schädel, M., Schimpf, E., Kirbach, U., Trautmann, N., Yakushev, A.B., Radiochim. Acta **89**, (2001) 737.
124. Düllmann, Ch.E., Eichler, B., Eichler, R., Gäggeler, H.W., Jost, D.T., Piguet, D., Türler, A.: Nucl. Instr. Meth. A **479**, (2002) 631.
125. Düllmann, Ch.E.: "Chemical Investigation of Hassium (Hs, Z=108)", PhD thesis, Bern University (2002).

Index

Chapter 5

Liquid-Phase Chemistry

J.V. Kratz

Institut für Kernchemie, Universität Mainz, Fritz-Straßmann-Weg 2, D-55128 Mainz, Germany

1. Introduction

In the liquid-phase chemistry of the transactinide elements and their lighter homologues, carrier-free radionuclides produced in a nuclear reaction are transported to a separation device by the gas-jet technique and are dissolved in an aqueous solution. In general, the latter contains suitable ligands for complex formation. The complexes are then chemically characterized by a partition method which can be liquid-liquid extraction, cation-exchange or anion-exchange chromatography, or reversed-phase extraction chromatography. The ultimate goal of the partition experiments is to determine the so-called distribution coefficient, the K_d value, as a function of ligand concentration. The K_d value is given in its simplest definition, which applies to liquid-liquid extraction, by the ratio of the activity in the organic phase to that in the aqueous phase. In a chromatography experiment, as we will see below, the distribution coefficient is closely related to the key observable, the retention time t_r.

Liquid-phase chemistry is performed mostly in a discontinuous, batch-wise manner. It is then necessary, in order to get statistically significant results, to repeat the same experiment several hundred or even several thousand times with a cycle time on the order of a minute. Recent discontinuous studies

M. Schädel (ed.), The Chemistry of Superheavy Elements, 159-203.

were either performed manually [1-6] or with the Automated Rapid Chemistry Apparatus (ARCA II) [7]. These discontinuous separations involve a rather time-consuming evaporation step to prepare weightless samples for alpha spectroscopy. This is avoided with the continuous ion-exchange chromatography with the multi-column technique (MCT), which was first used to study the fluoride complexation of rutherfordium [8-10]. The fast centrifuge system SISAK was coupled with on-line liquid-scintillation counting (LSC) and is heading for its first application in the study of transactinides [11,12]. This chapter gives an overview over the chemical separation and characterization experiments of the first three transactinide elements in liquid-phases, mainly aqueous solutions, the chemical properties which were obtained, and some perspectives for further studies; more technical details of these experiments are given in Chapter 4.

2. Rutherfordium (Rf, Element 104)

2.1 FIRST SURVEY EXPERIMENTS

According to the actinide concept [13], the 5f series in the Periodic Table ends with element 103, lawrencium (Lr), and a new 6d transition series is predicted to begin with element 104, rutherfordium (Rf)). After the discovery of a long-lived (65-s) α-particle emitting isotope, ^{261}Rf, by A. Ghiorso et al. [14] in 1970, R.J. Silva et al. [15] confirmed this placement of Rf in the Periodic Table by conducting the first aqueous-phase separations with a cation-exchange chromatography column and the chelating agent α-hydroxyisobutyric acid (α-HiB, 2-hydroxy-2-methyl-propionic acid). In this pioneering experiment, the ^{261}Rf was produced by irradiation of 47 μg of ^{248}Cm which was electrodeposited over 0.2 mm^2 area onto a Be foil with 92-MeV ^{18}O ions from the Berkeley Heavy Ion Linear Accelerator. A sketch of the system used is shown in Figure 1. The recoils from the target were stopped in He gas, were swept out of the recoil chamber through a nozzle, and were deposited on the surface of a rabbit which was coated with a thin layer of NH_4Cl. After fast transport of the rabbit to the chemistry apparatus, the ^{261}Rf was washed from the rabbit with 50 μL ammonium α-HiB (0.1 M, pH 4.0) onto the top of a 2 mm diameter by 2 cm long heated (~80°C) column of Dowex 50x12 cation-exchange resin. This solution was forced into the resin and, after adding more eluant, the washing was continued. The first two drops (free column volume) contained little or no activity and were discarded. The next four drops (taken in two-drop fractions) were collected on platinum discs, evaporated to dryness and heated to ~500°C to burn off any carbon residue. The discs were placed active side down directly over Si(Au) α-particle detectors. The number, energy and time distribution of α particles with energies between 6 and 12 MeV emitted by the sources were recorded. The average time from beam off to the start of counting was ~60 s. Approximately 100 atoms of ^{261}Rf were produced in several hundred experiments, however, only about 1/10 of this number of events were observed after chemistry due to decay, counting geometry and chemical losses. The overall yield of tracer quantities of the lighter homologues Zr and Hf recovered in drops 3-6 was about 50%.

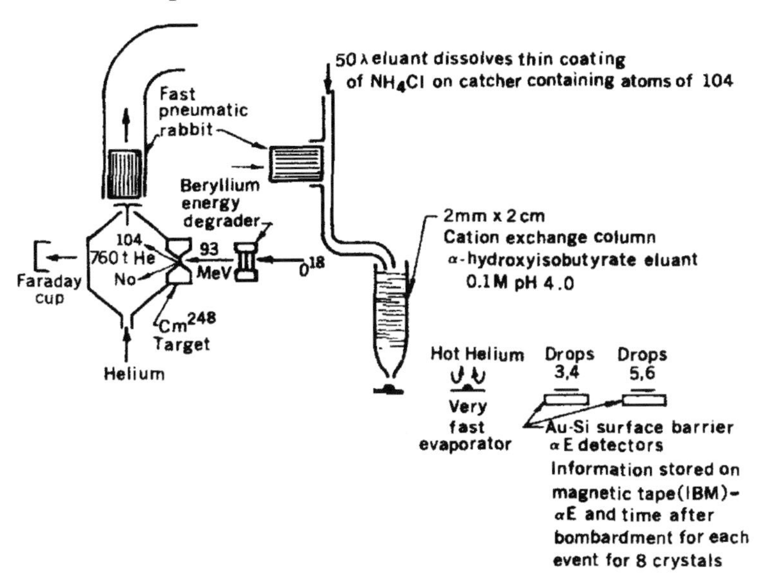

Fig. 1. Sketch of the chemical method used for the first liquid-phase separation of rutherfordium; taken from Ref. [16].

In contrast, trace quantities of the 3+ ions Tm, Cf, and Cm did not elute in over 100 column volumes. No^{2+}, as well as the alkaline earth elements, was retained even stronger on the resin. α-particle spectra were recorded for four sequential one-minute decay intervals from sources after chemical separation. These contained 17 α-particle events in the energy region 8.2 to 8.4 MeV. Approximately one half of these events are due to the decay of 26-s ^{257}No, the daughter of ^{261}Rf. In two experiments, two α-decay events occurred in the 8.2 to 8.4 MeV energy region within a time interval of one minute, representing a pair of correlated mother (^{261}Rf) and daughter (^{257}No) decays. This number of α-α mother-daughter correlations is consistent with the counting geometry used.

These data [15] show unambiguously that the chemical behavior of the activity assigned to ^{261}Rf is entirely different from that of trivalent and divalent actinides but is similar to that of Zr and Hf as one would expect for the next member of the Periodic Table following the actinide series.

The Rf experiment by R.J. Silva et al. [15], even though being "historical", is typical for many liquid-phase experiments performed later. It combines a fast transport system for transfer of the activities from the target-recoil chamber to the chemistry system with a discontinuously performed chemical separation that is repeated hundreds of times. It also includes the preparation of sources for α-particle spectroscopy by evaporation of the aqueous effluent from the column to dryness and it records energy and time of the α events and of αα correlations for unambiguous isotopic assignment of the activity. Above all, it compares the behavior of a transactinide element with that of its homologues under identical conditions. However, it has not yet been efficient enough to measure K_d values. This has only been achieved with improved techniques that have been developed more recently, i.e., within the last ~15 years.

Another first-generation experiment by E.K. Hulet et al. [17] testing the chloride complexation of Rf made use of computer automation to perform all chemical manipulations rapidly, to prepare α sources, and to do α spectroscopy. An extraction chromatographic method was chosen to investigate chloride complexing in high concentrations of HCl, which thereby avoided the hydrolysis reaction possible at lower acidities. The extraction columns contained an inert support loaded with trioctyl-methylammonium chloride, since anionic-chloride complexes formed in the aqueous phase are strongly extracted into this ammonium compound. Such complexes are formed in 12 M HCl by the group-4 elements and are extracted, whereas group-1, group-2, and group-3 elements, including the actinides, are not appreciably extracted. Thus, these latter activities were eluted from the column with 12 M HCl while Zr and Hf, and Rf were extracted and subsequently eluted with 6 M HCl, in which chloride complexation is less favored. Figure 2 shows the atoms of ^{261}Rf observed by α decay in three sequential elution fractions.

Only six ^{261}Rf events were observed in over one hundred experiments, one in the feed fraction (12 M HCl), two in elution fraction 2 (6 M HCl), and three in elution fraction 3 (6 M HCl). The percentage of Hf in these same fractions was 12%, 59%, and 29%. These results showed the chloride complexation of Rf is consistently stronger than that of the trivalent actinides and is similar to that of Hf. Again, no K_d value was determined in this "early" experiment.

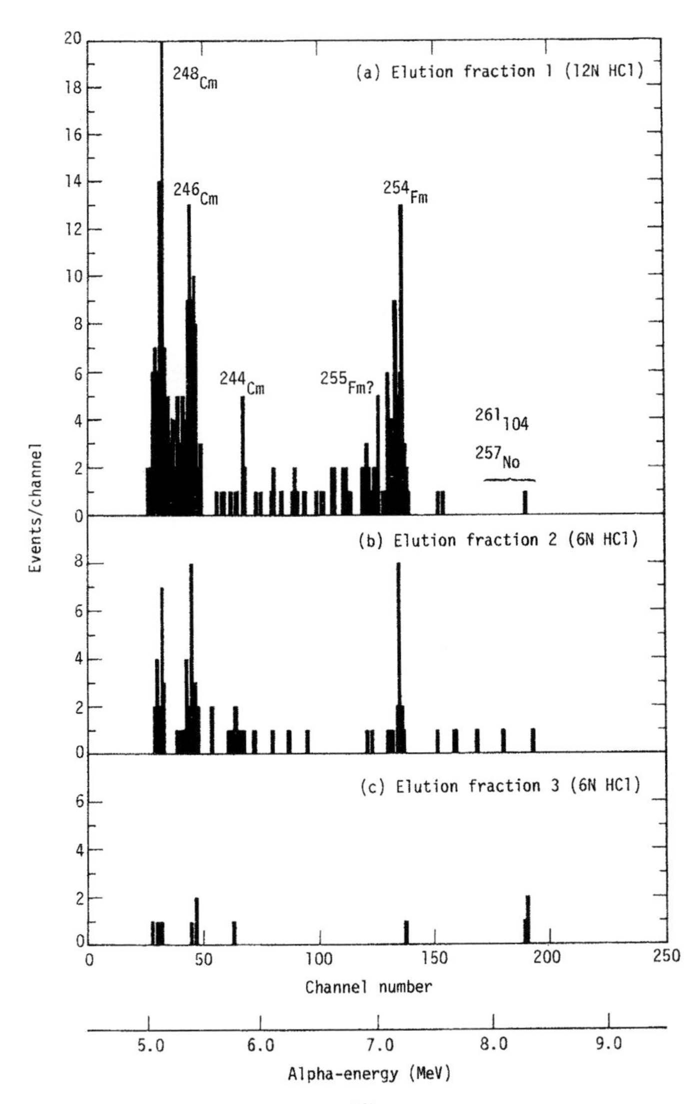

Fig. 2. α spectra showing the decay of a few ^{261}Rf atoms in three sequential elution fractions from a column containing trioctylmethylammonium chloride. Reprinted from [17] with the permission from Excerpta Medica Inc. (2001).

2.2 DETAILED STUDIES

A renewed interest in studying the chemical properties of the transactinide elements in more detail both experimentally and theoretically arose in the late 1980s, see [18-26] for recent reviews. A series of manually performed separations of ^{261}Rf in aqueous solutions was performed by the Berkeley group at the Berkeley 88-Inch Cyclotron [2-6,27]. Their experiments involving liquid-liquid extractions typically comprised the following steps:

A He(KCl)-jet transportation system was used for the transfer of the activities. The KCl aerosol with the reaction products was collected by impaction on a Pt or Teflon™ slip for 60 or 90 s, was picked up with 10μL of the aqueous phase and was transferred to a 1 mL centrifuge cone containing 20 μL of the organic phase. The phases were mixed ultrasonically for 5 s and were centrifuged for 10 s for phase separation. The organic phase was transferred to a glass cover slip, was evaporated to dryness on a hot plate, and was placed over a passivated ion-implanted planar silicon detector (PIPS). This procedure took about 1 min.

2.2.1 *Hydrolysis versus halide complexation – studies using liquid-liquid extraction techniques*

Liquid-liquid extractions with triisooctyl amine (TiOA) from 12 M HCl [2] confirmed the results of [17]. Cationic species were investigated [27] by extraction into thenoyltrifluoroacetone (TTA). A distribution coefficient for Rf between those of the tetravalent pseudo homologues Th and Pu indicated [27] that the hydrolysis of Rf is less than that for Zr, Hf, and Pu.

K.R. Czerwinski et al. [3] performed a series of liquid-liquid extractions with tributylphosphate (TBP) in benzene to study the effect of HCl, Cl^--, and H^+ -ion concentration between 8 and 12 M on the extraction of Zr^{4+}, Hf^{4+}, Th^{4+}, Pu^{4+}, and Rf^{4+}. It was found [3] that Rf extracts efficiently as the neutral tetrachloride into TBP from 12 M HCl like Zr, Th, and Pu while the extraction of Hf was relatively low and increased from 20 % to 60 % between 8 M and 12 M HCl. Extraction of Rf increased from 60 % to 100 % between 8 M and 12 M HCl, thus defining an extraction sequence Zr > Rf > Hf for the group-4 chlorides. Surprising results were obtained when the chloride concentration was varied at a constant H^+ concentration of 8 M. Above 10 M Cl^- concentration, the extraction of Rf decreased and behaved differently from Zr, Hf, and Th, and resembled that of Pu^{4+}. This was interpreted in terms of stronger chloride complexing in Rf than in Zr, Hf, and Th, leading to the formation of $RfCl_6^{2-}$ [18] which is not extracted into TBP. Extraction studies at a constant concentration of 12 M Cl^- showed that Rf extraction is sharply increasing with increasing H^+-concentration between 8 M and 12 M [3]. Such a behavior is not exhibited by Zr and Hf. As some of these extraction experiments suffered from differences in the details of the chemical procedures applied to the different elements, e.g. different contact times and volumes used, it was important to confirm these very interesting and somehow unexpected findings in experiments that establish identical conditions for all homologues elements including Rf.

C.D. Kacher et al. [5] performed some additional chloride extractions into TBP/benzene with Zr, Hf and Ti. The reported low extraction yields of Hf by

K.R. Czerwinski et al. [3] could not be reproduced by C.D. Kacher et al. [5] who reported that they observed that significant amounts of Hf (more than 50 % in some cases) stick to Teflon surfaces. (They actually conducted their subsequent experiments with polypropylene equipment because only negligible adsorption was observed with polypropylene surfaces.) The Hf results from the Czerwinski et al. experiments [3] were based on on-line data taken at the 88-Inch Cyclotron where the activity was collected on a teflon disc which according to [5] accounts for the seemingly low Hf extraction. Surprisingly, a similar loss of Rf due to adsorption in the Czerwinski et al. work [3] was not suspected by Kacher et al., and so the latter authors, based on their new Zr-, Hf-, and Ti-results and on the old [3] Rf results, suggested a revised, still questionable, sequence of extraction into TBP/benzene from around 8 M HCl as Zr > Hf > Rf > Ti. In a parallel study of liquid-liquid extractions into TBP/benzene from HBr solutions, extraction of Rf was found to be low and was only increased for bromide concentrations beyond 9 M [5]. The extraction behavior of the group-4 elements into TBP from both HCl and HBr solutions was primarily attributed to their different tendencies to hydrolyze [5].

The latter statement refers to concurrent work by A. Bilewicz et al. [4] who studied the sorption of Zr, Hf, Th, and Rf on cobalt ferrocyanide surfaces. These ferrocyanides are known to be selective sorbents for heavy univalent cations such as Fr^+, Cs^+, and Rb^+. However, some ferrocyanides such as Co ferrocyanide have been found to exhibit also particularly high affinities for tetravalent elements such as Zr^{4+}, Hf^{4+}, and Th^{4+} involving the formation of a new ferrocyanide phase between the 4^+ cation and the $Fe(CN)_6^{4-}$ anion.

The hydrolysis of a 4^+ cation is shown in the following reaction

$$M(H_2O)_x^{4+} \leftrightarrows M(H_2O)_{x-1}(OH)^{3+} + H^+ \quad (1)$$

On the left hand side of Equation 1, we have the hydrated 4^+ cation, on the right hand side the first hydrolysis product being a 3^+ cation.

As 3^+ cations are essentially not sorbed by ferrocyanide surfaces, the onset of hydrolysis at decreasing HCl concentration in the aqueous phase will be reflected by a rapid decrease of the sorbed activity. This decrease was observed by A. Bilewicz et al. [4] below 3 M HCl for Rf, below 1 M HCl for Zr, and below 0.5 M HCl for Hf, establishing seemingly a hydrolysis sequence Rf > Zr > Hf. As, in general, hydrolysis increases with decreasing radius of the cation, the stronger hydrolysis of the larger size Rf ion is very surprising and in conflict with the results presented in [27]. A. Bilewicz et al. suggested as an explanation that the coordination number CN (x in

Equation 1) for Zr^{4+} and Hf^{4+} is 8, and changes to CN = 6 for Rf due to relativistic effects making the $6d_{5/2}$ orbitals unavailable for ligand bonding of water molecules. However, R. Günther et al. [28] have shown that this does not withstand a critical examination. It is the author's opinion that some experimental deficiency and not the increased tendency of Rf to hydrolyze produced the surprising results [4]. For example, the contact time of the aqueous phase with the ferrocyanide surface was only 10 s in the Rf experiments. In a kinetic study, the authors found that Zr and Hf sorbed within 20 s and 40 s, respectively, while Th required more than 90 s to achieve nearly complete sorption [4]. It is conceivable that, within the 10 s interaction of the aqueous phase in the Rf experiments, no equilibrium was established thus making the Rf data meaningless.

A study of the extraction of fluoride complexes of Ti^{4+}, Zr^{4+}, Hf^{4+}, and Rf^{4+} into TiOA was also reported by C.D. Kacher et al. [6]. This work presents some evidence for extraction of ^{261}Rf into TiOA from 0.5 M HF, however no quantitative assessment of the extraction yield or K_d value is made. Consequently, the conclusion in [6] that the extraction into TiOA for the group-4 elements decreases in the order Ti > Zr ≈ Hf > Rf is not reproducible.

In view of the somewhat unsatisfactory situation with the conflicting Hf results in [3,5] and with the intention to establish an independent set of data characterizing the extraction sequence of Zr, Hf, and Rf from 8 M HCl into TBP, Günther et al. [28] have determined distribution coefficients of these elements from HCl solutions. In 8 M HCl, the K_d of Zr is 1180, that for Hf 64. This difference makes possible a chromatographic separation of Hf from Zr in ARCA II on 1.6 x 8 mm columns filled with TBP on an inert support. This separation was also studied with the short-lived ^{169}Hf from the Gd(^{18}O,xn) reaction yielding $K_d = 53^{+15}_{-13}$ in agreement with the above results from batch extraction experiments. 78-s ^{261}Rf was produced in the $^{248}Cm(^{18}O, 5n)$ reaction at the Philips Cyclotron of the Paul Scherrer Institute (PSI), Villigen/Switzerland, and from the distribution of α-events between the Hf and Zr fraction, a K_d value of 150 was determined for Rf in 8 M HCl. This gives the extraction sequence Zr > Rf > Hf. Such a sequence is expected from theoretical considerations [28,30] of complex formation and the concurrent hydrolysis of complexes

$$M(OH)_x Cl_{4-x} + x\, HCl \leftrightarrows MCl_4 + x\, H_2O \qquad (2)$$

To predict the equilibrium constant of Equation 2, one has to consider the difference in total energies of the (partially) hydrolyzed species on the left hand side of Equation 2 that are not extracted and the extractable MCl_4. (It is assumed that the OH^- containing species will not extract into the organic phase because of the strong hydrogen bonding interaction between OH^- and H_2O.) This can be done by quantum chemical calculations, presently using the DS DV method [26,29-31], which allows determining the differences in the Coulomb and the covalent parts of the binding energy separately. Calculations for Zr, Hf, and Rf are still to be done, but one can already draw qualitative conclusions based on a parallel study of the hydrolysis of chloro complexes of the group-5 elements Nb, Ta, Pa, and Db [30]. For group 5, the order of complex formation described by equilibria similar to the reaction (2) in 4-12 M HCl solutions was found to be Pa > Nb > Db > Ta in excellent agreement with experimental data [33] to be discussed in Section 3. Reasons for this are dominant differences in the Coulomb part of the free energy of reaction when OH^- groups are replaced by Cl^- anions. Earlier calculations [33] of MCl_4 (M = Zr, Hf, and Rf) have shown that the compounds are very similar which can also be supposed for $M(OH)_xCl_{4-x}$. Knowing the analogy in the electronic structure of the halides and oxyhalides of groups 4 and 5 [34,26], one can postulate the same order of complex formation Zr > Rf > Hf according to eq. (2) as it was found for the corresponding group-5 elements. For both group 4 and 5, such a sequence is in full agreement with experimental data for Zr, Hf, and Nb, Ta, respectively, showing that compounds of the 5d elements are more hydrolyzed than those of the 4d elements and are hence less extracted. For these elements, the hydrolysis of complexes [26,30] is of opposite order to the hydrolysis of cations [29]. This shows that the statements in [5,6] concerning the seemingly low extraction of Hf and Rf as being due to an increased tendency of their cations to hydrolyze are completely misleading.

2.2.2 *Hydrolysis versus halide complexation -- studies using ion-exchange resins*

Fluoride complexation of group-4 elements was studied by Z. Szeglowski et al. [8], by G. Pfrepper et al. [9,10], and by E. Strub et al. [35,36]. In the work of [8], ^{261}Rf was transported on line to the chemistry apparatus, was continuously dissolved in 0.2 M HF in a degassing unit, and the solution was passed through three ion-exchange columns. In the first cation-exchange column, actinides produced directly in the $^{18}O + ^{248}Cm$ reaction were removed from the solution. In the next anion-exchange column, ^{261}Rf was sorbed, presumably as RfF_6^{2-}, while the following cation-exchange column retained its cationic decay products. After the end of bombardment, the long-lived descendants ^{253}Fm and ^{253}Es were desorbed from the third column and detected off-line by α spectroscopy. Their detection was proof that Rf forms anionic fluoride complexes which are sorbed on an anion-exchange resin.

G. Pfrepper et al. [9,10] have developed this technique further, thereby making it a quantitative technique capable of measuring distribution coefficients K_d. In the conventional off-line chromatography as performed by ARCA II, the distribution coefficient is determined via the retention time (elution position) as

$$K_d = [t_r - t_o]\frac{V}{M} \quad (3)$$

with t_r = retention time, t_o = column hold-up time due to the free column volume, V = flow rate of the mobile phase (mL min^{-1}) and M = mass of the ion exchanger (g).

As in [8], the detection of the transactinide isotope itself, ^{261}Rf, is abandoned and replaced by the detection of its long-lived descendant, 20-d ^{253}Es. This way, one gains the possibility of a continuous on-line mode over many hours. The feeding of ^{261}Rf onto the anion-exchange column is performed under conditions in which the retention time t_r is on the order of the nuclear half-life $t_{1/2}$, i.e., K_d values on the order of 10-50 are selected [9]. Similarly to the principle used in the on-line isothermal gas chromatography, one is using the nuclear half-life as an internal clock. As in [8], three ion-exchange columns are used in series, first a cation-exchange column that retains the ^{253}Es from the continuously flowing feed solution. This is necessary as ^{253}Es could be produced directly by transfer reactions. It follows the true chromatographic column (C) filled with an anion-exchange resin. The long-lived decay products (D_1) that are formed by radioactive decay of ^{261}Rf during its retention time on the anion-exchange column are eluted from this column as cations and are fixed on the following cation-exchange column. The part of the ^{261}Rf that survives the retention time on the anion-exchange column is eluted from it and passes the following cation-exchange column to be subsequently collected in a reservoir in which it decays into the long lived decay products (D_2). D_1 and D_2 are isolated separately after the end of the on-line experiment and assayed off-line for α-activity of ^{253}Es. From the ratio of D_1 and D_2 and the nuclear half-life $t_{1/2}$ of ^{261}Rf, one obtains the distribution coefficient

$$K_d = \left[\frac{t_{1/2}}{\ln 2}\ \ln\frac{D_1 + D_2}{D_2} - t_o\right]\frac{V}{M} \quad (4)$$

G. Pfrepper et al. [9] have verified the equivalence of Equations 3 and 4. They compared K_d values for Hf isotopes obtained, firstly, in batch experiments and by conventional elution chromatography in the anion-exchange system HS36 (a resin with the quaternary triethyl-ammonium group) – 0.27 M HF/x M HNO_3 (x variable) with , secondly, K_d values obtained in the same system with the MCT technique. They used the short-lived Hf isotopes $^{165,166}Hf$ and detected the daughter nuclides $^{165,166}Tm$ and ^{166}Yb by γ-ray spectroscopy. In order to determine the ionic charge of the Hf fluoride complexes, the concentration of the counter ion NO_3^- was varied. K_d values were found to be 12.7 ± 1.8 mL/g with 0.2 M HNO_3 and 2.4 ± 0.8 mL/g with 0.5 M HNO_3 corresponding to an ionic charge of –1.9 ± 0.4 for the fluoride complex of Hf as determined from the slope of the plot of logK_d versus the logarithm of the concentration of the counter ion. From this it was concluded that Hf probably forms HfF_6^{2-}. This is evidenced also by Figure 3 in which the K_d values obtained by the on-line MCT technique are compared with those from a series of batch experiments performed in parallel with the same anion- exchange resin. From the good agreement, it was concluded that with the MCT technique, determination of the ionic charge of complexes of transactinides must be feasible.

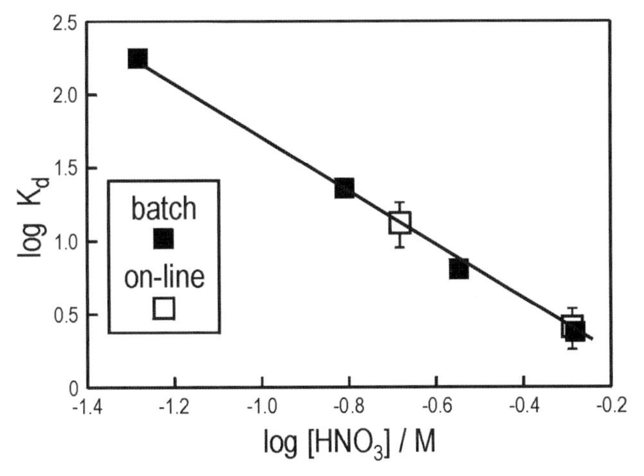

Fig. 3. logK_d of Hf in the system HS36-0.27 M HF-HNO_3 as a function of the HNO_3 concentration, measured in batch and in on-line experiments. Reprinted from [9] with the permission of Oldenbourg Verlag.

In the application of the MCT technique to ^{261}Rf [10], G. Pfrepper et al. used about 20 mg (11 mg) of the anion-exchange resin HS36 in the chromatographic column at an elution rate of approximately 0.35 mL min^{-1} of 0.27 M HF and 0.2 (0.1) M HNO_3. The cation-exchange column(s) that sorbed the descendants 23-s ^{257}No, 3-d ^{253}Fm, and 20-d ^{253} Es contained 50 mg of the strongly acidic cation exchanger Wofatit KPS. For the similar cation-exchange resin Dowex 50x12, G. Pfrepper et al. [9] had shown previously that, for HNO_3 concentrations <0.3 M, K_d values for trivalent rare earth cations exceed 10^4 mL/g resulting in a breakthrough volume in excess of 500 mL . Assuming the same conditions for their ^{261}Rf experiment [10], it was decided to renew the cation-exchange columns collecting the descendant fractions D_1 and D_2 every 12 h. This corresponded to eluent volumes of about 260 mL while the breakthrough volume was estimated to be about 10^3 mL. Thus, at first sight, breakthrough of the cationic “daughter nuclides” was safely prevented.

^{261}Rf and $^{165-169}$Hf were produced simultaneously by irradiation of a combined target of ^{248}Cm and ^{152,154,158}Gd with ^{18}O ions at the Dubna U-400 Cyclotron. In the γ-ray spectra of D_1 and D_2, the nuclides ^{165}Tm, ^{166}Yb, ^{167}Tm, ^{169}Lu, and ^{169}Yb were detected as the decay products of ^{165,166,167,169}Hf. K_d values on the order of 14.8 and 47 were determined in 0.27 M HF/0.2 M HNO_3 and 0.27 M HF/0.1 M HNO_3, respectively, compatible with an ionic charge of the fluoride complex of -1.85 ± 0.1. For ^{261}Rf, the α-decays of its descendant ^{253}Es were evaluated yielding K_d values of about 14 and 50 in 0.27 M HF/0.2 M HNO_3 and 0.27 M HF/0.1 M HNO_3, respectively, giving an ionic charge of -1.9 ± 0.2. It was concluded [10] that Rf has properties of a close homologue of Hf and forms RfF_6^{2-}, the stoichiometry characteristic of Hf and other group-4 elements.

A. Kronenberg et al. [36] have observed that the K_d values of trivalent lanthanide (Tb^{3+}) and actinide cations (^{241}Am, ^{250}Fm) in mixed HF/0.1 M HNO_3 solutions on the cation-exchange resin Dowex 50Wx8 exceed 10^4 mL/g only for HF concentrations below 10^{-2} M. For higher HF concentrations, the K_d values decrease to 2000 in 0.05 M HF/0.1 M HNO_3, 1300 in 0.1 M HF/0.1 M HNO_3, 700 in 0.5 M HF/0.1 M HNO_3, and 500 in 1 M HF/0.1 M HNO_3. A simultaneous rise of K_d values on the anion-exchange resin Dowex 1x8 is observed [36] for >0.05 M HF/0.1 M HNO_3 with values reaching 170 in 0.1 M HF/0.1 M HNO_3, 1600 in 0.5 M HF/0.1 M HNO_3, and 2500 in 1 M HF/0.1 M HNO_3 thus indicating that anionic fluoride complexes are formed. The K_d value in 0.27 M HF/0.1 M HNO_3 on a strongly acidic cation-exchange resin can be estimated to be on the order of 800. Thus, the breakthrough volume for lanthanides and actinides in the experiments by G. Pfrepper et al. [10] was not >500 mL but rather <40 mL, i.e., both the Hf and the Rf descendants were breaking through the cation-exchange columns in less than 2 h. It is impossible to reconstruct what this did to the K_d values in [10], however, the significant differences between the K_d values of Hf and Rf in mixed HF/HNO_3 solutions observed in [35,36] do cast doubt on the validity of the results in [10], where indistinguishable K_d values for Hf and Rf were reported.

E. Strub et al. [35] have investigated in batch experiments K_d values of the long-lived tracers ^{95}Zr, ^{175}Hf, and ^{233}Th on a strongly acidic cation-exchange resin, Aminex A5, and on a strongly basic anion-exchange resin (Riedel-de Häen) in 0.1 M HNO_3 solutions containing variable concentrations of HF. On the cation-exchange resin, below 10^{-3} M HF, the K_d values for Zr, Hf, and Th are $>10^3$ indicating the presence of cations. In the range 10^{-3}M < [HF] < 10^{-2} M, the K_d values of Zr and Hf decrease due to the formation of neutral or anionic fluoride complexes. The behavior of Zr and Hf is very similar (Hf sticking to the resin even at slightly higher HF concentrations). The elution of Th from the resin is observed at more than one order of magnitude higher HF concentrations, see Figure 4. These off-line data were compared with on-line data taken with ARCA II at the Gesellschaft für Schwerionenforschung (GSI), Darmstadt, UNILAC accelerator where short-lived Hf isotopes $^{166,167,168...}$Hf were produced in xn-reactions of a ^{12}C beam with an enriched ^{158}Dy target. As is shown by Figure 4, there is general agreement between both sets of data.

On an anion-exchange resin, as expected, there is little sorption of Zr, Hf, and Th at low HF concentrations. In the range 10^{-3} M < [HF] < 10^{-2} M, the K_d values of Zr and Hf in batch experiments rise simultaneously, i.e., in the same range as they decrease on the cation- exchange resin. This shows that the disappearance of the cationic species at low HF concentrations is followed immediately by the formation of anionic fluoride complexes. For Th, the sorption on the anion-exchange resin stays low for all HF concentrations indicating that Th does not form anionic fluoride complexes.

In the Rf experiments [35] performed at the PSI Philips Cyclotron, ^{261}Rf was produced in the ^{248}Cm(^{18}O, 5n) reaction at 100 MeV. The target contained 10% Gd enriched in ^{152}Gd to produce simultaneously short-lived Hf isotopes that were used to monitor the behavior of Hf and to perform yield checks by γ spectroscopy. Rf and Hf were transported by a He(KCl) gas jet and were collected for 90s by impaction inside ARCA II [7]. The deposit was dissolved in 200 µL 0.1 M HNO_3/x M HF (x variable) and was fed onto the 1.6x8 mm cation-exchange column at a flow rate of 1 mL min^{-1}. The effluent was evaporated to dryness as sample 1. In order to elute remaining Rf and Hf from the column, a second fraction of 200 µL 0.1 M HNO_3/0.1 M HF was collected to strip all group-4 elements from the column. The fraction was prepared as sample 2.

In the anion-exchange experiments, Rf and Hf were transported, collected and loaded onto the anion-exchange column as in the experiments with the cation-exchange columns. Again, the effluent was evaporated to dryness as sample 1. In order to elute remaining Rf and Hf from the column, a second fraction of 200 µL 5 M HNO_3/0.01 M HF was used. This fraction was prepared as sample 2.

K_d values are often calculated as follows:

$$K_d = \frac{A_s}{A_l} = \frac{A_{s0}\ V_l}{A_{l0}\ m_s} \qquad (5)$$

with A_s specific activity in the solid phase [Bq/g], A_l specific activity in the liquid phase [Bq/mL], A_{s0} activity in the solid phase [Bq], A_{l0} activity in the liquid phase [Bq], V_l volume of the solution [mL], and m_s mass of the resin

[g]. The K_d values using Equation 5 are obtained such that the activity of sample 1 is attributed to the liquid phase and the activity of sample 2 to the solid phase. However, in elution experiments Equation 5 yields only correct K_d values for the fortunate situation of equal activities in fraction 1 and 2, i.e., cuts between fraction 1 and 2 at the maximum position of the elution curve. To circumvent this deficiency, K_d values shown in Figure 4 were calculated by a computer program that simulates the elution process. The number of theoretical plates – obtained from the shape of tracer elution curves -, the free column volume, the volume of fraction 1 and an assumed Kd value are the initial input parameters. In an iteration process, the K_d value is varied until the experimentally observed activity in fraction 1 results.

The results of the on-line experiments with Rf and Hf on cation-exchange columns are given in Figure 4 together with the results of the off-line batch experiments. The on-line data for Hf are consistent with the off-line data.

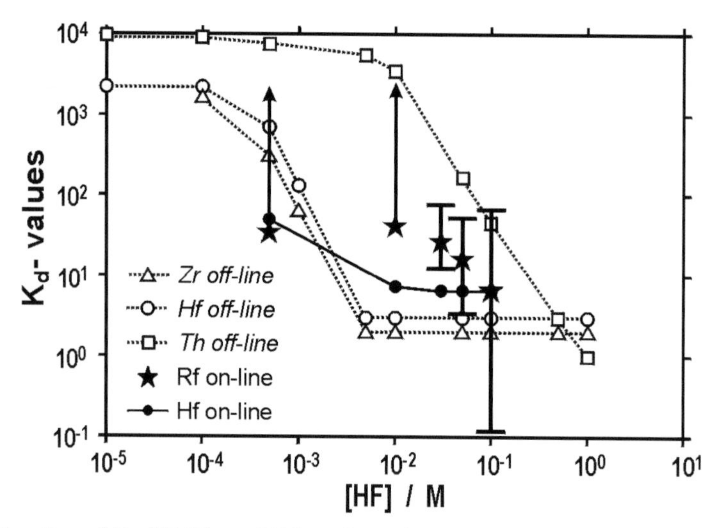

Fig. 4. Sorption of Zr, Hf, Th, and Rf on the cation-exchange resin Aminex A6 in 0.1 M HNO_3 at various HF concentrations. Off-line data are shown for Zr, Hf, and Th, and on-line data for Rf and Hf; re-evaluated data from [35].

The K_d values for ^{261}Rf at $5 \cdot 10^{-4}$ M HF and at 10^{-2} M HF are lower limits as no ^{261}Rf decays were detected in sample 1. It is seen that the decrease of the K_d values for Rf occurs between 0.01 M HF and 0.1 M HF, i.e., at one order of magnitude higher HF concentrations than for Zr and Hf. Under the given conditions, the behavior of Rf is intermediate between that of its pseudo-homologue Th and that of its group members Zr and Hf.

The results of the on-line experiments with the anion-exchange columns are given in Figure 5. While the off-line data for Hf clearly indicate that anionic fluoride complexes of Hf are formed for >10^{-3} M HF, one observes that the Hf data taken on-line are systematically lower. The reason for this discrepancy is unknown at this time. Seemingly, the behavior of Rf is different from that of Zr and Hf – if compared with the off-line data - as there is almost no sorption of Rf on the resin. Even for 1 M HF, which is about two orders of magnitude higher than the concentration from whereon maximum K_d values are observed for Zr and Hf, there is no indication of a rise in the K_d values for Rf.

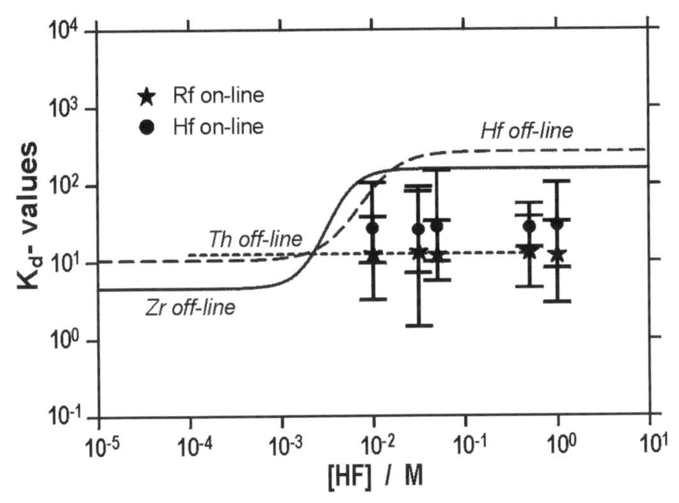

Fig. 5. Sorption of Zr, Hf, Th, and Rf on the anion-exchange resin (Riedel-de Häen) in 0.1 M HNO_3 at various HF concentrations. The data points of Hf and Rf are taken from the on-line experiments. The lines indicate the trends exhibited by the off-line data. Reprinted from [35] with the permission of Oldenbourg Verlag.

As it would be in contradiction to the results in [6,8,10] that there are no anionic fluoride complexes for Rf at any HF concentration, the concentration of HNO_3 was varied [35] between 0.1 M and 0.01 M at a constant HF concentration of 0.05 M thus varying the concentration of NO_3^- which acts as a counter ion competing for the binding sites on the anion exchanger. It was observed [35] that the K_d values of Hf and Rf rise with decreasing NO_3^- concentration indicating that negatively charged complexes are formed both for Hf and for Rf, but that the counter ion NO_3^- is much more effective in removing Rf from the binding sites on the anion exchanger than in removing Zr and Hf. A possible influence of the kinetics on the differences between on-line and off-line data remains speculative at the moment.

This was the starting point for a new series of experiments with anion-exchange chromatography with Aminex A27 in which the HNO_3 and the HF concentration were varied systematically [36]. The following observations were made:

- The dependence of the K_d values of Rf on the NO_3^- concentration at fixed HF concentration (0.05 M) is much more pronounced than that for the K_d values of Hf.
- In the absence of NO_3^- , the K_d values for Hf and Rf are equally high at $>10^{-2}$ M HF.

The latter observation suggested that the counter ion NO_3^- acts with a remarkable selectivity for the Rf-fluoride complex. A second counter ion discussed in [36] is the HF_2^- -ion. If K_d values of Hf and Rf are plotted versus the ratio of concentrations $[F^-]/([NO_3^-]+[HF_2^-])$ one observes the following: The K_d values of Hf rise between 10^{-4} and 10^{-3} and reach a plateau between 10^{-3} and 10^{-1}. There is a second rise between 10^{-1} and 10^{0} followed by a second plateau. One might speculate that the two steps reflect the subsequent formation and anion exchange of HfF_5^- and HfF_6^{2-}. The K_d values of Rf stay low until 10^{-1} from where on they rise and reach the same level as the K_d values of Hf in the second plateau. One might speculate that the RfF_5^-, if formed in the same concentration range as HfF_5^- (10^{-3}–10^{-1}), is very effectively replaced by the counter ions, and that appreciable anion exchange occurs only when the hexafluoride anion RfF_6^{2-} is formed.

We note that in [36] there were also two MCT experiments performed with ^{261}Rf using 0.5 M HF/0.1 M HNO_3 and 0.01 M HF, respectively. In the first case, the cation-exchange columns F and D were filled with 330 mg of Dowex 50Wx8 and were used for 3 h to prevent breakthrough of the cationic descendants. Column C was filled with 50 mg of the anion exchange resin Dowex 1x8 in the nitrate form. In the second case, columns F and D were filled with 68 mg Dowex 50Wx8 and used for 4 h, and column C with 17 mg of Dowex 1x8 in the fluoride form. Each experiment was run for 24 h at an average beam intensity of $3 \cdot 10^{12}$ s^{-1}.

The first experiment with 0.5 M HF/0.1 M HNO_3 was performed to corroborate the low K_d values obtained in [35] with ARCA II on an anion-exchange resin, see Figure 5. A total of 80 α decays were observed attributable to ^{253}Es in the sample 1 representing the liquid phase and no event in the strip fraction representing the solid phase. Taking zero to be compatible with 3 events at 95% confidence level in Poisson statistics, leads to an upper limit for the K_d value of Rf of <3 which is rather consistent with the data in [35], but not with the value in 0.27 M HF/0.1 M HNO_3 in [10]. The insensitivity of the K_d values of Rf to the actual HF concentration in 0.1 M HNO_3, see Figure 5, makes this statement possible.

In the second experiment, performed in pure 0.01 M HF without any HNO_3, 90 events attributable to ^{253}Es were observed in the strip fraction representing the solid phase and zero events in the liquid phase. This leads to a lower limit for the K_d value on the order of 300 compatible with the one measured with ARCA II in pure 0.01 M HF.

In summary, one can state that the K_d values for Rf obtained with the MCT, i.e., low in 0.5 M HF/0.1 M HNO_3 and high, and comparable to that of Hf, in 0.01 M HF, are consistent with the ones determined with ARCA II. They corroborate the so far unexplained finding that sorption of the Rf fluoride anions shows an extreme sensitivity to the presence of HNO_3,, much more pronounced than the respective sensitivity of Hf .

3. Dubnium (Db, Element 105)

3.1 FIRST SURVEY EXPERIMENTS

First studies of the aqueous-phase chemistry of element 105, dubnium (Db), were conducted manually in 1987 by K.E. Gregorich et al. [37]. The isotope 34-s ^{262}Db was used which is produced in the ^{249}Bk(^{18}O,5n) reaction [38,39] at around 100 MeV with a cross section of a few nb. As it was known that group-5 elements sorb on glass surfaces from strong nitric acid [40], the following adsorption experiment was carried out: The activity bearing aerosol from a He(KCl) gas-jet was collected on a glass plate. At the end of the 60-s collection time, the glass plate was removed from the collection site and was placed on a hot plate. The potassium chloride spot on the glass plate was fumed with 3 μL of 15 M nitric acid. After this nitric acid dried, a second fuming was performed with 7 μL of 15 M nitric acid. When the second drop of nitric acid had dried, the potassium nitrate and the actinide activities on the glass plate were removed by washing the plate with 1.5 M nitric acid from a squirt bottle. Any remaining dilute nitric acid was removed by washing the glass plate with acetone from a second squirt bottle. The glass plate was immediately dried in a stream of hot air from an electric 'heat gun' and placed over one of the Si(Au) surface barrier detectors. The average time from the end of accumulating the aerosol on a glass plate to the beginning of counting for α and spontaneous fission (SF) decays was 51 s.

In the ^{18}O + ^{249}Bk bombardments, 801 adsorption experiments were performed in [37]. The decontamination from actinides was very good. The decay rates of $^{252\text{-}255}$Fm indicated that only 0.25% of the fermium activities remained on the glass. A total of 26 α events in the energy range from 8.42 to 8.70 MeV were observed during the first 140 s of counting (~ 4 half-lives of ^{262}Db). By looking at the time distribution of α events in this energy range out to 500 s, it was estimated that the time interval from 0 s to 140 s contained ~2 background events from longer-lived activities, leaving ~24 α events due to the decay of ^{262}Db or its daughter ^{258}Lr. A spectrum containing the α data from the first 30 s of counting for all 801 samples is presented in Figure 6. For comparison, a summed spectrum of unseparated products from this reaction, taken with a rotating wheel system for a 20-s interval starting 60 s after the end of collection of aerosols, is presented in Figure 7 for

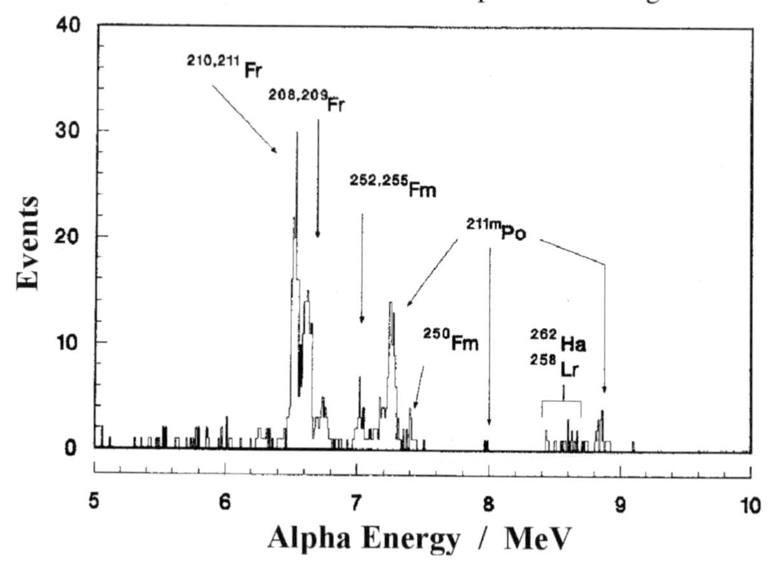

Fig. 6. A summed spectrum containing all of the α data from the first 30 s of counting from the 801 experiments involving the adsorption of ^{262}Db on glass from concentrated nitric acid solution. Reprinted from [37] with the permission of Oldenbourg Verlag.

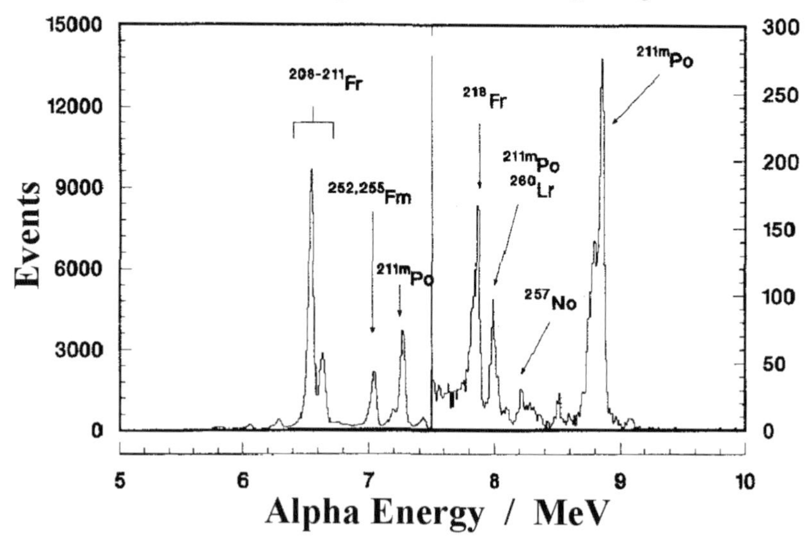

Fig. 7. A representative spectrum of the unseparated product mixture from 101 MeV ^{18}O + ^{249}Bk bombardments taken on a rotating wheel system. These data were recorded over a 20-s time interval starting 60 s after the end of collection. One should note the change of vertical scale at the center of the spectrum. Reprinted from [37] with the permission of Oldenbourg Verlag.

comparison. It appears that in Figure 6 there is little fractionation between the francium-polonium activities, which are produced by interaction of the ^{18}O beam with a small lead impurity in the target, and the fermium activities. There is, however, a large relative enhancement of the activities due to ^{262}Db and its daughter, ^{258}Lr, in the adsorption experiments. Figures 6 and 7 demonstrate that insufficient decontamination from polonium can seriously hamper the detection of transactinides.

The 26 α events in Figure 6 attributed to ^{262}Db and ^{258}Lr contain 5 time-correlated parent-daughter pairs, as indicated in Table 1, as well as ~14 uncorrelated events and ~2 background events. A maximum likelihood fit to the half-life of the 16 α singles events which occurred at times shorter than 140 s from the beginning of counting together with the 5 parent events of the mother-daughter correlations gives a half-life of 28^{+7}_{-5} s, consistent with the accepted ^{262}Db half-life of 34 s. From a detector geometry of 30% of 4π, one would expect the ratio of correlated pairs to uncorrelated events to be 0.214. The measured ratio of 0.36 is within the one sigma uncertainty of 0.18 consistent with the expected ratio. A maximum likelihood fit to the ^{258}Lr half-life based on the time intervals between parent and daughter events gives $2.5^{+1.9}_{-1.0}$ s which is also consistent with the accepted value of 4.3 ± 0.5 s.

Table 1. α-α parent-daughter correlations in [37].

#	Parent energy (keV)	Parent time[a] (s)	Daughter energy (keV)	Daughter time since parent (s)
1	8640	10.03	8752	4.45
2	8533	131.87	8636	0.08
3	8681	10.82	8661	3.40
4	8437	4.12	8603	2.02
5	8681	2.55	8611	7.88

[a] Time from the beginning of counting to the α event

In the adsorption experiments, K.E. Gregorich et al. also observed 26 SF decays in the first 140 s of which 23 were assigned to the decay of ^{262}Db giving a half-life of 32^{+8}_{-6} s, again consistent with the known 34 s half-life. From this, an α-decay branch of 51 ± 14% was deduced for ^{262}Db, with the remainder of the decay being either by SF or by electron capture to ^{262}Rf which then decays by SF. An overall production cross section of 3.2 ± 0.5 nb was estimated based on an estimated adsorption yield of 80%. In view of a more recent determination of that cross section [41], it appears that the adsorption yield was probably overestimated by about a factor of 2.

The other chemical separation attempted in [37] involved the extraction of anionic fluoride species into methyl isobutyl ketone (MIBK). The extraction system with 3.8 M HNO_3 and 1.1 M HF as the aqueous phase, and MIBK as the organic phase was chosen because MIBK had been found to be an ideal solvent for the rapid preparation of α sources by evaporation. Under these conditions, tantalum is extracted into MIBK nearly quantitatively, while niobium is extracted to only a small extent. It was expected that this trend would continue and dubnium would be extracted quantitatively.

In these experiments, the activity-bearing aerosols were collected for 90 s on Pt foils. The KCl and the activities were dissolved in 20 μL of 3.8 M HNO_3/1.1 M HF. This solution was then placed in a centrifuge cone containing 20 μL of MIBK. The phases were mixed ultrasonically for 2 s and separated in a centrifuge within 4 s. After centrifuging, the MIBK upper phase was pipetted onto a Ni disc which was heated from the edge. After drying, the Ni disc was placed over one of the Si(Au) surface barrier detectors. The average time from the end of the aerosol collection to the start of α and SF counting was 50 s. In tests, performed with ^{172}Ta, the chemical yield was found to be 75%.

In the dubnium experiments, 335 extractions were performed. No α particles in the energy range 8.4 – 8.7 MeV nor any fission were observed within the first two minutes of counting, demonstrating conclusively that, under these conditions, dubnium is not behaving chemically like its lighter homologue tantalum. The non Ta-like behavior of Db might indicate that Db forms polynegative anions like DbF_7^{2-} under the chosen conditions. The higher charge would then prevent extraction even into solvents with a relatively high dielectric constant such as MIBK.

3.2 DETAILED STUDIES

3.2.1 *Liquid-liquid extraction studies from mixed halide solutions*

While polynegative species cannot be extracted into ketones, it is possible to extract them by anion exchange into high molecular-weight ammonium salts. Amine extractions have another advantage for chemical studies of species that are produced as single atoms such as ^{262}Db, so that their chemical behavior must be studied "one-atom-at-a time". Due to their high viscosity, high molecular-weight ammonium salts are best suited as stationary phase on inert column support materials for High Performance Liquid Chromatography (HPLC). Then the principle of chromatography can be applied to single atoms or ions, and the many adsorption-desorption cycles along the column in the elution of that ion ensure a statistical behavior, so that one can be reasonably certain that the observation represents the 'true' chemistry of the element. Thus, in 1988 J.V. Kratz et al. [42] brought their ARCA II equipment to the Berkeley 88-Inch Cyclotron to perform a large number of automated reversed-phase extraction chromatography separations with the 1.6x8 mm columns filled with tri-iso-octyl amine (TiOA) on an inert support (Voltalef™, 32 – 63 μm, weight ratio 1:5).

The performance of ARCA II was studied with tracers of Zr, Nb, Hf, Ta, Pa, and some lanthanides produced on line and transported to ARCA II with a He(KCl) jet. To this end, the effluents from the TiOA mini columns were collected in fractions of three drops in small test tubes and were assayed for γ-ray activities using two Ge detectors. In agreement with earlier tracer studies, chemical yields for Nb, Ta, and Pa were consistently found to be 85 ± 5%.

In the beginning of the ^{262}Db experiments, it was most important to verify that the dubnium halide complexes were extracted into the TiOA phase under the same conditions as niobium, tantalum, and protactinium which were known to extract either from 12 M HCl/0.02 M HF or from 10 M HCl. ^{262}Db was produced in the ^{249}Bk(^{18}O,5n) reaction at 99 MeV with beam currents of 0.4 to 0.5 particle microamperes. Its extraction was verified by feeding the activities to the columns, and by stripping the amine (along with the extracted activities) from the columns by dissolving the TiOA in acetone. The acetone strip fraction was evaporated to dryness, and an assay of the samples for α-particle and SF decay started about 45 s after the end of the 60-s collection time. Elution curves for actinides, Nb, Ta, and Pa under the same conditions, with carrier-free activities produced on-line, are shown in the top section of Figure 8. In a first series of 207 experiments, extraction into the column material occurred from 12 M HCl/0.02 M HF, and 9 ^{262}Db/^{258}Lr events including 1 correlated pair, and 12 SF decays were recorded, indicating that ^{262}Db forms anionic complexes that are extracted by TiOA.

In a second series of 133 experiments, extraction into the amine phase occurred from 10 M HCl. This time, two ^{262}Db/^{258}Lr α events and four SF decays were detected in the acetone strip fraction. This demonstrates that dubnium behaves similar to Nb, Ta, and Pa also in 10 M HCl.

In the next series of 721 collection and separation cycles, the activity was adsorbed from 12 M HCl/0.02 M HF on the TiOA column as before, a Nb,Pa fraction was removed from the column with 4 M HCl/0.02 M HF, followed by the stripping of Ta from the column with 6 M HNO_3/0.015 M HF, as shown in the middle part of Figure 8. On the average, there was a tailing of about 10% of the Nb,Pa activity into the Ta fraction. 88% of the ^{262}Db/^{258}Lr α activity (38 events including 4 α-α correlations) were found in the Nb,Pa fraction, and 12% (5 events, with no correlated pair) in the Ta fraction. This distribution is identical with that of Nb and Pa, and distinctly different from

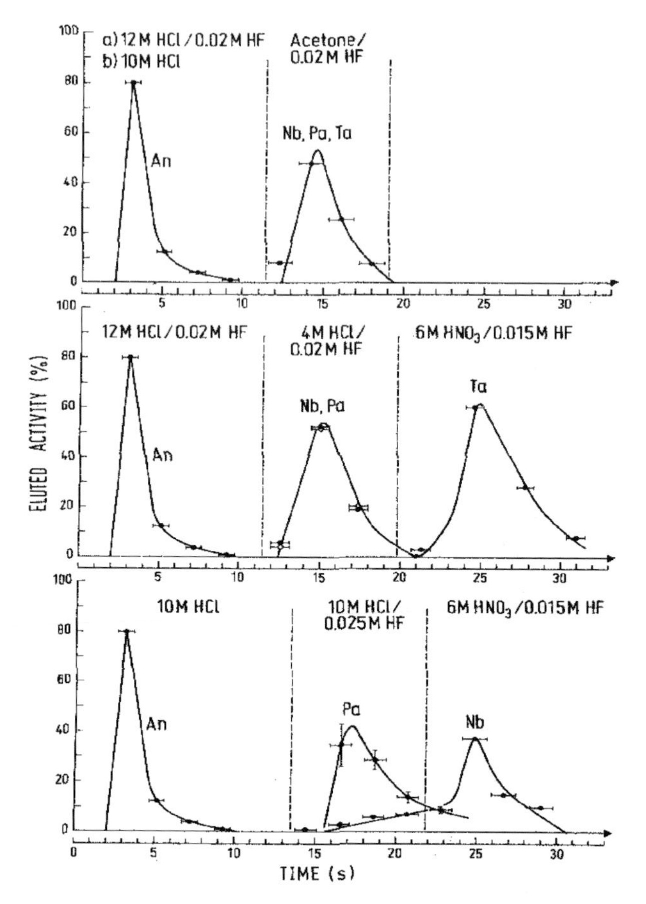

Fig. 8. Elution curves for carrier-free actinides (An), Nb, Ta, and Pa from TiOA/Voltalef™ columns (1.6x8 mm) under the same conditions as in the dubnium experiments. The horizontal error bars are associated with uncertainties in converting drop numbers into effluent volumes, i.e., times. In the upper part, the ^{262}Db is sorbed on the column from a) 12 M HCl/0.02 M HF, b) 10 M HCl, and is stripped along with the TiOA from the column in acetone/0.02 M HF. In the middle part, the activity is extracted as in a), followed by separate elutions of a Nb,Pa fraction, and a Ta fraction. In the lower part, the activity is extracted as in b), followed by a Pa, and then a Nb fraction. The number of ^{262}Db decays observed in a given fraction is given in the text. Reproduced from [42] with the permission of Oldenbourg Verlag.

that of Ta. This shows, in contrast to simple extrapolations, that the trend in the chemical properties in group 5 from Nb to Ta does not continue, but is *reversed* in going from Ta to Db. The distribution of SF decays between the Nb,Pa fraction (39) and the Ta fraction (10) corroborates the above finding.

In a last series of 536 experiments, a separation of Pa from Nb was performed (see bottom part of Figure 8). After feeding of the activities in 10 M HCl onto the column, Pa was eluted first with 10 M HCl/0.025 M HF. Under these conditions, a fraction of the Nb activity begins to break through.

The change of the eluent to 6 M HNO_3/0.015 M HF was timed such that the Pa fraction contained 80% of the Pa and 20% of the Nb, while the Nb fraction contained the remaining 20% of Pa and 80% of Nb. The ^{262}Db decays were equally divided between the Pa and Nb fractions: There were 25 α events, including 5 correlated pairs, in the Pa fraction, and 27 α events, including 5 correlated pairs, in the Nb fraction. The Pa/Nb ratio of SF decays was 25:19. These results indicate that the halide complexing strength in dubnium is between that for Nb and for Pa.

The half-lives deduced from the 106 α singles, 15 α-α correlations and 109 SF decays were all compatible with the 34-s half-life of ^{262}Db and the 4.3-s half-life of ^{258}Lr. The total production cross section of ^{262}Db at 99 MeV was 8.3 ± 2.4 nb.

In the discussion of the rather unexpected chemical results [42], it was suggested "that the chemical properties of the heaviest elements cannot reliably be predicted by simple extrapolations of trends within a group of elements", and that "relativistic, quantum chemical calculations for compounds of Nb, Ta, Pa, and Db are needed to understand in detail the differences in the halide complexing of the group-5 elements".

The TiOA column experiments were continued [43] by feeding the activities onto the columns in 12 M HCl/0.01 M HF and eluting a Pa fraction in 0.5 M HCl/0.01 M HF followed by elution of a Nb fraction in 4 M HCl/0.02 M HF. It was found that ^{262}Db elutes earlier than the bulk of the Nb activity, i.e., dubnium shows a more Pa-like behavior. From the distribution of the ^{262}Db events between fractions 2 and 3 in [42,43] K_d values, or the fractional extraction, %p, of ^{262}Db were estimated in [43] for 10 M HCl/0.025 M Hf, 4 M HCl/0.01 M HF, and the newly investigated 0.5 M HCl/0.01 M HF. The results are shown in Figure 9.

It is seen that Db shows a striking non Ta-like behavior and that it follows, at all HCl concentrations, the behavior of its lighter homologue Nb and that of its pseudo-homologue Pa. From this similarity, oxygen containing structures such as known for Nb and Pa, e.g., $NbOCl_4^-$, $PaOCl_4^-$, or $Pa(OH)_2Cl_4^-$, were discussed also for Db [43], and the preferential formation of oxygen containing complexes of Db was also predicted theoretically[44]. However, in [44], the extraction sequence Pa > Db > Nb was predicted which is the inverse sequence of that observed experimentally, see Figure 9.

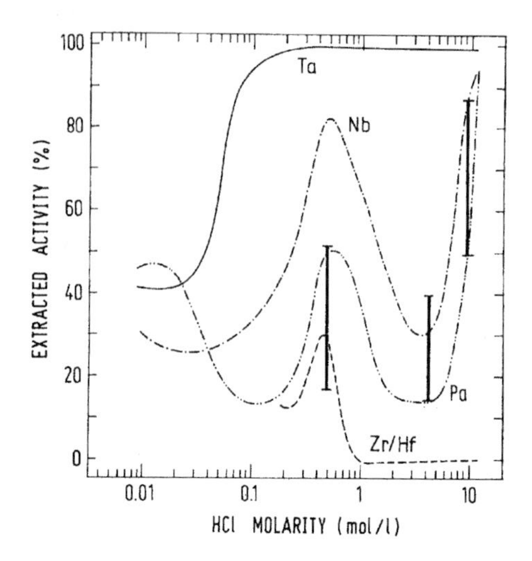

Fig. 9. Fractional extraction, %p, of Nb, Ta, Pa, and Zr/Hf vs. HCl molarity in the system TiOA-HCl/HF. The bold bars encompass the upper and lower limits of %p deduced from the Db elution positions. The bar for the extraction of Db from 12 M HCl/0.02 M HF is not included in the figure for clarity. The figure suggests that the element with the unusual behavior is Ta. Reproduced from [43] with the permission of Oldenbourg Verlag.

Due to the complicated situation in mixed HCl/HF solutions, with possibilities to form mixed chloride/fluoride complexes or even pure fluoride complexes, it was recommended in [44] to repeat the experiments in the pure HCl system. The original decision of the experimenters to add small amounts of HF to the HCl solutions was recommended in the literature [45] to prevent hydrolysis and "to maintain reproducible solution chemistry" of the group-5 elements.

The proximity in the Db behavior in the TiOA experiments to that of Pa was the reason for performing a series of extractions of Db into diisobutyl-carbinol (DIBC), a secondary alcohol which is a very specific extractant for Pa. The extraction from concentrated HBr solution in ARCA II was followed by the elution of a Nb fraction in 6 M HCl/0.0002 M HF, and of a Pa fraction in 0.5 M HCl. The number of ^{262}Db decays observed in the Nb fraction indicated that less than 45% of the Db was extracted into DIBC, and the extraction sequence Db < Nb < Pa was established [46]. This suggests that the seeming similarity between Db and Pa concluded from the TiOA experiments is not a general phenomenon.

3.2.2 *Liquid-liquid extraction studies from pure halide solutions*

As suggested in [44], the amine extractions of the group-5 elements were systematically revisited by W. Paulus et al. [32,49] in pure HF, HCl, and HBr solutions. V. Pershina [29-31], on the basis of an improved model of hydrolysis [29], and by explicitly considering the competition between chloride-complex formation and hydrolysis of the chloride complexes [30] has predicted the extraction sequence Pa >> Nb ≥ Db > Ta in the pure chloride system. She also predicted the extraction sequence fluorides >> chlorides > bromides for the halide complexes of the group-5 elements [31]. For tracer activities of Pa, Nb, and Ta, these predictions were verified experimentally for a number of amines [32,49] in batch extractions. Figure 10 shows an example for the quaternary ammonium salt Aliquat336/$Cl^-$and aqueous HCl solutions.

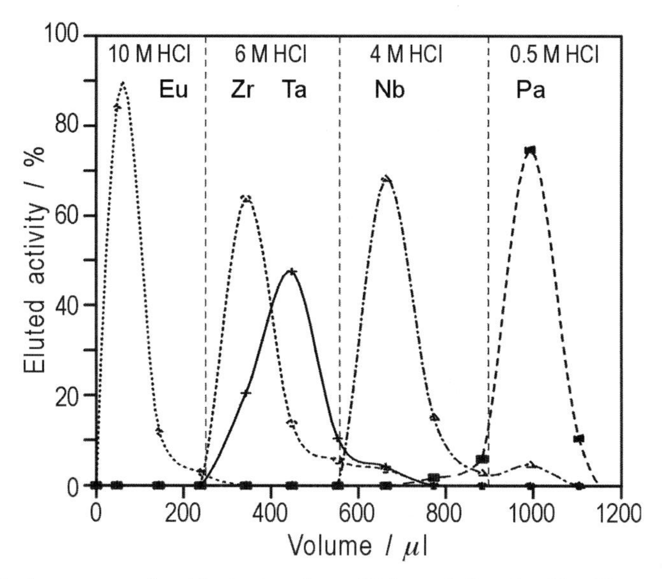

Fig. 10. Elution curves for trivalent cations (Eu), and for Zr, Ta, Nb, and Pa from Aliquat336/Cl^- -Voltalef™ (1:5) columns (1.6x8 mm) in ARCA II. The activities are fed onto the column in 10 M HCl. This is followed by separate elutions of a Ta fraction (6 M HCl), a Nb fraction (4 M HCl), and a Pa fraction (0.5 M HCl). Reproduced from [32] with the permission of Oldenbourg Verlag.

Studies with tracer activities of the lighter homologues. Based on these results, new chromatographic column separations with ARCA II were elaborated [32]. In these experiments, most of the amines tested in the HCl system showed slow kinetics for back extraction into the aqueous phase, resulting in elution peaks with an unacceptable tailing of the activities into the subsequent fraction, a feature that was not observed in the mixed HCl/HF system [42]. An acceptable chromatographic separation was only achieved with the quaternary ammonium salt Aliquat336/Cl^- at a flow rate of the mobile aqueous phase of 1 mL/min. Carrier-free tracers of Eu, Nb, Ta, Pa, Zr, and Hf were fed onto the column in 10 M HCl. Zr, Hf, and Ta were eluted in 6 M HCl, Nb in 4 M HCl, and Pa in 0.5 M HCl, see Figure 10. Thus, this system provides conditions for extracting all relevant elements from one HCl solution in a first step and for differentiating between Ta, Nb, and Pa in subsequent elutions.

Similarly, new partition experiments and chromatographic separations were performed with the fluoride salt of Aliquat336 in HF solutions and with the bromide salt of Aliquat336 in HBr solutions. The K_d values in the HBr system show the same sequence as in the HCl system: Pa > Nb > Ta. However, the threshold HBr concentrations above which an appreciable extraction is observed are shifted to higher HBr molarities, i.e., to 6 M HBr for Pa, 9 M for Nb, while Ta is not extracted even from 12 M HBr. Chromatographic separations were performed by loading the activities onto the 1.6x8 mm column in 12 M HBr and by eluting Nb in 7 M HBr and Pa in 2 M HBr. The fact that Ta did not extract from HBr solutions made this system the least attractive for an application to element 105.

With the fluoride salt of Aliquat336, K_d values on the order of $\geq 10^3$ were observed for all tracer activities (Pa, Nb, Ta) and somewhat lower ones for Zr, Hf even at low HF concentrations. The same extraction sequence Pa > Nb > Ta as in the HCl- and HBr-systems was observed at 0.5 M HF. For increasing HF concentrations, the K_d values stay high for Nb and Ta up to 12 M HF while they decrease for Pa, Zr, and Hf due to the formation of polynegative fluoro complexes such as PaF_7^{2-} and MF_6^{2-} (M=Zr, Hf), respectively. In chromatographic separations, after feeding the activities onto the 1.6x8 mm column in 0.5 M HF, a Pa fraction was eluted in 4 M HF and Nb and Ta were subsequently stripped from the column with 6 M HNO_3/0.015 M HF.

Studies of the dubnium behavior. 930 experiments were conducted with ^{262}Db produced at the Berkeley 88-Inch Cyclotron in the $^{249}Bk(^{18}O,5n)$ reaction. Extractions were performed in the Aliquat336/Cl^- - HCl system with a 50 s cyclic collection time of the KCl aerosol on a Kel-F™ slider in ARCA II. The reaction products were fed onto the column in 10 M HCl followed by the elution of a Ta fraction in 6 M HCl and a Nb,Pa fraction in 6 M HNO_3/0.015 M HF. The effluents were continuously sprayed through a 60 μm nozzle onto hot Ta discs on which they were evaporated to dryness by hot He gas and infrared light. Start of measurement of the activities was 60 s (Ta fraction) and 76 s (Nb, Pa fraction) after the end of collection, respectively.

6 α singles in the Ta fraction and 12 in the Nb,Pa fraction were registered with life times compatible with the 34-s half-life of ^{262}Db. In addition, 3 pairs of correlated mother-daughter decays were registered, one in the Ta fraction and two in the Nb,Pa fraction. From this distribution, a K_d value of ~440 for element 105 in 6 M HCl was deduced [32], see Figure 11. This

value is much larger than that of Ta (22), smaller or equal to that of Nb (680), and smaller than that of Pa (1440). From this, the extraction sequence Pa >> Nb ≥ Db > Ta was deduced [32] which is in agreement with the theoretical prediction [30].

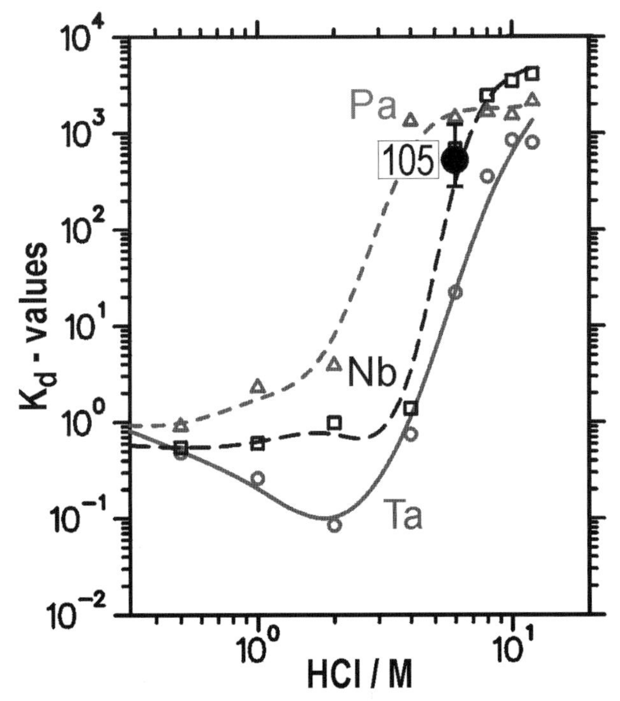

Fig. 11. Extraction behavior of Pa, Nb, and Ta from HCl solutions into Aliquat336/Cl^-. The K_d of element 105 in 6 M HCl (with error bars encompassing the 68% confidence limit) is indicated by the bold dot. Reproduced from [32] with the permission of Oldenbourg Verlag.

In the Aliquat336/F^--HF system, 377 experiments were performed with ARCA II. The collection time of the He(KF) gas-jet was 50 s. The dissolved reaction products were loaded onto the column in 0.5 M HF. In elutions with 4 M HF (Pa fraction) and 6 M HNO_3/0.015 M HF (Nb,Ta fraction), 4 α-α correlations were detected in the Nb,Ta fraction and none in the Pa fraction. (Due to a Po contamination, α singles were not evaluated.) The K_d value resulting from the probability distributions for zero (Pa fraction) and 4 (Nb,Ta fraction) observed correlated events is >570 in 4 M HF which is close to that for Nb and Ta ($\geq 10^3$) and differs markedly from that for Pa (~19), see Figure 12.

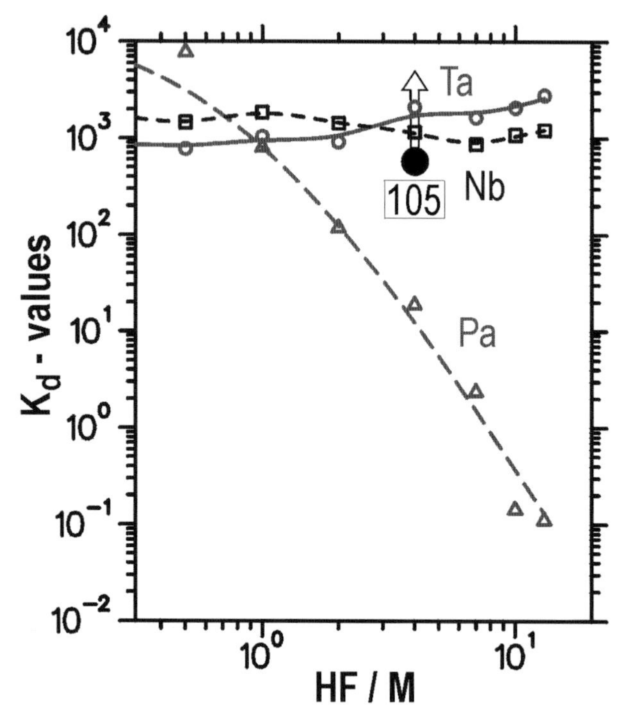

Fig. 12. Extraction behavior of Pa, Nb and Ta from HF solutions in Aliquat336/F^-. The lower limit for the K_d value of element 105 in 4 M HF (representing the 68% confidence limit) is indicated by the bold dot with the arrow. Reproduced from [32] with the permission of Oldenbourg Verlag.

To conclude, the amine extraction behavior of dubnium halide complexes is always close to that of its lighter homologue Nb in agreement with the predicted *inversion of the trend of the properties when going from the 5d elements to the 6d elements* [30,31]. In pure HF solutions, it differs mostly from the behavior of Pa. In pure HCl solutions, it differs considerably from both Pa and Ta. In mixed HCl/HF solutions, it differs markedly from the behavior of Ta. The studies of the halide complexing of the group-5 elements both theoretically [30,31] and experimentally [32,42,43] demonstrate the progress that has been made in understanding details of the chemical properties of the transactinides.

3.2.3 *Studies using ion-exchange resins and an organic complexing agent -- chemical and nuclear results*

Another series of experiments with Db used its complexation with the α-hydroxyisobutyrate (α-HiB), $(CH_3)_2COH\ COO^-$, that had been used in the first liquid-phase rutherfordium experiment [15]. Chelating by α-HiB depends strongly on the charge of the metal ion [41]. Elutions from cation exchange columns in ARCA II with unbuffered dilute α-HiB show that Nb, Ta, and Pa are eluted promptly while tetravalent and trivalent metal ions are strongly retained on the resin [41]. This is shown in Figure 13 for manually performed separations of pentavalent Nb and Pa from tetravalent Zr using 0.075 M α-HiB and a 1.7x25 mm column filled with Aminex A6 (particle size 17.5 ± 2 μm). The tracer activities are fed onto the column through a sample loop. After the dead volume of 130 μL (7 s), both activities appear in the effluent and are eluted completely in another 150 μL (9 s). The first activity of the tetravalent Zr is detected after the elution of another 3150 μL (190 s). After the elution of Nb and Pa in 0.075 M α-HIB, Zr can be eluted in 210 μL of 0.5 M α-HiB, see Figure 13.

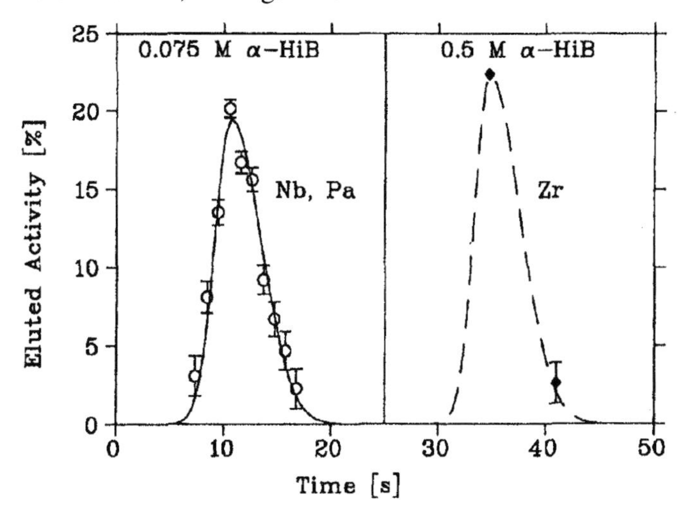

Fig. 13. Left frame: Manual HPLC elution (1 mL/min) of ^{95}Nb (open circles) from a 1.7x25 mm Aminex A6 column in 0.075 M α-HiB. ^{233}Pa (not shown) elutes in an identical manner. Right frame: Elution of ^{95}Zr from the same column in 0.5 M α-HiB. Only two fractions were counted (black diamonds). The dashed curve has the same shape as the elution curve in the left frame. Reproduced from [41] with the permission of Oldenbourg Verlag.

In experiments at the UNILAC accelerator at GSI, Darmstadt, the tantalum isotopes $^{168\text{-}170}$Ta were transported by the He(KCl) gas-jet and deposited on a polyethylene frit in ARCA II. The dissolution of the collected tantalum activity from the frit was investigated as a function of the α-HiB concentration. Because of the smaller column size in ARCA II (1.6x8 mm) which might cause an earlier breakthrough of the tetravalent and trivalent metal ions it was desirable to decrease the α-HiB concentration. Even with 34 μL of 0.025 M α-HiB, dissolution of >75% of the tantalum activity was achieved in 2 s. The time required for the complete elution of Ta from the column was about 4 s. Similar experiments performed at the Mainz TRIGA reactor with ^{99m}Nb confirm this result as shown in Figure 14.

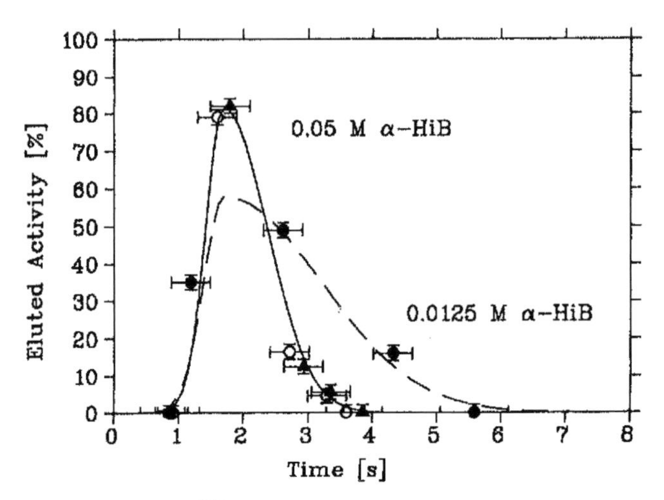

Fig. 14. Automated elution of ^{99m}Nb from 1.6x8 mm Aminex A6 columns in ARCA II at a flow rate of 1 mL/min (16.7 μL/s). Open circles and closed triangles, connected by the solid line, refer to series of elutions with 0.05 M α-HiB. Black circles and dashed curve: Elution in 0.0125 M α-HiB. Reproduced from [41] with the permission of Oldenbourg Verlag.

The data for 0.05 M α-HiB show that the elution position and the width are quite reproducible. As the elution obviously takes more time with 0.0125 M α-HiB, see Figure 14, 0.05 M α-HiB was finally selected for the dubnium experiment.

The time sequence for the Db separations was as follows: After collection of the transported activities on the polyethylene frit in ARCA II for 1 minute, the frit was washed with 100 μL of unbuffered 0.05 M α-HiB solution (6.5 s). This solution was passed through one of 38 cation exchange columns contained in two magazines and was collected on a Ta disc and quickly evaporated to dryness. After flaming and cooling, the Ta disc was inserted into one of ten detector stations for α-particle and SF detection. Start of counting was 39 s after the end of collection. After three separations each, a new column was positioned below the collection frit. Thus, 114 continuous collection and separation cycles were conducted before the program was stopped; the used magazines were removed, and two new magazines were introduced for the next 114 1-min cycles.

In the production runs with ^{262}Db at the 88-Inch Cyclotron of the Lawrence Berkeley Laboratory, the beam energy in the ^{249}Bk target was 99 MeV. The target originally consisted of 0.54 mg/cm^2 of 330-d ^{249}Bk and decreased to 0.51 mg/cm^2 with the remainder of the total thickness being its ^{249}Cf daughter. The activity was transported by a He(KCl) gas-jet (2 L/min) over 5 m to the collection site in ARCA II and was collected on polyethylene frits. The He(KCl) gas-jet efficiency was measured frequently during the experiments by dissolving the activity in 0.05 M α-HiB and eluting it through an empty column directly onto a Ta disc. After evaporation to dryness and flaming, the production rate of the $^{252\text{-}255}$Fm transfer products was determined by α-pulse height analysis. For normalization, separate bombardments of the Bk target were performed in which all of the products recoiling from the target were caught in a Au-catcher foil located directly behind the target. After one hour of irradiation, the foil was dissolved in aqua regia to which an aliquot of ^{241}Am had been added to trace the chemical yield of actinides. The gold was removed on an anion-exchange column. The actinide fraction, which passed through the column, was collected and dried on a Pt disc for α-particle spectroscopy. By comparing the apparent production rates measured after transport through the He(KCl) gas-jet with the absolute production rates from the Au-catcher experiments, the He-transport efficiency was determined. In these experiments, the transport efficiency varied between 40 and 53%. The jet-transport yields are based on the measurement of 50-100 counts, so they are only accurate to 10 or 20%.

In 525 collection and elution cycles at 99 MeV, the α-particle spectrum shown in Figure 15 was obtained. Apart from ^{249}Cf contamination sputtered from the target (this material is not dissolved by the weekly acidic α-HiB solution and is washed mechanically through the column), and from small amounts of Bi and Po activities produced by transfer reactions on a Pb impurity in the target, the spectrum is very clean.

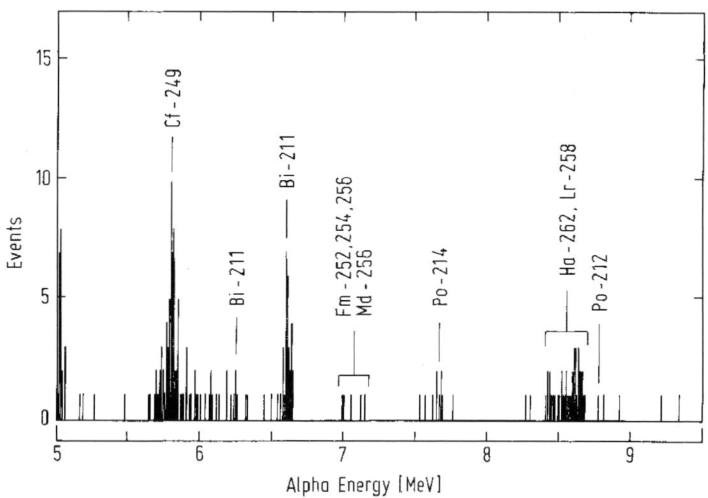

Fig. 15. α-energy spectrum of Db fractions from α-HiB separations observed in the bombardment of a ^{249}Bk target with 99 MeV ^{18}O ions. Reproduced from [47]; Copyright (2001) by the American Physical Society.

The spectrum contains 41 α events attributable to element 105, among them 9 pairs of ^{262}Db-^{258}Lr parent-daughter decays. The half-lives of $35.7^{+6.9}_{-5.4}$ s for

^{262}Db and $4.2^{+1.5}_{-1.1}$ s for ^{258}Lr were in good agreement with the previously determined values [42]. Also, 23 fissions attributable to Db decays were detected. These event rates are consistent with the detection efficiency and the known production rate at 99 MeV [42]. From a chemical point of view, this means that the data are consistent with a high chemical yield of Db in these separations and with the fact that Db resumes the pentavalent oxydation state under the present conditions [41].

The α-HiB procedure was also applied to the products of a ^{249}Bk bombardment with ^{18}O ions at the lower bombarding energy of 93 MeV in the target in an attempt to discover the unknown isotope ^{263}Db produced in the 4n reaction [41]. In a total of 374 experiments, 9 α particles with energies between 8.3 and 8.5 MeV, see Figure 16, as well as 18 SF events were registered.

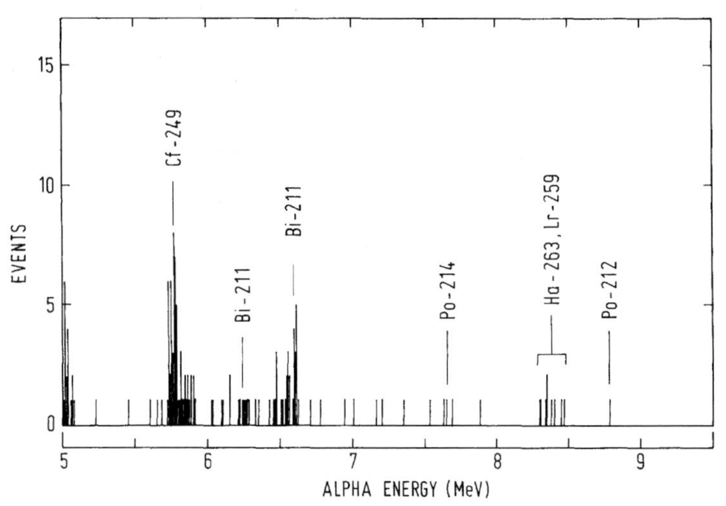

Fig. 16. α-particle energy spectrum of Db fractions from α-HiB separations observed in the bombardment of a ^{249}Bk target with 93 MeV ^{18}O ions. Reproduced from [47]; Copyright (2001) by the American Physical Society.

The absence of α particles above 8.5 MeV indicates that the ^{262}Db/^{258}Lr pair is no longer present at the lower bombarding energy (σ_{5n} < 1.3 nb [47]) in agreement with evaporation calculations. Instead, α groups at 8.35 and 8.45 MeV are detected that are assigned to the new isotope ^{263}Db and its daughter ^{259}Lr, respectively [41,47]. ^{263}Db, decays with a half-life of 27^{+10}_{-7} s and has a SF branch of 57^{+13}_{-15}%. The cross section of 10±6 nb at 93 MeV is on the same order of magnitude as the 5n cross section at 99 MeV in agreement with evaporation calculations. The mass assignment is supported by the observation of the 8.45 MeV α particles of the daughter, ^{259}Lr, in the chemically separated Db fractions and by the observation of two events [47] in which the α decay of ^{263}Db was followed by the SF decay [48] of the ^{259}Lr daughter.

Based on these results, the cross section for production of ^{262}Db at 99 MeV and the SF branch in ^{262}Db as reported earlier [42] had to be revised as it was clear now that both isotopes, 34-s ^{262}Db and 27-s ^{263}Db, are produced at this energy. The new value for the SF branch in ^{262}Db [47] resulted as 33 %, and the cross sections for production of ^{262}Db and ^{263}Db at 99 MeV resulted as 6±3 nb and 2±1 nb, respectively [47]. This shows that chemical separations of transactinides can also be useful in obtaining new data on nuclear properties of their isotopes.

4. Seaborgium (Sg, Element 106)

Tests of liquid-phase separations suitable for the group-6 elements Mo, W, including sometimes the pseudo-homologue U, were first performed in [50] with the fission products ^{93}Y, ^{97}Zr, and ^{99}Mo and with W isotopes produced in the ^{20}Ne + ^{152}Gd reaction at the PSI Philips Cyclotron. $5 \cdot 10^{-2}$ M α-HiB solutions, pH = 2.65 and pH = 5 were found to elute W in a one-stage separation from cation-exchange columns and they provided good separations from Hf and Lu. Likewise, solutions with 0.1 M HCl and various HF concentrations between 10^{-4} and 10^{-2} M were eluting W rapidly while Hf was retained on the cation-exchange resin. Hf was observed to be partly eluted for $\geq 2.8 \cdot 10^{-3}$ M in 0.1 M HCl. As the sorption of Hf on the cation-exchange resin was even higher in 0.1 M $HNO_3/5 \cdot 10^{-4}$ M HF, the decision was made [51] to elute a seaborgium fraction from cation columns using this solution. This would also avoid the formation of mixed chloride-fluoride complexes which are difficult to model. $MO_2F_3(H_2O)^-$ is a likely form of the complexes that are eluted under such conditions but neutral species such as $MO_2F_2(H_2O)_2$ cannot be excluded. Some problems were encountered with adsorption of the activities on the slider in ARCA II. Among the various materials tested, titanium showed the lowest losses of W and Hf due to adsorption. Figure 17 shows the elution curve for short-lived tungsten isotopes from the reaction of ^{20}Ne with enriched ^{152}Gd.

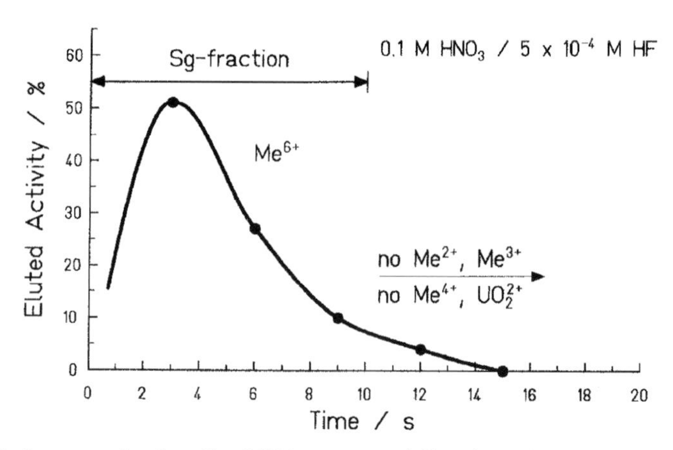

Fig. 17. Elution curve for short-lived W isotopes modeling the seaborgium separation [52] in ARCA II using a solution of 0.1 M $HNO_3/5\cdot10^{-4}$ M HF with a flow rate of 1 mL/min. The 1.6x8 mm columns are filled with the cation-exchange resin Aminex A6. Reproduced from [52] with the permission of Oldenbourg Verlag.

The activity was transported to ARCA II with a He(KCl) gas-jet within about 3 s. After deposition on a titanium slider it was dissolved and washed through the 1.6x8 mm column (filled with the cation-exchange resin Aminex A6, 17.5±2 µm) at a flow rate of 1 mL/min with 0.1 M $HNO_3/5\cdot10^{-4}$ M HF. 85% of the W elute within 10 s. Neither divalent or trivalent metal ions nor group-4 ions are eluted within the first 15 s. Also the pseudo-homologue uranium, in the form of UO_2^{2+}, is completely retained on the column.

In the experiments [52] with element 106, seaborgium (Sg), at the GSI UNILAC a 950 µg/cm^2 ^{248}Cm target was bombarded with $3\cdot10^{12}$ ^{22}Ne ions s^{-1} at 121 MeV to produce the isotope ^{265}Sg with a half-life of $7.4^{+3.3}_{-2.7}$ s [53]. 3900 identical separations were conducted with a collection and cycle time of 45 s and a total dose of $4.58\cdot10^{17}$ Ne ions. The transport efficiency of the He(KCl) gas-jet was 45%. On the average, counting of the samples started 38 s after the end of collection. The overall chemical yield for W was 80%. Three correlated α-α mother-daughter decays were observed that were assigned to the decay of ^{261}Rf and ^{257}No as the descendants of ^{265}Sg, see Figure 18. These three correlations have to be compared with an expectation value of 0.27 random correlations. This gives a probability of 0.24% that the three events are random correlations. As the mother decays were not observed, Figure 18, it is important to note that ^{261}Rf and ^{257}No can only be observed if ^{265}Sg passes through the column because group-4 elements and No are strongly retained on the cation exchange columns in ARCA II [35], see Figure 4. Most likely, the decay of 7-s ^{265}Sg was not observed because it occurred in the time interval between the end-of-separation and the start-of-measurement which is equivalent to four half-lives.

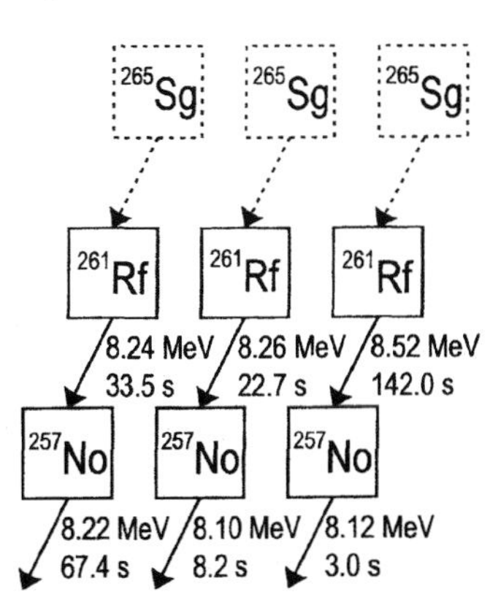

Fig. 18. Nuclear decay chains originating from ^{265}Sg after chemical separation with ARCA II [52]. The α-decay energies are given in MeV and the observed lifetimes in seconds. Reproduced from [54] with the permission of Nature & Macmillan Publishers Ltd.

From the observation of the three correlated α-decay chains of ^{265}Sg daughters, it could thus be concluded that seaborgium shows a behavior typical of a group-6 element and different from that of the pseudo-group-6 element uranium which is fixed as UO_2^{2+} on the cation-exchange column. Presumably, Sg formed $SgO_2F_3(H_2O)^-$ or the neutral species SgO_2F_2, but due to the low fluoride concentration used, the anionic SgO_4^{2-} ("seaborgate" in analogy to molybdate, MoO_4^{2-}, or tungstate, WO_4^{2-}) could not be excluded from these experiments; see below for results from further investigations.

Thus, it appeared to be most attractive to test what would happen if no fluoride ions were provided in otherwise identical experiments, i.e., in attempts to elute Sg from cation-exchange columns in 0.1 M HNO_3. To this end, a 690 μg/cm^2 ^{248}Cm target containing 22 μg/cm^2 enriched ^{152}Gd was bombarded with 123 MeV ^{22}Ne ions [55]. ^{169}W produced in the $^{152}Gd(^{22}Ne,5n)$ reaction served as a yield monitor. 45-s cycles were run in which the effluent from the Aminex A6 columns was evaporated on thin (~500 μg/cm^2) Ti foils mounted on Al frames. The Ti foils were thin enough to be counted in close geometry by pairs of Si detectors thus increasing the efficiency for α-α-correlations by a factor of four as compared to the preceding experiment. A beam dose of $4.32 \cdot 10^{17}$ ^{22}Ne ions was collected in 4575 separations. Only one α-α-correlation attributable to the ^{261}Rf – ^{257}No pair was observed. With an expected number of random correlations of 0.5 this is likely (with a probability of 30%) to be a random correlation. From the beam integral and the overall yield, as measured simultaneously for ^{169}W (27% including jet transportation and chemical yield), a total of 5 correlated events was to be expected. The absence of these events is interpreted [55] as to indicate that, in the absence of fluoride ion, there is sorption of seaborgium on the cation-exchange resin.

This obviously non-tungsten like behavior of seaborgium under the given conditions is attributed to its weaker tendency to hydrolyze:

$$M(H_2O)_6^{6+} \leftrightarrow M(OH)(H_2O)_5^{5+} + H^+ \quad (6)$$

$$\vdots$$

$$MO(OH)_3(H_2O)_2^{+} \leftrightarrow MO_2(OH)_2(H_2O)_2 + H^+ \quad (7)$$

$$MO_2(OH)_2(H_2O)_2 \leftrightarrow MO_3(OH)^- + 2\,H_2O + H^+ \quad (8)$$

$$MO_3(OH) \leftrightarrow MO_4^{2-} + H^+ \quad (9)$$

The measured equilibrium constants for this stepwise deprotonation scheme for Mo and W have been collected from the literature in [56]. They show that Mo is more hydrolyzed than W, and that the deprotonation sequence for Mo and W at pH = 1 reaches the neutral species $MO_2(OH)_2(H_2O)_2$. Assuming the deprotonation processes for the Sg compounds to be similar to those of Mo and W, Equations (6-9), V. Pershina and J.V. Kratz performed fully relativistic density-functional calculations of the electronic structure of the hydrated and hydrolyzed structures for Mo, W, and Sg [56]. By use of the electronic density distribution data, relative values of the free energy changes and by use of the hydrolysis model [29,30], constants of hydrolysis reactions (6-9) were defined [56]. These results show hydrolysis of the cationic species to the neutral species to decrease in the order Mo>W>Sg which is in agreement with the experimental data on hydrolysis of Mo and W, and on Sg [55] for which the deprotonation sequence may end earlier with a cationic species such as $SgO(OH)_3(H_2O)_2^+$ that is sorbed on the cation-exchange resin.

Looking back to the experiments in [52] where fluoride ions were present having a strong tendency to replace OH^- ligands, it appears plausible that, in this experiment, neutral or anionic species were formed:

$$MO_2(OH)_2(H_2O)_2 + 2HF \leftrightarrow MO_2F_2(H_2O)_2 + 2\,H_2O \quad (10)$$

$$MO_2F_2(H_2O)_2 + F^- \leftrightarrow MO_2F_3(H_2O)^- + H_2O \quad (11)$$

Thus, the presence of fluoride ions seems to be an important prerequisite for planned experiments in which the K_d value of Sg on an anion-exchange resin is to be determined with the MCT technique [36,57], see Section 5.

5. Perspectives

For rutherfordium and for dubnium, many more interesting chemical investigations in the liquid phase are intelligible.

For seaborgium, it is planned to determine the K_d value on an anion-exchange resin (Dowex 1x8) in 0.1 M HNO_3/5·10^{-3} M HF. A large amount of work has been invested to prepare this important experiment by measuring K_d values in batch experiments for tracer activities of Zr, Hf, Mo, Ta, W, Th with various resins of different functionality and aqueous solutions containing HCl, HNO_3, HCl/HF, and HNO_3/HF [36,57]. In [57], an anion separation scheme in mixed HNO_3/HF solutions was elaborated for a W/Ta separation on an anion-exchange resin with the quaternary ammonium group $-N^+(C_2H_5)_3$ for which relatively low K_d values for W and high K_d values for Ta are observed yielding a W/Ta separation factor of 2270. With this resin, batch and MCT experiments were performed yielding mutually consistent K_d values. By varying the concentration of the counter ion NO_3^- competing for the binding sites on the resin, a slope of –1 in a plot of log K_d vs. log c(HNO_3) was verified indicating that W, in the presence of HF, forms anions with an electric charge of –1, e.g., $WO_2F_3(H_2O)^-$. Thus the feasibility of a Sg experiment with the MCT technique was demonstrated.

As the α-decay daughters of ^{265}Sg, i.e., ^{261}Rf, ^{257}No, ^{253}Fm, and ^{253}Es, are cations in HNO_3/HF solutions at sufficiently low HF concentrations, a separation scheme for the planned Sg experiment is to use cation-exchange resins for the filter column F and the daughter column D_1, while the chromatographic column C contains the anion-exchange resin Dowex 1x8. In 0.1 M HNO_3/5·10^{-3} M HF, ^{261}Rf is still present as a cation [35], and Mo and W have K_d values on Dowex 1x8 of 82 and 11.2, respectively [36]. As a continuous dissolution of the KCl aerosol of the transport jet in a degassing unit of the type used in [9,10] would result in a hold-up time too long for 7-s ^{265}Sg, a new apparatus was constructed. Here, the activity-bearing aerosol is deposited for 2 s on a Ta disc by impaction Thereafter, the Ta disc with the active spot is rotated into a position where the activity is dissolved within 2 s in 0.1 M HNO_3/5· 10^{-3} M HF and is fed into the 3-column system quasi continuously. This MCT experiment has been tested successfully with short-lived W isotopes and awaits its application to ^{265}Sg.

Since the fast centrifuge system SISAK is equipped with liquid scintillation counting LSC [12,58], it is in principle capable of investigating short-lived α-decaying nuclides of the transactinides. β/γ pulses and α pulses are distinguished by pulse-shape discrimination PSD and pile-up pulses are rejected by a pile-up rejection system PUR. This analog electronics proved to result in insufficient background suppression. Thus, two new approaches are being pursued: K. Eberhardt et al. [59] have coupled the existing PSD/PUR electronics with a fast transient recorder which records the LSC pulses in 500 2-ns steps thus facilitating the recognition of background pulses in the off-line analysis. These digitally recorded pulses will be subject of analysis by a neural network that has been trained with α events, β events, and β/γ pile-up events in order to recognize true α events [60].

Another perspective for SISAK is to suppress the interfering β/γ background by an electromagnetic separator. J.P. Omtvedt et al. [61] performed recently first successful experiments with ^{261}Rf by coupling SISAK to the Berkeley Gas Filled Separator (BGS); see Chapter 4 for more details.

Probably, a combination of both i) digitally recording of LSC pulses plus their analysis by a neural network and ii) reducing the amount of reaction products entering the SISAK system with a pre-separator gives the most promising perspective. Various liquid-liquid extraction schemes for the transactinides have already been elaborated. An example is the separation procedure proposed for liquid-phase studies of element 107, bohrium (Bh) [62].

The Flerov Laboratory of Nuclear Reactions (FLNR) in Dubna, Russia, has recently announced the observation of relatively long-lived isotopes of elements 108, 110, 112, 114, and 116 [63-66] confirming the over 30 years old theoretical prediction of an "island of stability" of spherical superheavy elements. Due to the half-lives of the observed isotopes in the range of seconds to minutes, chemical investigations of these heaviest elements in the Periodic Table appear now to be feasible. The chemistry of these elements should be extremely interesting due to the predicted dramatic influence of relativistic effects [67]. In addition, the chemical identification of the newly discovered superheavy elements is highly desirable as the observed decay chains [63-66] cannot be linked to known nuclides which has been heavily criticized [68,69].

Elements 108 – 116 are homologues of Os through Po and are expected to be partially very noble metals. Thus it is obvious that their electrochemical deposition could be an attractive method for their separation from aqueous solutions. It is known that the potential associated with the electrochemical deposition of radionuclides in metallic form from solutions of extremely small concentration is strongly influenced by the electrode material. This is reproduced in a macroscopic model [70], in which the interaction between the microcomponent A and the electrode material B is described by the partial molar adsorption enthalpy and adsorption entropy. By combination with the thermodynamic description of the electrode process, a potential is calculated that characterizes the process at 50% deposition:

$$E_{50\%} = E^0 - \Delta H(A\text{–}B)/nF + T\Delta S_{vibr}(A\text{–}B)/nF - (RT/nF)\ln(A_m/1000) \qquad (12)$$

Here, ΔH(A-B) is the partial molar net adsorption enthalpy associated with the transformation of 1 mol of the pure metal A in its standard state into the state of "zero coverage" on the surface of the electrode material B, ΔS_{vibr} is the difference in the vibrational entropies in the above states, n is the number of electrons involved in the electrode process, F the Faraday constant, and A_m the surface of 1 mol of A as a mono layer on the electrode metal B [70]. For the calculation of the thermodynamic functions in (12), a number of models were used in [70] and calculations were performed for Ni-, Cu-, Pd-, Ag-, Pt-, and Au-electrodes and the micro components Hg, Tl, Pb, Bi, and Po, confirming the decisive influence of the choice of the electrode material on the deposition potential. For Pd and Pt, particularly large, positive values of $E_{50\%}$ were calculated, larger than the standard electrode potentials tabulated for these elements. This makes these electrode materials the prime choice for practical applications. An application of the same model to the superheavy elements still needs to be done, but one can anticipate that the preference for Pd and Pt will persist. The latter are metals in which, due to the formation of the metallic bond, almost or completely filled d orbitals are broken up, such that these metals tend in an extreme way towards the formation of intermetallic compounds with sp-metals. The perspective is to make use of the Pd or Pt in form of a tape on which the tracer activities are electrodeposited and the deposition zone is subsequently stepped between pairs of Si detectors for α-spectroscopy and SF measurements.

References

1. Gregorich, K.E., Henderson, R.A., Lee, D.M., Nurmia, M.J., Chasteler, R.M., Hall, H.L., Bennett, D.A., Gannett, C.M., Chadwick, R.B., Leyba, J.D., Hoffman, D.C., Herrmann, G.: Radiochim. Acta **43**, (1988) 223.
2. Czerwinski, K.R. , Gregorich, K.E., Hannink, N.J., Kacher, C.D., Kadkhodayan, B., Kreek, S.A., Lee, D.M., Nurmia, M.J., Türler, A., Seaborg, G.T., Hoffman, D.C.: Radiochim. Acta **64**, (1994) 23.
3. Czerwinski, K.R., Kacher, C.D., Gregorich, K.E., Hamilton, T.M., Hannink, N.J., Kadkhodayan, B., Kreek, S.A., Lee, D.M., Nurmia, M.J., Türler, A., Seaborg, G.T., Hoffman, D.C.: Radiochim. Acta **64**, (1994) 29.
4. Bilewicz, A., Siekierski, S., Kacher, C.D., Gregorich, K.E., Lee, D.M., Stoyer, N.J., Kadkhodayan, B., Kreek, S.A., Lane, M.R., Sylwester, E.R., Neu, M.P., Mohar, M.F., Hoffman, D.C.: Radiochim. Acta **75**, (1996) 121.
5. Kacher, C.D., Gregorich, K.E., Lee, D.M., Watanabe, Y., Kadkhodayan, B., Wierczinski, B., Lane, M.R., Sylwester, E.R., Keeney, D.A., Hendriks, M., Stoyer, N.J., Yang, J., Hsu, M., Hoffman, D.C., Bilewicz, A.: Radiochim. Acta **75,** (1996) 127.
6. Kacher, C.D., Gregorich, K.E., Lee, D.M., Watanabe, Y., Kadkhodayan, B., Wierczinski, B., Lane, M.R., Sylwester, E.R., Keeney, D.A., Hendriks, M., Hoffman, D.C., Bilewicz, A.: Radiochim. Acta **75,** (1996) 135.
7. Schädel, M., Brüchle, W., Jäger, E., Schimpf, E., Kratz, J.V., Scherer, U.W., Zimmermann, H.P.: Radiochim. Acta **48,** (1989) 171.
8. Szeglowski, Z., Bruchertseifer, H., Domanov, V.P., Gleisberg, B., Guseva, L.J., Hussonnois, M., Tikhomirova, G.S., Zvara, I., Oganessian, Yu.Ts.: Radiochim. Acta **51,** (1990) 71.
9. Pfrepper, G., Pfrepper, R., Yakushev, A.B., Timokhin, S.N., Zvara, I.: Radiochim. Acta **77,** (1997) 201.
10. Pfrepper, G., Pfrepper, R., Krauss, D., Yakushev, A.B., Timokhin, S.N., Zvara, I.: Radiochim. Acta **80,** (1998) 7.
11. Persson, H., Skarnemark, G. Skålberg, M., Alstad, J., Liljenzin, J.O., Bauer, G., Haberberger, F., Kaffrell, N., Rogowski, J., Trautmann, N.: Radiochim. Acta **48,** (1989) 177, and Alstad, J., Skarnemark, G., Haberberger, F., Herrmann, G., Nähler, A., Pense-Maskow, M., Trautmann, N.: J. Radioanal. Nucl. Chem. **189,** (1995) 133.
12. Wierczinski, B., Eberhardt, K., Herrmann, G., Kratz, J.V., Mendel, M., Nähler, A., Rocker, F., Tharun, U., Trautmann, N., Weiner, K., Wiehl, N., Alstad, J., Skarnemark, G.: Nucl. Instr. and Meth. in Phys. Res. **A370,** (1996), 532.
13. Seaborg, G.T., Loveland, W.D., *The Elements Beyond Uranium*, John Wiley and Sons, Inc., New York (1990).
14. Ghiorso, A., Nurmia, M., Eskola, K., Eskola, P.: Phys. Lett. B **32,** (1970) 95.
15. Silva, R.J., Harris, J., Nurmia, M., Eskola, K., Ghiorso, A.: Inorg. Nucl. Chem. Lett. **6,** (1970) 871.
16. Ghiroso, A.: Report XBL 702-6151 (1970).
17. Hulet, E.K., Lougheed, R.W., Wild, J.F., Landrum, J.H., Nitschke, J.M., Ghiorso, A.: J. Inorg. Nucl. Chem. **42,** (1980), 79.
18. Hoffman, D.C.: "Chemistry of the Transactinide Elements". In: Proceedings of "The Robert A. Welch Foundation Conference on Chemical Research XXXIV Fifty Years with Transuranium Elements", Houston, Texas, 22-23 October 1990, pp. 255-276.
19. Hoffman, D.C.: Radiochim. Acta **61,** (1993) 123.
20. Kratz, J.V.: Proc. Int. Conf. on Actinides, Santa Fe, New Mexico, 19-24 September 1993, Elsevier, Amsterdam (1994), and J. Alloys Comp. **213, 214,** (1994) 20.
21. Kratz, J.V.: Chemie in unserer Zeit, 29. Jahrg. (1995) Nr. 4, 194.
22. Hoffman, D.C.: Chem. Eng. News **72 (18),** (1994) 24.
23. Schädel, M.: Radiochim. Acta **70/71,** (1995) 207.

24. Kratz, J.V.: "Chemistry of Seaborgium and Prospects for Chemical Studies of Bohrium and Hassium". In: Proceedings of "The Robert A. Welch Foundation Conference on Chemical Research XXXXI The Transactinide Elements", Houston, Texas, 27-28 October 1997, pp. 65-93.
25. Kratz, J.V.: "Chemical Properties of the Transactinide Elements". In: *Heavy Elements and Related New Phenomena*, Eds. Greiner, W., Gupta, R.K., World Scientific, Singapore, (1999) 129-193.
26. Pershina, V.: "Electronic Structure and Chemistry of the Heaviest Elements". In: *Heavy Elements and Related New Phenomena*, Eds. Greiner, W., Gupta, R.K., World Scientific, Singapore, (1999) 194-262.
27. Czerwinski, K.R.: Ph.D. thesis, University of California, Lawrence Berkeley Laboratory Report LBL-32233 (1992).
28. Günther, R., Paulus, W., Kratz, J.V, Seibert, A., Thörle, P., Zauner, S., Brüchle, W., Jäger, E., Pershina, V., Schädel, M., Schausten, B., Schumann, D., Eichler, B., Gäggeler, H.W., Jost, D.T., Türler, A.: Radiochim. Acta **80,** (1998) 121.
29. Pershina, V.: Radiochim. Acta **80**, (1998) 65.
30. Pershina, V.: Radiochim. Acta **80**, (1998) 75.
31. Pershina, V.: Radiochim. Acta **84**, (1999) 79.
32. Paulus, W., Kratz, J.V., Strub, E., Zauner, S., Brüchle, W., Pershina, V., Schädel, M., Schausten, B., Adams, J.L., Gregorich, K.E., Hoffman, D.C., Lane, M.R., Laue, C., Lee, D.M., McGrath, C.A., Shaughnessy, D.K., Strellis, D.A., Sylwester, E.R.: Radiochim. Acta **80**, (1999) 69.
33. Pershina, V., Fricke, B.: J. Phys. Chem. **98**, (1994) 6468.
34. Pershina, V.: Chem. Rev. **96**, (1996) 1977.
35. Strub, E., Kratz, J.V., Kronenberg, A., Nähler, A., Thörle, P., Zauner, S., Brüchle, W., Jäger, E., Schädel, M., Schausten, B., Schimpf, E., Li Zongwei, Kirbach, U., Schumann, D., Jost, D., Türler, A., Asai, M., Nagame, Y., Sakama, M., Tsukada, K., Gäggeler, H.W., Glatz, J.P.: Radiochim. Acta **88**, (2000) 265.
36. Kronenberg, A., Kratz, J.V., Mohapatra, P.K., Nähler, A., Rieth, U., Strub, E., Thörle, P., Zauner, S., Brüchle, W., Jäger, E., Pershina, V., Schädel, M., Schausten, B., Schimpf, E., Jost, D., Türler, A., Gäggeler, H.W.: Radiochim. Acta, to be submitted.
37. Gregorich, K.E., Henderson, R.A., Lee, D.M., Nurmia, M.J., Chasteler, R.M., Hall, H.L., Bennett, D.A., Gannett, C.M., Chadwick, R.B., Leyba, J.D., Hoffman, D.C., Herrmann, G.: Radiochim. Acta **43**, (1988) 223.
38. Ghiorso, A., Nurmia, M., Eskola, K., Eskola, P.: Phys. Rev. C **4**, (1971) 1850.
39. Bemis, C.E., Ferguson, R.L., Plasil, F., Silva, R.J., O'Kelley, G.D., Keifer, M.L., Hahn, R.L., Hensley, D.C.: Phys. Rev. Lett. **39**, (1977) 1246.
40. Weis, M., Ahrens, H., Denschlag, H.O., Fariwar, M., Herrmann, G., Trautmann, N.: Radiochim. Acta **42,** (1987) 201.
41. Schädel, M., Brüchle, W., Schimpf, E., Zimmermann, H.P., Gober, M.K., Kratz, J.V., Trautmann, N., Gäggeler, H., Jost, D., Kovacs, J., Scherer, U.W., Weber, A., Gregorich, K.E., Türler, A., Czerwinski, K.R., Hannink, N.J., Kadkhodayan, B., Lee, D.M., Nurmia, M.J., Hoffman, D.C.: Radiochim. Acta **57**, (1992) 85.
42. Kratz, J.V., Zimmermann, H.P., Scherer, U.W., Schädel, M., Brüchle, W., Gregorich, K.E., Gannett, C.M., Hall, H.L., Henderson, R.A., Lee, D.M., Leyba, J.D., Nurmia, M.J., Hoffman, D.C., Gäggeler, H., Jost, D., Baltensperger, U., Nai-Qi, Ya, Türler, A., Lienert, Ch.: Radiochim. Acta **48**, (1989) 121.
43. Zimmermann, H.P., Gober, M.K., Kratz, J.V., Schädel, M., Brüchle, W., Schimpf, E., Gregorich, K.E., Türler, A., Czerwinski, K.R., Hannink, N.J., Kadkhodayan, B., Lee, D.M., Nurmia, M.J., Hoffman, D.C., Gäggeler, H., Jost, D., Kovacs, J., Scherer, U.W., Weber, A.: Radiochim. **60**, (1993) 11.
44. Pershina, V., Fricke, B., Kratz, J.V., Ionova, G.V.: Radiochim. Acta **64**, (1994) 37.

45. Korkisch, J., *Handbook of Ion Exchange Resins: Their Application to Inorganic Analytical Chemistry, Vol. IV,* CRC Press, Boca Raton, Florida (1989) 257.
46. Gober, M.K., Kratz, J.V., Zimmermann, H.P., Schädel, M., Brüchle, W., Schimpf, E., Gregorich, K.E., Türler, A., Hannink, N.J., Czerwinski, K.R., Kadkhodayan, B., Lee, D.M., Nurmia, M.J., Hoffman, D.C., Gäggeler, H., Jost, D., Kovacs, J., Scherer, U.W., Weber A.: Radiochim. Acta **57,** (1992) 77.
47. Kratz, J.V., Gober, M.K., Zimmermann, H.P., Schädel, M., Brüchle, W., Schimpf, E., Gregorich K.E., Türler, A., Hannink, N.J., Czerwinski, K.R., Kadkhodayan, B., Lee, D.M., Nurmia, M.J., Hoffman, D.C., Gäggeler, H., Jost, D., Kovacs, J., Scherer, U.W., Weber, A.: Phys. Rev. C **45**, (1992) 1064.
48. Gregorich, K.E., Hall, H.L., Henderson, R.A., Leyba, J.D., Czerwinski, K.R., Kreek, S.A., Kadkhodayan, B., Nurmia, M.J., Lee, D.M., Hoffman, D.C.: Phys. Rev. C **45**, (1992) 1058.
49. Paulus, W., Kratz, J.V., Strub, E., Zauner, S., Brüchle, W., Pershina, V., Schädel, M., Schausten, B., Adams, J.L., Gregorich, K.E., Hoffman, D.C., Lane, M.R., Laue, C., Lee, D.M., McGrath, C.A., Shaughnessy, D.K., Strellis, D.A., Sylwester, E.R.: Proc. Int. Conf. on Actinides, Baden-Baden, 21-26 September 1997, J. Alloys Comp. **271-273**, (1998) 292.
50. Brüchle, W., Schausten, B., Jäger, E., Schimpf, E., Schädel, M., Kratz, J.V., Trautmann, N., Zimmermann, H.P., Bruchertseifer, H., Heller, W.: GSI Scientific Report 1991, **GSI 92-1**, (1992) 315.
51. Günther, R., Paulus, W., Posledni, A., Kratz, J.V., Schädel, M., Brüchle, W., Jäger, E., Schimpf, E., Schausten, B., Schumann, D., Binder, R.: Institut für Kernchemie, Univ. Mainz, Jahresbericht 1994, **IKMz 95-1**, (1995) 2.
52. Schädel, M., Brüchle, W., Schausten, B., Schimpf, E., Jäger, E., Wirth, G., Günther, R., Kratz, J.V., Paulus, W., Seibert, A., Thörle, P., Trautmann, N., Zauner, S., Schumann, D., Andrassy, M., Misiak, R., Gregorich, K.E., Hoffman, D.C., Lee, D.M., Sylwester, E.R., Nagame, Y., Oura, Y.: Radiochim. Acta **77**, (1997) 149.
53. Türler, A., Brüchle, W., Dressler, R., Eichler, B., Gäggeler, H.W., Gregorich, K.E., Jost, D.T., Trautmann, N., Schädel, M.: Phys. Rev. C **57**, (1998) 1648.
54. Schädel, M., Brüchle, W., Dressler, R., Eichler, B., Gäggeler, H.W., Günther, R., Gregorich, K.E., Hoffman, D.C., Hübener, S., Jost, D.T., Kratz, J.V., Paulus, W., Schumann, D., Timokhin, S., Trautmann, N., Türler, A., Wirth, G., Yakushev, A.: Nature **388,** (1997) 55.
55. Schädel, M., Brüchle, W., Jäger, E., Schausten, B., Wirth, G., Paulus, W., Günther, R., Eberhardt, K., Kratz, J.V., Seibert, A., Strub, E., Thörle, P., Trautmann, N., Waldek, A., Zauner, S., Schumann, D., Kirbach, U., Kubica, B., Misiak, R., Nagame, Y., Gregorich, K.E.: Radiochim. Acta **83**, (1998) 163.
56. Pershina, V., Kratz, J.V.: Inorg. Chem. **40**, (2001) 776.
57. Pfrepper, G., Pfrepper, R., Kronenberg, A., Kratz, J.V., Nähler, A., Brüchle, W., Schädel, M.: Radiochim. Acta **88**, (2000) 273.
58. Wierczinski, B., Alstad, J., Eberhardt, K., Kratz, J.V., Mendel, M., Nähler, A., Omtvedt, J.P., Skarnemark, G., Trautmann, N., Wiehl, N.: J. Radioanalyt. Nucl. Chem. **236**, (1998) 193.
59. Eberhardt, K., Alstad, J., Kling, H.-O., Kratz, J.V., Langrock, G., Omtvedt, J.P., Skarnemark, G., Stavsetra, L., Tharun, U., Trautmann, N., Wiehl, N., Wierczinski, B.: Contribution to the LSC2001 International Conference on Liquid Scintillation Spectrometry, Karlsruhe, Germany, 7-11 May 2001.
60. Langrock, G., Messerschmidt. M., Wiehl, N., Kling, H.O., Mendel, M., Nähler, A., Tharun, U., Eberhardt, K., Trautmann, N., Kratz, J.V., Skarnemark, G., Omtvedt, J.P., Johansson, M., Hult, E.A., Dyve, J.E., Breivik, H., Alstad, J.: Contribution to the LSC2001 International Conference on Advances in Liquid Scintillation Spectrometry, Karlsruhe, Germany, 7-11 May 2001.

61. Omtvedt, J.P., Alstad, J., Eberhardt, K., Folden, C.M., Johansson, M., Kirbach, U.W., Lee, D.M., Mendel, M., Nähler, A., Patin, J.B., Skarnemark, G., Stavsetra, L., Sudowe, R., Wiehl, N., Zielinski, R., Trautmann, N., Nitsche, H., Hoffman, D.C.: Abstract submitted to Actinides 2001, Hayama, Japan.
62. Malmbeck, R., Skarnemark, G., Alstad, J., Fure, K., Johansson, M., Omtvedt, J.P.: J. Radioanal. Nucl. Chem. **246**, (2000) 349.
63. Oganessian, Yu.Ts., Yeremin, A.V., Gulbekian, G.G., Bogomolov, S.L., Chepigin, V.I., Gikal, B.N., Gorschkov, V.A., Itkis, M.G., Kabachenko, A.P., Kutner, V.B., Lavrentev, Y.Yu., Malyshev, O.N., Popeko, A.G., Rohac, J., Sagaidak, R.N., Hofmann, S., Münzenberg, G., Veselsky, M., Saro, S., Iwasa, N., Morita, K.: Eur. Phys. J. A **5**, (1999) 63.
64. Oganessian, Yu.Ts, Yeremin, A.V., Popeko, A.G., Bogomolov, S.L., Buklanov, G.V., Chelnokov, M.L., Chepigin, V.I., Gikal, B.N., Gorschkov, V.A., Gulbekian, G.G., Itkis, M.G., Kabachenko, A.P., Lavrentev, A.Yu., Malyshev, O.N., Rohac, J., Sgaidak, R.N., Hofmann, S., Saro, S., Giardina, G., Morita, K.: Nature **400**, (1999) 242.
65. Oganessian, Yu.Ts., Utyonkov, V.K., Lobanov, Yu.V., Abdullin, F.Sh., Polyakov, A.N., Shirokovsky, I.V., Isyganov, Yu. S., Gulbekian, G.G., Bogomolov, S.L., Gigkal, B.N., Mezentsev, A.N., Iliev, S., Subbotin, V.G., Sukhov, A.M., Ivanov, O.V., Buklanov, G.V., Subotin, K., Itkis, M.G., Moody, K.J., Wild, J.F., Stoyer, N.J., Stoyer, M.A., Lougheed, R.W.: Phys. Rev. C **62**, (2000) 04160.
66. Oganessian, Yu.Ts., Utyonkov, V.K., Lobanov, Yu.V., Abdullin, F.Sh., Polyakov, A.N., Shirokovsky, I.V., Tsyganov, Yu.S., Gulbekian, G.G., Bogomolov, S.L., Gikal, B.N., Mezentsev, A.N., Iliev, S., Subbotin, V.G., Sukhov, A.M., Ivanov, O.V., Buklanov, G.V., Subotic, K., Itkis, M.G., Moody, K.J., Wild, J.F., Stoyer, N.J., Stoyer, M.A., Lougheed, R.W., Laue, C.A., Karelin, Ye.A., Tatarinov, A.N.: Phys. Rev. C **63**, (2001) 011301.
67. Schwerdtfeger, P., Seth, M.: "Relativistic Effects of the Superheavy Elements". In: *Encyclopedia of Computational Chemistry,* Eds. von R. Schleyer, P., Schreiner, P.R., Allinger, N.L., Clark, T., Gasteiger, J., Kollman, P.A., Schaefer III, H.F., Vol. 4, Wiley, New York, (1998) 2480.
68. Armbruster, P.: Eur. Phys. J. A **7**, (2000) 23.
69. Armbruster, P.: Annu. Rev. Nucl. Part. Sci. **50**, (2000) 411.
70. Fricke, B.: Structure and Bonding **21**, (1975) 89.
71. Eichler, B., Kratz, J.V.: Radiochim. Acta **88**, (2000) 475.

Index

Chapter 6

Gas-Phase Adsorption Chromatographic Determination of Thermochemical Data and Empirical Methods for their Estimation

B. Eichler[a], R. Eichler[b]

[a] *Paul Scherrer Institut, CH-5232 Villigen PSI, Switzerland*
[b] *Gesellschaft für Schwerionenforschung, Planckstr. 1, D-64291 Darmstadt, Germany*

Part I: Basic principles of the determination of adsorption properties using gas-phase adsorption chromatographic methods

1. Introduction

Volatilization processes, combined with gas adsorption chromatographic investigations, are well established methods in nuclear chemistry. Fast reactions and high transport and separation velocities are crucial advantages of these methods. In addition, the fast sample preparation for α-spectroscopy and spontaneous fission measurements directly after the gas-phase separation is a very advantageous feature. Formation probabilities of defined chemical compounds and their volatility can be investigated on the basis of experimentally determined and of predicted thermochemical data, the latter are discussed in Part II of this chapter.

Volatile elements as well as a large variety of volatile chemical compounds can be investigated by using a broad assortment of reactive carrier gases. Moreover, different stationary phase materials, available for gas adsorption chromatography, are further broadening the areas of application, see Table 1.

M. Schädel (ed.), The Chemistry of Superheavy Elements, 205-236.

Apart from that, the separation quality can be influenced by the size of the chromatographic surface, e.g., by fillings in the chromatography column. Experimental investigations were carried out in the temperature range from 2400 K [1] to 85 K [2]. Suitable chemical states and stationary phases exist for almost all elements to allow their investigation in adsorption chromatography.

Table 1. Chemical states and applicable reactive carrier gases and stationary phases

Chemical state	Typical reactive carrier gases	Typical stationary phases
Element (atomic state)	H_2	Fused silica, glass, metals, alumina, graphite, ice
Halides	Halogens, hydrogen halides, $SOCl_2$, CCl_4	Fused silica, glass, metals, alumina, alkaline chlorides, alkaline earth chlorides
Oxyhalides	Halogens + O_2 hydrogen halides + O_2	Fused silica, glass, metals, alumina, alkaline chlorides, alkaline earth chlorides
Oxides	O_2	Fused silica, glass, alumina
Oxyhydroxides/ Hydroxides	$O_2 + H_2O$	Fused silica, glass, alumina
Complex compounds	Beta diketones, CO, $AlCl_3$	Fused silica, glass

As the quantities of elements applied in nuclear chemistry are often small, down to one-atom-at-a-time, deposition and volatilization are predominately related to adsorption and desorption phenomena, respectively. Practically, pure condensed phases do not occur. The volatilization and the gas phase transport through a chromatography column depend on

- experimental parameters, as the flow rate and the type of carrier gases,
- the design of the chromatography columns,
- the temperature, applied to the column, and
- the interaction of the gaseous atoms and molecules with the column surface (the stationary phase).

The entire gas phase transport is called gas-adsorption chromatography.

This chapter describes basic physico-chemical relations between the gas phase transport of atoms and molecules and their thermochemical properties, which are related to the adsorption-desorption equilibrium. These methods can either be used to predict the behavior of the adsorbates in the chromatographic processes, in order to design experiments, or to characterize the absorbate from its experimentally observed behavior in a process. While Part I of this chapter is devoted to basic principles of the process, the derivation of thermochemical data is discussed in Part II. Symbols used in the following sections of Part I are described in Section 5. For results, which were obtained applying the described evaluation methods in gas-adsorption chromatography, see Chapters 4 and 7 of this book.

2. Thermochemical Description of the Transport of Substances in Gas-adsorption Chromatography Processes

2.1 THE ADSORPTION EQUILIBRIUM

This description is based on the differential equation for the transport velocity of a substance in the ideal linear gas chromatography [3]:

$$\frac{dy}{dt} = \frac{u}{1+k_i} \tag{1}$$

k_i represents the partition coefficient corrected by the ratio of both phases, the solid and the gas phase:

$$k_i = \frac{s}{v} \cdot K_{ads} \tag{2}$$

Here we consider the simple adsorption-desorption reaction equilibrium of a reversible mobile adsorption process without any chemical reactions:

$$A_{gas} \leftrightarrow A_{ads} \tag{3}$$

For the formulation of a dimensionless adsorption equilibrium constant the definition of a standard state is crucial. The standard state is freely selectable, regardless of the possibility of its physical realization. It is defined according to its expediency. The standard state of adsorption is assumed to be the ratio of a standard molar volume to the standard molar surface [4].

$$\frac{V}{A} = \frac{c^o_{ads}}{c^o_{gas}} = 1\,\text{cm} \rightarrow \frac{V}{100 \cdot A} = \frac{c^o_{ads}}{c^o_{gas}} = 1\,\text{m} \tag{4}$$

$$K = K_{ads} \cdot \frac{100 \cdot A}{V} = \frac{c_{ads}}{c_{gas}} \cdot \frac{c^o_{gas}}{c^o_{ads}} \tag{5}$$

The dimensionless equilibrium constant of a simple reversible adsorption reaction is related to thermodynamic standard quantities – the standard adsorption enthalpy and the standard entropy of adsorption which are assumed to be temperature independent:

$$\Delta G^0_{ads} = -R \cdot T \cdot \ln(K) = \Delta H^0_{ads} - T \cdot \Delta S^0_{ads} \tag{6}$$

For a localized adsorption the concentration of the adsorption sites has to be taken into account [4]. In addition, a reversible change of the chemical state of the adsorbate in the chromatography process has to be considered, e.g. dissociative and substitutive adsorptions as described in Part II of this chapter (Part II, Section 2.3, Equations 52-54). The reaction enthalpy and entropy has to be introduced into the calculations [5-8] as well as the equilibrium constant for the chemical reaction with its standard states of the occurring chemical states of:

$$K = K_r \cdot K_{ads} \cdot \frac{100 \cdot A}{V} \quad (7)$$

$$\Delta G^o_{r,ads} = -R \cdot T \cdot \ln(K) = (\Delta H^o_r + \Delta H^o_{ads}) - T \cdot (\Delta S^o_r + \Delta S^o_{ads}) \quad (8)$$

The entropy of a mobile adsorption process can be determined from the model given in [4]. It is based on the assumption that during the adsorption process a species in the gas phase, where it has three degrees of freedom (translation), is transferred into the adsorbed state with two translational degrees of freedom parallel to the surface and one vibration degree of freedom vertical to the surface. From statistical thermodynamics the following equation for the calculation of the adsorption entropy is derived:

$$\Delta S^o_{ads} = R \cdot \left(\ln(\frac{100 \cdot A}{V \cdot \nu_b}) \cdot \sqrt{\frac{R \cdot T}{2 \cdot \pi \cdot M_a}} + \frac{1}{2} \right) \quad (9)$$

As a good approximation it is assumed, that the adsorbed species are vibrating in resonance with the lattice phonon vibrations of the solid stationary phase. The phonon frequency can be evaluated from phonon spectra, from the standard entropy of solid metals, from the Debye temperatures or from the Lindemann equation [9].

2.2 ISOTHERMAL GAS CHROMATOGRAPHY

Combining Equations 1-5, the temperature dependency of the carrier gas flow rate, and integrating over the column length the following relation for the retention time in isothermal gas chromatography is obtained [10]:

$$t_r = \frac{L \cdot T_o \cdot \varnothing}{\overline{V}_o \cdot T_{iso}} \cdot \left[1 + \frac{s}{v} \cdot \frac{V}{100 \cdot A} \cdot \exp\left(-\frac{\Delta H^o_{ads}}{R \cdot T_{iso}} \right) \cdot \exp\left(\frac{\Delta S^o_{ads}}{R} \right) \right] \quad (10)$$

The first addend in Equation 10 can be neglected. Thus the retention volume is calculated as:

$$\ln\left(t_r \cdot \overline{V}_o \cdot \frac{T_{iso}}{T_o} \right) = \ln\left(\frac{L \cdot \varnothing \cdot s \cdot V}{v \cdot 100 \cdot A} \right) + \left(-\frac{\Delta H^o_{ads}}{R \cdot T_{iso}} \right) + \left(\frac{\Delta S^o_{ads}}{R} \right) \quad (11)$$

In experiments with long-lived nuclides retention times equal the time of the experimental duration. However, for short-lived nuclides, at the temperature, where 50 % of the nuclides pass the isothermal chromatography column, the retention time equals the half-life of the species:

$$t_r = T_{1/2} \quad (12)$$

ΔH_{ads} can be evaluated from the slope of an Arrhenius-type plot of the retention volume (first term in Equation 11) against $1/T_{iso}$. From the intersection with the ordinate axis ΔS_{ads} can be evaluated.

2.3 THERMOCHROMATOGRAPHY

2.3.1 *Gas thermochromatography*

The given description of the gas phase transport in a tube with a temperature gradient is only valid for adsorption equilibria of reversible mobile adsorptions without any superimposed chemical reaction. The temperature profile along the chromatography column is approximated to be linear by:

$$T = T_s - a \cdot y \tag{13}$$

The substitution of the corrected partition coefficient and the introduction of the standard state (eq.1-8) lead to:

$$t_r = -\frac{1}{a} \cdot \int_{T_s}^{T_{dep}} \frac{\left(1 + \frac{s}{v} \cdot \frac{V}{100 \cdot A} \cdot K(T)\right)}{u(T)} \cdot dT \tag{14}$$

With $u(T) = u_o \cdot \dfrac{T}{T_o}$ (15) and $\overline{V}_o = u_o \cdot \varnothing$ (16)

the integration yields:

$$t_r = -\frac{T_o}{a \cdot u_o} \cdot \ln\left(\frac{T_{dep}}{T_s}\right) + \frac{s \cdot \varnothing \cdot T_o \cdot \dfrac{V}{100 \cdot A} \cdot \exp\left(\dfrac{\Delta S^o_{ads}}{R}\right)}{\overline{V}_o \cdot a \cdot v} \cdot \left[Ei^*\left(-\frac{\Delta H^o_{ads}}{R \cdot T_{dep}}\right) - Ei^*\left(-\frac{\Delta H^o_{ads}}{R \cdot T_s}\right)\right] \tag{17}$$

Ei*(x) is the exponential integral function of x. For x »1 this function can be estimated by:

$$Ei^*(x) = \frac{\exp(x)}{x} \tag{18}$$

For $T_{dep} \ll T_s$, which is mostly the case, the first addend in Equation 15 is negligible and

$$Ei^*\left(-\frac{\Delta H^o_{ads}}{R \cdot T_{dep}}\right) \gg Ei^*\left(-\frac{\Delta H^o_{ads}}{R \cdot T_s}\right) \tag{19}$$

Hence, the following simplification [11] is conceivable:

$$\frac{a \cdot t_r \cdot \overline{V}_o}{s \cdot T_{dep} \cdot \frac{V}{100 \cdot A}} = \left(\frac{R \cdot T_o}{-\Delta H^o_{ads}} \right) \cdot \exp\left(-\frac{\Delta H^o_{ads}}{R \cdot T_{dep}} \right) \cdot \exp\left(\frac{\Delta S^o_{ads}}{R} \right) \tag{20}$$

Thus, the deposition temperature and the thermodynamic state function of the adsorption are combined and they can easily be determined from each other. The retention time for a short-lived radioactive species is calculated as the radioactive lifetime of the nuclide:

$$t_r = \frac{T_{1/2}}{\ln(2)} \tag{21}$$

or equals the duration of the experiment for a long-lived species. Note, that Equations 17 and 20 are only valid, if the chemical state of the investigated species does not change along the entire temperature range. The simplification of Equation 17 is not possible, if very short-lived nuclides ($t_{1/2}$<1 s) are used.

2.3.2 *Vacuum thermochromatography*

The model of ideal linear gas chromatography (Equation 1) [12] also describes the transport of a chemical species in the temperature gradient of a vacuum tube. At molecular flow conditions the linear velocity of the carrier gas, which is identical to the transport velocity of the adsorbate in the gas phase, has to be substituted by the fraction of the column length over the average retention time of the species in the column:

$$u = \frac{y}{t_r} = \frac{4 \cdot d_i \cdot \sqrt{\frac{2 \cdot \pi \cdot k \cdot T}{m}}}{3 \cdot y} \tag{22}$$

The deposition site of a species in a defined temperature region, under conditions where the retention time in the adsorbed state exceeds the duration of the experiment, is assumed to be similar to a corresponding effective length of the column at which this species would have exited. This length is the distance of the deposition region with T_{dep} from the end of the column with T_s. Thus, the average retention time is a function of the column volume and of the conductivity of the column for inert gas phase species at molecular flow conditions. The probability of residence of the species in the adsorbed state is mainly dependent on the temperature dependent equilibrium constant and thus, on the thermochemical constants of the adsorption reaction. An increasing adsorption probability with lowering temperatures leads to a decrease of the migration velocity of the adsorbate zone. According to a limited experiment duration and radioactive life-time of the species this leads to a deposition zone of the adsorbate in the column.

Applying Equation 1 and substituting the corrected partition coefficient by a gas phase kinetic formulation of the adsorption equilibrium constant

$$k_i = \frac{4}{d_i} \cdot \frac{h}{k \cdot T} \cdot \sqrt{\frac{k \cdot T}{2 \cdot \pi \cdot m}} \cdot \exp\left(-\frac{\Delta H^o_{ads}}{R \cdot T}\right) \tag{23}$$

the following functional dependency between the retention time, the experimental parameters, T_{dep} and ΔH_{ads} is obtained:

$$\begin{aligned} t_r = {} & 2 \cdot A \cdot \left(\frac{1}{3} \cdot \sqrt{T^3_{dep}} - T_s \cdot \sqrt{T_{dep}} + \frac{2}{3} \cdot \sqrt{T^3_s}\right) + \\ & + B \cdot \left[T_{dep} \cdot \exp\left(-\frac{\Delta H^o_{ads}}{R \cdot T_{dep}}\right) - T_s \cdot \left(-\frac{\Delta H^o_{ads}}{R \cdot T_{dep}}\right)\right] + \\ & + B \cdot \left(T_s - \left(-\frac{\Delta H^o_{ads}}{R}\right)\right) \cdot \left[Ei^*\left(-\frac{\Delta H^o_{ads}}{R \cdot T_{dep}}\right) - Ei^*\left(-\frac{\Delta H^o_{ads}}{R \cdot T_s}\right)\right] \end{aligned} \tag{24}$$

$$A = \frac{3}{4 \cdot a^2 \cdot d_i \cdot \sqrt{\frac{2 \cdot \pi \cdot k}{m}}} \tag{25}$$

$$B = \frac{3 \cdot h}{2 \cdot a^2 \cdot d_i^2 \cdot k} \tag{26}$$

For practical applications some simplifications [13] similar to Equations 18 and 19 lead to:

$$t_r = \frac{R \cdot T_s \cdot T_{dep} \cdot B}{-\Delta H^o_{ads}} \cdot \exp\left(-\frac{\Delta H^o_{ads}}{R \cdot T_{dep}}\right) \tag{27}$$

Hence, from the experimental parameters and from the measured deposition temperature ΔH^o_{ads} can be determined.

Again this simplification holds only for $t_r \gg 1$ s [14,15]. For experiments with very short-lived nuclides the Equations 12, 13, 22, and 23 have to be combined and the resulting integral has to be solved numerically.

Introducing the mobile adsorption model into this derivation [16] k_i changes to:

$$k_i = \frac{4}{d_i} \cdot \frac{V}{100 \cdot A} \exp\left(\frac{\Delta S^o_{ads}}{R}\right) \cdot \exp\left(-\frac{\Delta H^o_{ads}}{R \cdot T}\right) \tag{28}$$

Then Equation 12, 13, 22, and 28 have to be combined and, again, the resulting integral has to be solved numerically.

3. Microscopic Kinetic Description of the Adsorption Process in Gas Chromatography - Monte Carlo Methods

3.1 GAS CHROMATOGRAPHY WITH CARRIER GASES

This section explains the basic principles of Monte-Carlo models, which are successfully applied for the evaluation of adsorption enthalpies in gas-phase adsorption chromatography investigations of the heaviest elements and their lighter homologues, see Chapters 4 and 7. The application of Monte-Carlo simulation methods in gas adsorption chromatography is based on ideas given in [17]. These models open the way for a microscopic description of the chromatographic adsorption-desorption process on an atomic scale. Hence, they are kinetic models of gas adsorption chromatography.

To determine ΔH^{o}_{ads} of the adsorbate on the stationary phase the experimental parameters and the physical data of the carrier gases and of the adsorbate are used as an input into these models. For a set of ΔH^{o}_{ads} values the model yields chromatograms which are compared to an actual experimental result. The ΔH^{o}_{ads} value is obtained from the best agreement between model and experiment.
A flow chart of the Monte Carlo model is presented in Figure 1.

The formalism can easily be adapted to a PC-based program for an evaluation of the ΔH^{o}_{ads} from gas adsorption chromatographic results and, vice versa, to predict the behavior of an adsorbate in gas chromatography from its known adsorption data. The formulas used in the PC-based simulation are given here. For their derivations see [17].

The Monte-Carlo principle uses random numbers in the region $0 \le \text{random} < 1$. The main randomized values in the Monte-Carlo simulation, see Figure 1, of a gas-phase adsorption process for radioactive species are:

1. The radioactive lifetime.
For every atom a random lifetime, which is distributed logarithmically according to the radioactive decay law with the half-life, $T_{1/2}$, of the nuclide is calculated:

$$t_{\lambda} = -\frac{\ln(2)}{T_{1/2}} \cdot \ln(1 - \text{random}) \qquad (29)$$

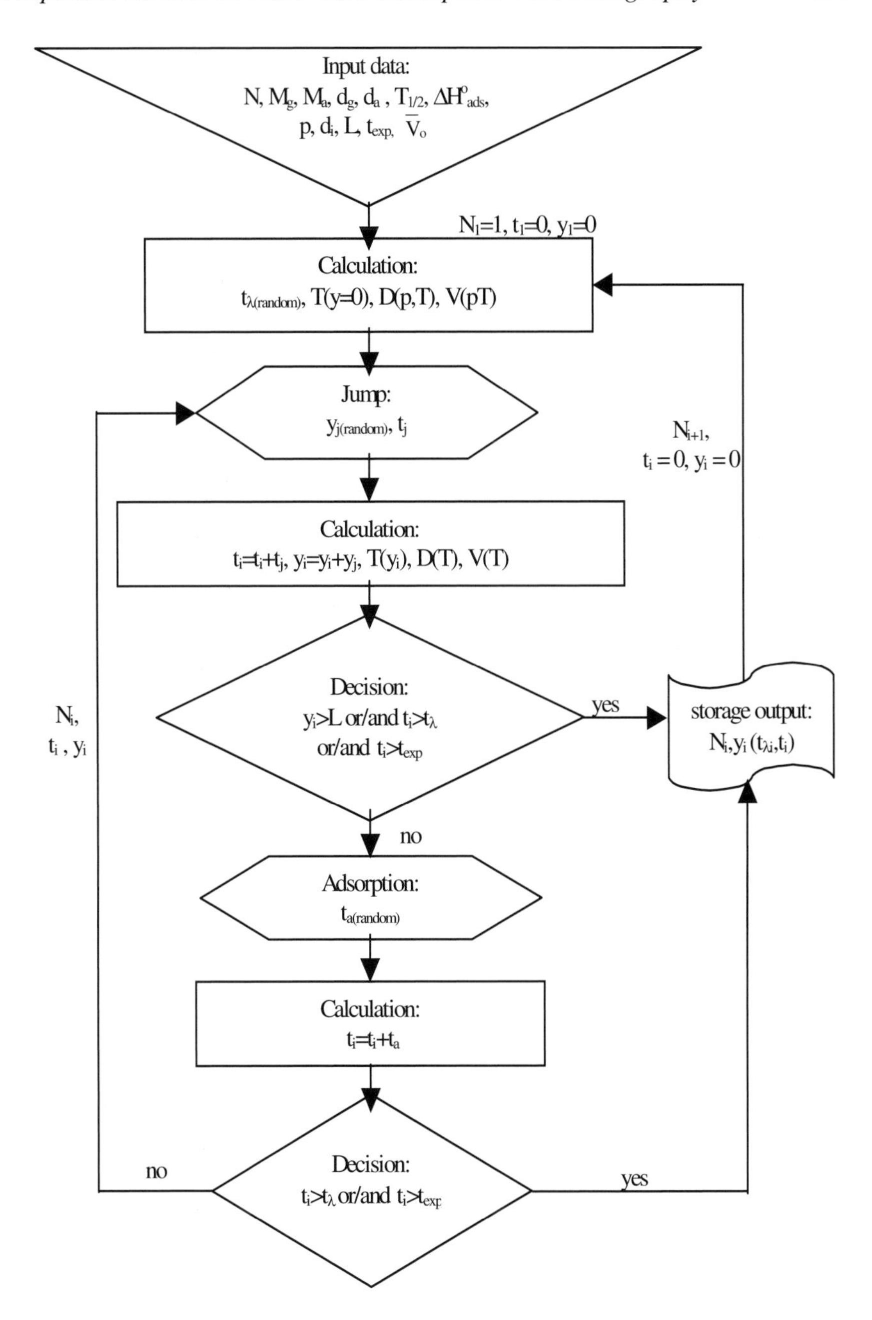

Fig. 1. Flow chart of the PC-based Monte-Carlo simulation of gas chromatography. The indices i and j stand for the number of the atom and of the individual jump, respectively.

2. The jump length and the time of flight in the gas phase.

The mean jump length of the adsorbate can be approximated by:

$$Y_{jm} = \frac{11 \cdot \overline{V}(p,T)}{48 \cdot \pi \cdot D(p,T)} \tag{30}$$

The pressure and temperature dependent diffusion coefficient of the adsorbate in the carrier gas is approximated according to Gilliland [18]:

$$D = 0.01378 \cdot 10^{-7} \cdot T_o^{1.5} \cdot \sqrt{\frac{1}{M_g} + \frac{1}{M_a}} \cdot p^{-1} \cdot \left(\sqrt[3]{\frac{d_g}{M_g}} + \sqrt[3]{\frac{d_a}{M_a}} \right)^{-2} \cdot \left(\frac{T}{T_o} \right)^{1.75} \tag{31}$$

The gas flow also depends on the temperature and on the pressure. Assuming an ideal gas behavior of the carrier gas it can be calculated from the flow rate at standard conditions:

$$\overline{V}(p,T) = \overline{V}_o \cdot \frac{T \cdot p_o}{T_o \cdot p} \tag{32}$$

Subsequently, the random jump length, which is distributed logarithmically, is calculated by:

$$y_j = -Y_{jm} \cdot \ln(1 - \text{random}) \tag{33}$$

Thus, the time the atom remains in the gas phase can be derived:

$$t_j = \frac{y_j \cdot \pi \cdot d_i^2}{4 \cdot \overline{V}(p,T)} \tag{34}$$

3. The residence time.

The mean number of wall collisions along Y_{jm} is

$$N_m = Y_{jm} \cdot \frac{d_i}{2 \cdot \overline{V}(p,T)} \cdot \sqrt{\frac{2 \cdot \pi \cdot R \cdot T}{M_a}} \tag{35}$$

The randomized number of wall collisions during the travel of the atom is distributed logarithmically:

$$N_a = -N_m \cdot \ln(1 - \text{random}) \tag{36}$$

The mean residence time of the atom in the adsorbed state is calculated by a Frenkel-type equation:

$$t_{am} = \frac{1}{\nu_b} \cdot \exp\left(-\frac{\Delta H^o_{ads}}{R \cdot T}\right) \tag{37}$$

Hence the residence time of the atom in the adsorbed state, which is distributed logarithmically too, can be calculated:

$$t_a = -N_a \cdot t_{am} \cdot \ln(1 - \text{random}) \tag{38}$$

3.2 GAS CHROMATOGRAPHY IN VACUUM

A Monte-Carlo simulation of gas chromatography at vacuum conditions follows the same scheme as of Figure 1 [16]. The lifetime of each atom or molecule is calculated using Equation 28. The mean residence time of the atom in the adsorbed state at a defined temperature is calculated using Equation 34. The residence time of the atom in the adsorbed state, which is distributed logarithmically, can be calculated:

$$t_a = -t_{am} \cdot \ln(1 - \text{random}) \qquad (39)$$

The transport of species along the chromatography column is assumed to be dependent only on the solid angle of the desorption direction and on the dimensions of the column. The solid angles of desorption from a plane surface into the vacuum are calculated according to Knudsen [19] assuming a cosine law:

$$\alpha = 2 \cdot \text{random} - 1 \qquad \varphi = \arcsin(\alpha) \qquad (40)$$

$$\beta = 2 \cdot \text{random} - 1 \qquad \theta = \arcsin(\beta) \qquad (41)$$

For the cylindrical column this approach leads to the following picture:

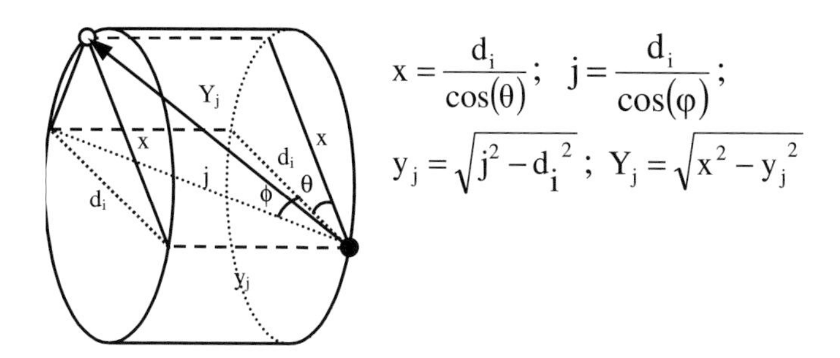

Fig 2. Trigonometry of an adsorbate jump at molecular flow conditions in a cylindrical tube

The velocity of the atoms can be set as the arithmetic mean velocity $\bar{v}$ from the Maxwell-Boltzmann distribution. It only dependents on the temperature and their molecular weight:

$$\bar{v} = \sqrt{\frac{8 \cdot R \cdot T}{\pi \cdot M_a}} \qquad (42)$$

Hence, the time the atom is in flight can be calculated as:

$$t_j = \frac{Y_j}{\bar{v}} \qquad (43)$$

The column is assumed to be "closed" at both ends.

4. Summary

At this point we would like to state some advantages and disadvantages of the thermodynamic and the kinetic approach.

Advantages of the thermodynamic model:

- It is less time consuming.
- Experiments, where complex stationary phases have been used (e.g. column fillings) can be evaluated.

Disadvantages of the thermodynamic model:

- The a linear temperature gradient has to be assumed.
- "Long jumps" cannot be described in the model of linear gas chromatography.
- The radioactive decay is approximated through an average lifetime.

Advantages of the kinetic Monte-Carlo model:

- The description of the microscopic chromatographic process at realistic temperature conditions at the surfaces (real temperature gradient) is possible.
- The probability distribution of the radioactive decay of the adsorbate is included.
- The probability that the adsorbate is transported along the column by "long jumps" is taken into account, which is important to describe the chromatographic separation especially at high gas flow rates.

Disadvantages of the kinetic Monte-Carlo model:

- The evaluation is time consuming.
- The surface geometry has to be simple.

5. Symbols

A	...	inner surface per 1m column length, m^2
a	...	negative temperature gradient, K/m
c_{ads}	...	surface concentration of the adsorbate, particles/m^2
c^o_{ads}	...	standard surface concentration of the adsorbate, $2.679 \cdot 10^{23}$ particles/m^2
c_{gas}	...	gas concentration of the adsorbate, particles/m^3
c^o_{gas}	...	standard gas concentration of the adsorbate, $2.679 \cdot 10^{25}$ particles/m^3 (ideal gas at STP)
d_a	...	density of the adsorbate at its melting point, kg/m^3
d_g	...	density carrier gas at its melting point, kg/m^3
d_i	...	inner diameter of the column, m
D	...	diffusion coefficient of the adsorbate in the carrier gas, m^2/s
$\varnothing$	...	free open cross section area of the column, m^2
h	...	Planck´s constant, J·s
ΔG^o_{ads}	...	free standard adsorption enthalpy of adsorption, J/mol
$\Delta G^o_{r,ads}$	...	free standard adsorption enthalpy with chemical reaction, J/mol
ΔH^o_{ads}	...	standard adsorption enthalpy at zero surface coverage, J/mol
ΔH^o_r	...	standard enthalpy of reaction, J/mol
K	...	equilibrium constantof adsorption, dimensionless
K_{ads}	...	distribution constant, m
k_i	...	partition coefficient corrected by the phases, dimensionless
K_r	...	reaction equilibrium constant, dimensionless
k	...	Boltzmann constant, J/K
L	...	length of the column, m
m	...	atomic or molecular mass of the adsorbate, kg
M_a	...	molar weight adsorbate, kg/mol
M_g	...	molar weight carrier gas, kg/mol
N	...	number of atoms
N_a	...	number of wall collisions
N_m	...	mean number of wall collisions
ν_b	...	maximum lattice phonon vibration frequency, 1/s
p	...	pressure, Pa
p_o	...	standard pressure, 101325 Pa
random	...	random number
R	...	gas constant, 8.31441 J/mol·K
s	...	open surface of the column per 1 m column length, m^2

ΔS^{o}_{ads}	…	standard entropy of adsorption at zero surface coverage, J/mol·K
t	…	time, s
t_a	…	time the adsorbate keeps adsorbed, s
t_{am}	…	mean time the adsorbate keeps adsorbed, s
t_{exp}	…	duration of the experiment, s
t_j	…	transport time, s
t_λ	…	lifetime, s
t_r	…	retention time, s
T	…	temperature, K
T_{dep}	…	deposition temperature, K
T_{iso}		isothermal temperature, K
T_o	…	standard temperature, 298.15 K
$T_{1/2}$	…	half-life, s
T_s	…	upper (start) temperature of the gradient, K
u	…	carrier gas velocity, m/s
u_o	…	carrier gas velocity at STP, m/s
V	…	inner volume of the column, m^3
v	…	open volume of the column per 1 m column length, m^3
$\overline{V}$	…	carrier gas flow, m^3/s
$\overline{V}_o$	…	carrier gas flow at STP, m^3/s
y	…	coordinate longitudinal to the column, m
y_j	…	jump length, m
Y_j	…	path length, m
Y_{jm}	…	mean jump length, m

Part II: Derivation of thermochemical data with respect to the periodicity of properties and their interrelations

1. Introduction

The discovery of new chemical elements - the transactinides or superheavy elements - stimulated the work on theoretical predictions of their chemical properties. Our intention is to present empirical methods [20-35], which can be used to predict chemical properties and which are relevant to gas phase chemical studies of transactinides.

Gas chemical methods are successfully applied for the chemical characterization of transactinides. They provide a fast separation of transactinides from a wide variety of by-products, which are instantly produced in the nuclear formation reactions of transactinides. These methods are mainly based on the measurement of adsorption properties of the atomic or molecular state of transactinides on different stationary phases.

Due to the extremely low production rates of transactinides in the nuclear fusion reactions, the chemical characterizations are carried out on a single atom level. The chemical reaction products are characterized on the basis of their behavior in the separation process or, exactly speaking, in the gas adsorption chromatographic process. In this process the formation probability of defined chemical states of transactinides and the subsequent interaction of the formed molecules with a solid state surface are studied.

The stability and the volatility are chemical properties, which define the behavior of the transactinide element or its compound in the gas adsorption chromatographic process. Therefore, the predictions of these properties are instrumental for the design of experiments and indispensable for the interpretation of experimental results. "Empirical" and "exact" methods are both of high importance for the prediction of properties of transactinide elements. The "exact" ab-initio methods, which are described in Chapter 2 of this book, yield the atomic ground state configuration, ionization potentials, atomic and ionic radii as well as binding energies in isolated single gaseous molecules of defined chemical compounds with satisfactory accuracy. It is in principle impossible to calculate the behavior of a single atom or molecule in an experimental setup "exactly". For the ab-initio calculation of the stability of pure solid phases, which is necessary for the evaluation of the volatility of compounds, the accuracy of the methods is not sufficient.

Up to now, only empirical methods can predict the volatility of compounds. These predictions represent an important part in the multiple step process of the chemical characterization of transactinide elements. The main steps in this process are:

I. The analysis of thermochemical data of chemical compounds in the solid and in the gaseous phase.

II. The calculation of the volatility of the elements and of their compounds.

III. The experimental determination of empirical correlations between the volatility of the pure macroscopic phase and the adsorption behavior of single atoms or molecules on defined surfaces.

IV. The prediction of the adsorption behavior of transactinides or their compounds at zero surface coverage from the predicted volatility (II.) applying the empirical correlations (III.).

V. The physico-chemical description of the gas adsorption chromatographic process.

VI. Model experiments using short-lived nuclides of lighter homologues for the determination of reaction rates, retention times and decontamination.

VII. Design of an experimental set-up for experiments with transactinides with respect to the predicted adsorption properties and its half-life using III. and V..

VIII. The interpretation of the results according to:
- the formation of compounds with similar volatility as known from their homologues,
- the relative volatility of equivalent chemical compounds compared with the homologues,
- the adsorption enthalpies and the volatilities of fictive macroscopic phases,
- the confirmation of predictions of the chemical reaction behavior and of the volatility, and
- the periodicity of chemical properties and trends along the groups of the periodic table.

The contribution to this chapter focuses on the steps I-IV. More details concerning step V and step VI are outlined in Chapter 2 and 4, respectively. Steps VII and VIII are discussed in Chapter 4 and 7.

2. Thermochemical Data

2.1 EXTRAPOLATIVE DEDUCTION AND ANALYSIS OF THERMO-CHEMICAL DATA FOR TRANSACTINIDES AND THEIR COMPOUNDS

It is assumed that the stability of a chemical state can be expressed in terms of the standard formation enthalpies of gaseous and solid compounds $\Delta H^{o}_{298}(g,s)$ and of their atomic standard formation enthalpies $\Delta H^{*}_{298}(g,s)$. Both values differ only in the elemental state: The first value stands for the standard state of the element and the second one for the gaseous monatomic state of the elements at standard conditions. These values are always relative stability measures in combination with a competing chemical state. Thermochemical state functions normally describe the behavior of a large number of atoms or molecules. For the description of single atoms we use these values as a measure of the formation probability of different chemical states and to quantify the binding energy in molecules. Gas-adsorption chromatography is a multi-step process, during which a single atom or molecule changes its chemical state in thousands of adsorption-desorption steps. Therefore, the resulting information on the volatility of the investigated chemical species is as reliable as results from experiments with a very large number of atoms or molecules from one evaporation-deposition experiment. For more detailed information on this topic see Chapter 3 of this book.

Trends of chemical properties of the elements and their compounds exist in the groups of the periodic table. Therefore, assuming that all transactinides systematically belong to defined groups rough estimates about their chemical properties are feasible. For these estimates shell effects and relativistic effects, which depend on the atomic number, and which lead to discontinuous variations of properties along the groups, have to be considered [36-38]. Examples for such effects are: the half occupancy of electron shells, which act especially on the chemical properties of elements with lower atomic numbers, and the sub-shell closure of the p-, d-, and f-electron levels with an increasingly strong spin-orbit splitting for elements at the highest row numbers - the transactinides - due to relativistic effects [21,25,28,33-38]. In general, elements with a shell closure or a half-filled electron shell in their ground state configuration form relatively volatile condensed phases and less stable chemical compounds compared to elements in the same group, which do not exhibit such effects. This creates a dilemma for extrapolative predictions of thermochemical data of transactinides along their groups as these configurations change with atomic or row number. Anyhow, no direct functional dependence exists between thermochemical state functions and electron configurations.

In the following part methods are outlined, which can be used to master the described dilemma. More trustworthy extrapolative predictions are possible by correlating thermochemical state functions mutually. In this case electron shell effects in the homologues are already implemented. To estimate standard formation data of transactinide compounds in their solid and in their gaseous states, the standard enthalpies of monatomic gaseous elements $\Delta H^{o}_{298}M(g)$ are of fundamental importance. They are used as primary input values for extrapolative predictions and they are mostly equal to the standard sublimation enthalpy of these elements. They can be estimated empirically for the transactinides by considering shell effects in the region of their homologues and by extrapolations along the groups and along the rows of the periodic table (Method ***1***) [28]. As an example, data for s- an p- elements are presented in Figure 3.

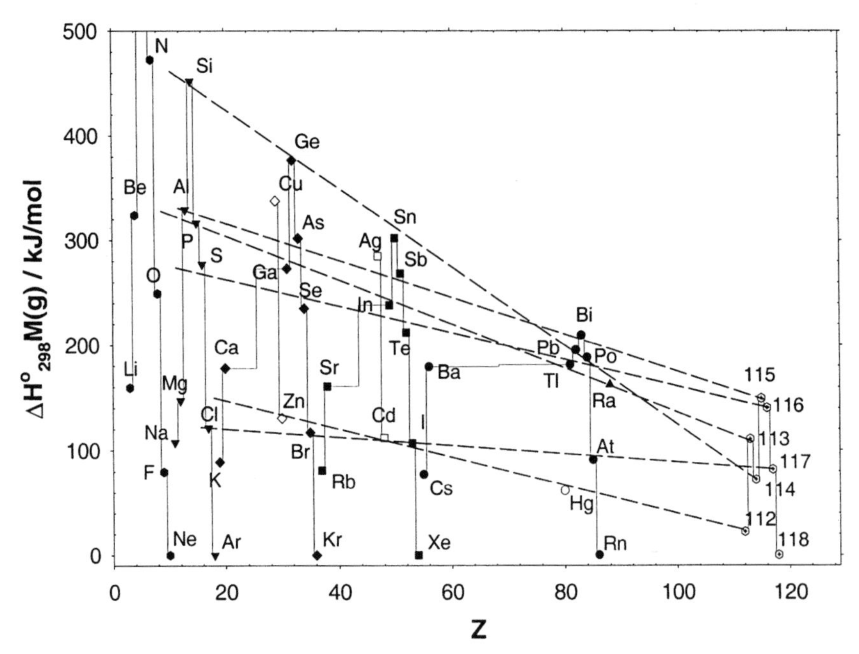

Fig. 3. Extrapolation of the standard enthalpies of monatomic gaseous elements $\Delta H^{o}_{298}M(g)$ along the groups of the periodic table based on the atomic number Z (Method ***1***).

In an illustrative way, the standard enthalpy of monatomic gaseous elements can be seen as the dissociation enthalpy of a (macromolecular) elemental crystal [39,40]. This value is a substantial constituent of the binding enthalpy of compounds [41]. Therefore, a coupling between the standard or the atomic standard formation enthalpies of solid and gaseous compounds and the standard enthalpy of monatomic gaseous elements can be expected and is certainly observed, e.g., see Figure 4.

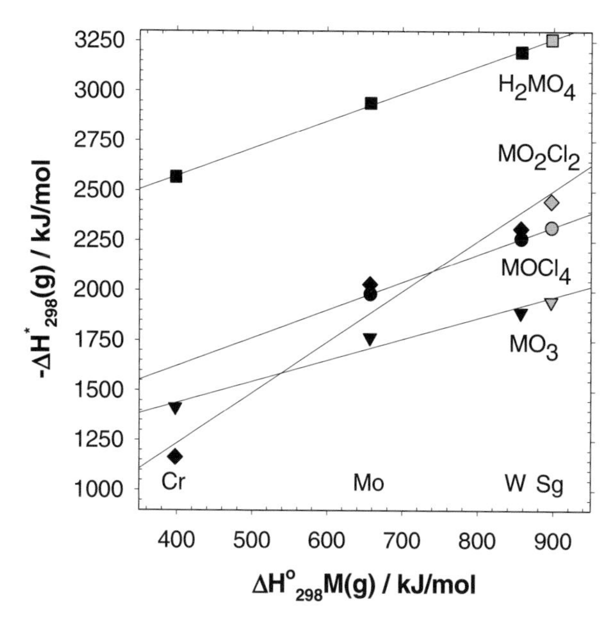

Fig. 4. Extrapolation of the atomic standard formation enthalpies $\Delta H^*_{298}(g)$ of gaseous compounds based on the standard enthalpies of monatomic gaseous elements $\Delta H^o_{298}M(g)$ for group 6 compounds with M = Cr, Mo, W, and Sg.

The enthalpies of formation obtained by this extrapolative Method ***2*** are listed in Table 2. Two results are given for the formation enthalpies, wherever different values of the standard enthalpies of monatomic gaseous elements are predicted.

Recently, correlations between the standard formation enthalpies or the atomic standard formation enthalpies of the solid state versus the corresponding values of the gaseous state have been used (Method ***3***) [42,43]. For similar types of compounds of elements along one group with equivalent oxidation states linear correlations can be observed. More generally, this type of correlation is observed for different compounds of transition elements in their highest achievable oxidation state, see Figure 5.

Table 2. Thermochemical data obtained using Methods ***1-8*** in kJ/mol

{***Method***}	ΔH^o_{298} $M_{(g)}$ {***1***}	$-\Delta H^*_{298}$ (s) {***2***}	$-\Delta H^o_{298}$ (s) {***2***}	$-\Delta H^*_{298}$ (g) {***2***}	$-\Delta H^o_{298}$ (g) {***2***}	ΔH^o_{subl} {***5-7***}	$-\Delta H^o_{ads}$ {***8***}	Lit.
Rf	648							[28]
$RfCl_4$			1031		909	122	95±8	[44]
						138	104±9	[44]
						126	97±8	[44]
$RfOCl_2$		2204		1745		459	297±17	[45]
Db	700							[46]
	835							[28]
$DbCl_5$			899		805	94	78±8	[44]
$DbOCl_3$			969		789	180.0	136±10	[44]
$DbBr_5$			541		423	118		[46]
			635		526	109		[46]
Sg	897							[28]
	1030							[42]
$SgCl_6$		2230		2125		105	85±8	[42]
		2400		2279		121	94±8	[42]
$SgOCl_4$		2413		2318		95	79±8	[42]
		2612		2504		108	86±8	[42]
SgO_2Cl_2		2592		2367		125	97±8	[42]
		2695		2551		144	108±9	[42]
SgO_3		2517		1967		549	380±23	[42]
		2727		2108		519	427±25	[42]
Bh	850							[46]
BhO_3			671		385	358	250±18	[43]
BhO_3Cl			650		561	89	75±7	[43]
BhO_3OH			837		710	127	93±11	[43]
Hs	790							[28]
HsO_4		2186		2130		57±1	46±9	[47]

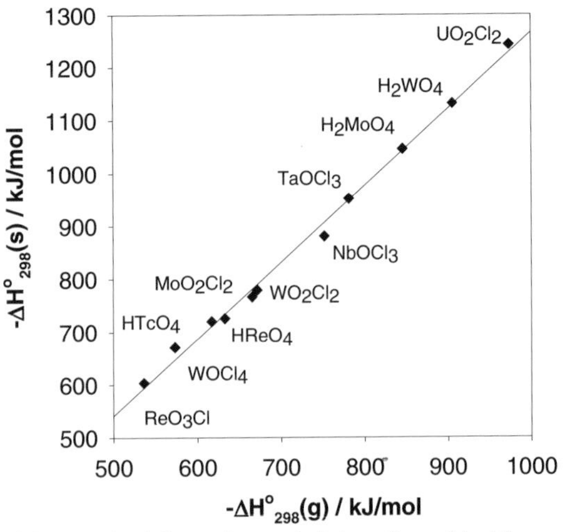

Fig. 5. Correlation of the standard formation enthalpies of oxychlorides and oxyhydroxides in their gaseous (ΔH^o_{298}(g)) and solid state (ΔH^o_{298}(s)), Method ***3***.

Interesting linear relations were obtained by Golutvin [48] correlating the standard formation enthalpies of solid and gaseous compounds of elements ($\Delta H^{o}_{eq.}(g,s)$, see Equation 44), which are normalized to the oxidation state of the metal ion in this compounds (w) and to the number of metal ions (metal equivalents) in the compounds (eq_{metal}),with the logarithm of the oxidation state of the metal ion in this compounds, see Figure 6.

$$\Delta H^{o}_{eq.}(g,s) = \frac{\Delta H^{o}_{298}(g,s)}{w \cdot eq_{metal}} \tag{44}$$

Thus, just by assuming the oxidation state of a compound, the formation enthalpy of a compound in its solid or gaseous state can be predicted (Method ***4***).

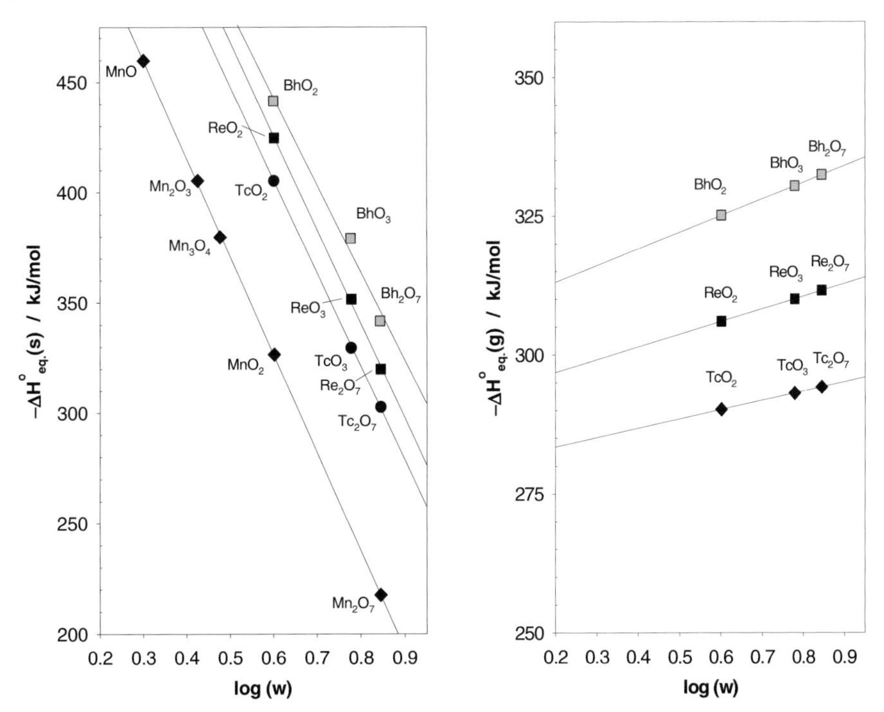

Fig. 6. Correlations between the normalized standard formation enthalpies $\Delta H^{o}_{eq.}(g,s)$ of gaseous and solid compounds of group 7 elements and the logarithm of the oxidation state (w) of the metal ion in these compounds (Method ***4***).

2.2 VOLATILITY

The temperature dependent vapor pressure of pure elements or compounds is frequently used to define volatility. For the adsorbed state the relevant quantity is the desorption pressure, which depends on the temperature and on the surface coverage. The individual crystal lattices with their characteristic binding properties and thus, their standard entropy of the pure solid phase, may largely influence the vapor pressure [42]. Thus, we define the standard sublimation enthalpy (ΔH^{o}_{subl}) as a measure for the volatility of a compound.

The prediction of the adsorption behavior of a transactinide compound starts with the calculation of the sublimation enthalpy (ΔH^{o}_{subl}) of the pure compound. Depending on the availability of data, the standard sublimation enthalpy can be calculated using the following methods:

- Method ***5***: Calculations employing the difference between the standard formation enthalpies of compounds in the gaseous and in the solid state:

$$\Delta H^{o,(*)}_{298}(g) - \Delta H^{o,(*)}_{298}(s) = \Delta H^{o}_{subl} \quad (45)$$

- Method ***6***: Correlations between the standard formation enthalpies of the gaseous and of the solid state with the linear regression coefficients a and b; e.g., see Figure 4:

$$-\Delta H^{o,(*)}_{298}(s) = a \cdot \Delta H^{o,(*)}_{298}(g) + b \quad (46)$$

The standard sublimation enthalpy is then calculated by:

$$(1-a) \cdot \Delta H^{o,(*)}_{298}(g) + b = \Delta H^{o}_{subl} \quad (47)$$

- Method *7*: Significant correlations exist between ionic radii in the solid state -or radii of electron orbitals of ions -and the sublimation enthalpies for various compound classes, especially for the halides. It is possible to calculate the standard sublimation enthalpy of transactinide compounds from their calculated ionic radii by making use of a radii-volatility correlation and of corresponding radii of the homologues [44]. However, in this procedure it is important to use consistent sets of radii for all homologues, see e.g. Figure 7.

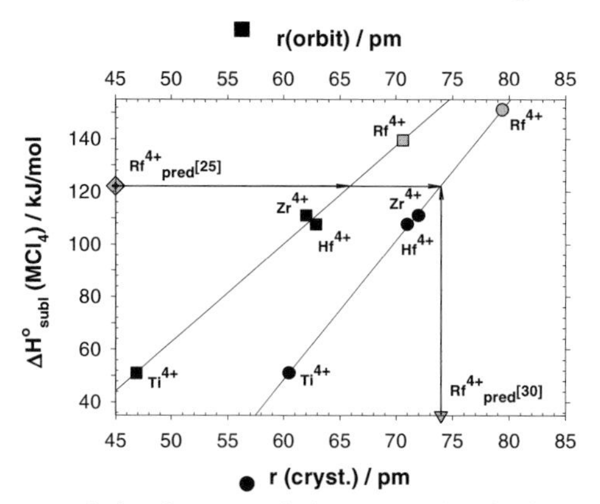

Fig. 7. Radii-volatility correlation for group 4 elements and rutherfordium. The filled squares represent the radii of the outer orbital (r(orbit), top axis) the filled circles show the ionic radii in solid crystals (r(cryst), bottom axis). Included are predicted data for ΔH^{o}_{subl} ($Rf_{pred.}$) from [44] (diamond) and r(cryst)Rf^{4+}_{pred} (triangle) from [49].

Sublimation enthalpies derived from Method ***5***, ***6***, and ***7*** are included in Table 2.

2.3 EMPIRICAL CORRELATIONS BETWEEN ADSORPTION PROPERTIES OF SINGLE ATOMS OR MOLECULES AND THE VOLATILITY OF PURE SUBSTANCES

Empirical extrapolative predictions regarding the stability and the volatility of transactinides and of their compounds are, in the next step, followed by the determination of empirical correlations between adsorption properties of extremely small amounts of these elements or compounds and the volatility of the pure macroscopic solid phase of this substance, respectively.
It is assumed that the binding energy of an adsorbed single molecule to the surface approximately equals its partial molar adsorption enthalpy at zero surface coverage. In the adsorbed state at zero surface coverage the individual variations of the entropy are partly but not completely suppressed. Hence, it is expected that this adsorption enthalpy is proportional to the standard sublimation enthalpy, which characterizes the volatility properties of pure solid phases as an integral value,:

$$-\Delta H^{o}_{ads} \propto \Delta H^{o}_{subl} \tag{48}$$

The experimental proof of such correlations for defined classes of pure substances is essentially for the prediction of adsorption properties of transactinides and their compounds. Therefore, a variety of gas adsorption chromatographic experiments were carried out with carrier free amounts of different radioisotopes using selected modified surfaces as stationary phases. The use of carrier free amounts is necessary to experimentally obtain adsorption conditions at nearly zero surface coverage.
The selection of proper elements and compounds needs to take into account the very complex situation. The chemical state of the radioisotopes in the reactive carrier gas and the standard sublimation enthalpy of this chemical state must be known. In principle all compounds, which are unstable under the selected conditions, have been rejected as well as compounds, which undergo diffusion processes or irreversible reactions with the stationary phase. The following correlations (Method ***8***) were obtained experimentally for elements and selected compound classes, see also Figure 8 A-C:
For elements (with H_2, (H_2O)) [50]

$$-\Delta H^{o}_{ads} = (-5.61 \pm 8.8) + (0.806 \pm 0.042) \cdot \Delta H^{o}_{298} M(g), \tag{49}$$

for chlorides and oxychlorides (with Cl_2, $SOCl_2$, HCl, (O_2)) [42]

$$-\Delta H^{o}_{ads} = (21.5 \pm 5.2) + (0.600 \pm 0.025) \cdot \Delta H^{o}_{subl}, \tag{50}$$

and for oxides and oxyhydroxides (with O_2, (H_2O)) [51]

$$-\Delta H^{o}_{ads} = (6.27 \pm 7.78) + (0.680 \pm 0.028) \cdot \Delta H^{o}_{subl} \tag{51}$$

The results of these empirical correlations clearly prove the postulated proportionality. These correlations suggest a similarity between the bond (with lower coordination) of the adsorbed particles to the modified surface and the bond to the surface of the pure macroscopic phase of the compound, which is relevant for the desublimation process.

The adsorption behavior of atoms and compounds for most of the experiments used in the described correlations were evaluated using differently defined standard adsorption entropies [28,52-57]. Adsorption data from more recent experimental results were evaluated applying the model of mobile adsorption [4]. In addition, data from previous experiments were re-evaluated using this model.

Correlations based on estimated standard sublimation enthalpies allow predictions of adsorption enthalpies for selected compounds for the case of zero surface coverage. These results are only valid for experimental conditions using the same reactive gases and thus, obtaining similarly modified stationary surfaces. For gas chemical studies we assume, in the simplest case, that under a given experimental condition the most stable chemical state is formed and that this states exists during the entire experiment. Pure and very reactive gas mixtures are used at high concentrations to obtain this chemical state. During gas adsorption processes the reactive carrier gas determines the chemical state of the investigated elements and modifies the surface of the stationary chromatographic phase in a characteristic way. Thus, in the case of dissociation reactions, which may occur with very high reaction rates, especially at high temperatures, the restoration of the most stable chemical state is possible and is realized very fast.

All deposition and volatilization processes of single atoms or single molecules (the nearest case to zero surface coverage) are basically adsorption and desorption processes, respectively. Two fundamentally different types of reversible processes can occur in the gas adsorption chromatography [7]:

(i) Firstly, an adsorption and desorption of molecules without any change in the oxidation state and no change in the ligands. For example, the simple adsorption of a group 5 oxychloride can be described by the following equation:

$$MOCl_3(g) \leftrightarrow MOCl_3(ads) \qquad (52)$$

If the interaction of the molecule with the surface is weak (Van der Waals forces), the molecule remains unchanged regarding its binding structure, oxidation state and number of ligands. This is typical for physisorption. Adsorption processes of molecules with non saturated coordinations show much stronger interactions. These adsorption processes are accompanied by changes of the coordination number of the central atom and, in some cases,

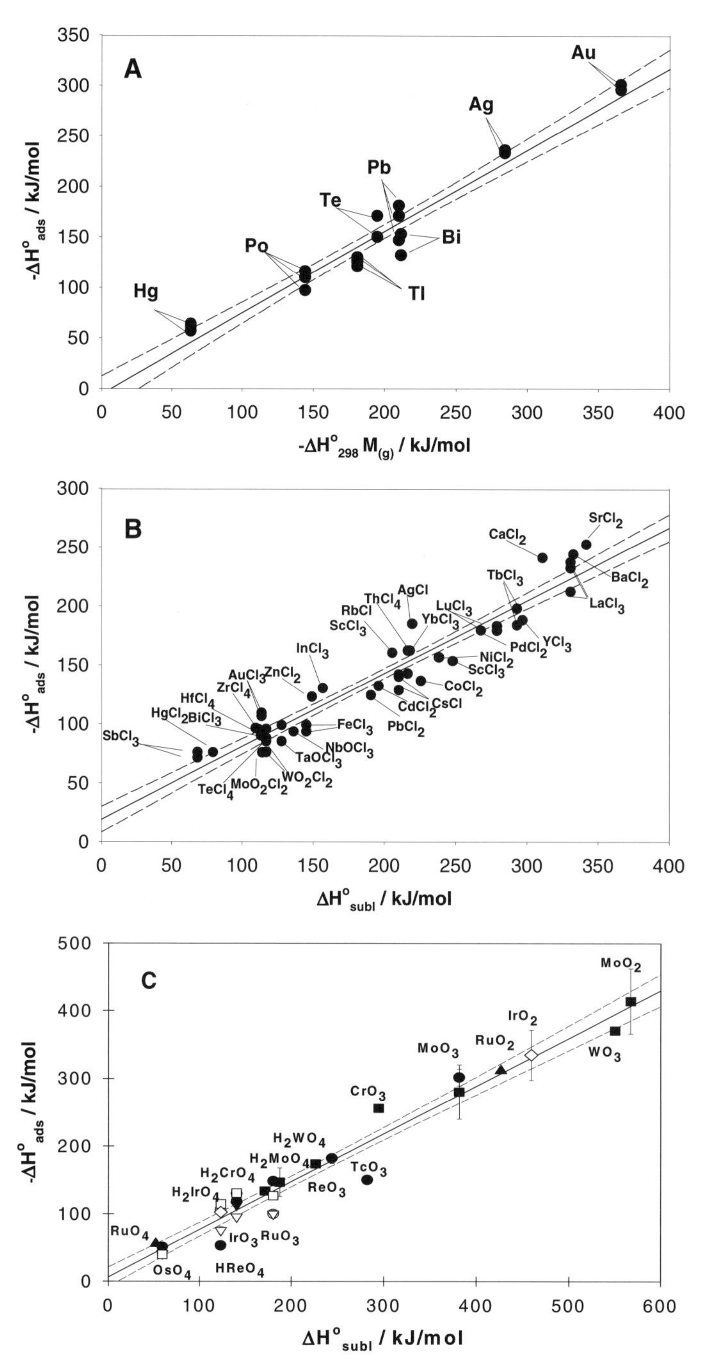

Fig. 8. Correlation of the molecular property adsorption enthalpy ΔH^o_{ads} with the property of the macroscopic solid phase sublimation enthalpy ΔH^o_{subl} for different gas phase chemical systems (Method ***8***): Panel **A** for elements in H_2, **B** for chlorides and oxychlorides in Cl_2, HCl, CCl_4 (O_2), and **C** for oxides and oxyhydroxides in O_2 (H_2O).

by a changes of the type of binding (e.g. break of a double bond). The oxidation state remains unchanged. This is a typical case of chemisorption.

(ii) Secondly, in a different kind of a chemisorption process a reversible change of the oxidation state or a reversible change of the number of ligands or even the nature of ligands may occur during the adsorption and the desorption of a molecule. For example, for the oxyhydroxide compound of a metal ion in the oxidation state 6+ the dissociative adsorption can be described by [7]:

$$H_2MO_4(g) \leftrightarrow MO_3(ads) + H_2O(g) \quad (53)$$

As another example, the substitutive adsorption of a group 4 tetrachloride is described by:

$$MCl_4(g) + 1/2O_2(g) \leftrightarrow MOCl_2(ads) + Cl_2(g) \quad (54)$$

The knowledge of the type of the adsorption process is crucial for the determination of the adsorption enthalpy from experimental results. One experimental approach to gain this information is by varying the partial pressure of the reactive carrier gas, which is involved in the mechanism of the adsorption reaction [45,8].

3. Empirical Calculation Methods for Adsorption Enthalpies

3.1 VAN DER WAALS INTERACTION

Depending on the type of interaction between an adsorbed particle and a solid state surface there are cases, where adsorption enthalpies can be calculated using empirical and semi-empirical relations. In the case of atoms with a noble-gas like ground-state configuration and of symmetrical molecules the binding energy (E_B) to a solid surface can be calculated as a function of the polarizability (α), the ionization potential (I_P), the distance (R) between the adsorbed atom or molecule and the surface, and the relative dielectric constants (ε) (Method ***9***) [58-61]:

$$E_B = 1.57 \cdot \alpha \cdot \frac{(\varepsilon - 1)}{(\varepsilon + 1)} \cdot \frac{(I_{p1} \cdot I_{P2})}{(I_{p1} + I_{P2})} \cdot \frac{1}{8 \cdot R^3} \quad (55)$$

The binding energy E_B approximately equals the adsorption enthalpy.

If the elements 112 and 114 have a noble-gas like character [62], then, in a fictitious solid state, they would form non conducting colorless crystals. A physisorptive type of adsorption may occur and their adsorption properties, for example on quartz, can be calculated with this method [61], see Table 3. For physisorbed noble gas atoms a roughly uniform distance to different surfaces of about 2.47±0.2 Å was deduced from experimental results [63].

Table 3. Physical data used in Method ***9*** together with the results of the calculation for the elements 112, 114, 118 [61]

Element	$\alpha * 10^{24}$ [cm^3]	ε (SiO_2)	R [nm]	I_{P1}, [eV]	I_{P2}, [eV]	$E_B \approx \Delta H^o_{ads}$ [kJ/mol]
112	5.33	3.75	0.25	11.7	11.4	24
114	6.47	3.75	0.25	11.7	8.5	25
118	8.31	3.75	0.25	11.7	8.7	33

3.2 ADSORPTION OF METALS ON METAL SURFACES

If the contact of a transactinide atom with a metallic surface leads to the formation of metallic bonds, then the periodic order scheme describing the adsorption of metallic elements will also accommodate this transactinide element. One general rule of the periodic table is that the metallic character of the elements increases with increasing atomic number Z along the groups of s-elements and of p-elements.

The relative difference between the dissociation enthalpy of two-atomic elemental molecules given by Gurevic [64] and the sublimation enthalpy of these elements [65] from Method ***1*** can be used to estimate the metallic character (m) of elements (Method ***10***) [28], see Figure 9:

$$m = \left(\Delta H^o_{298}(g) - 0.5 \cdot \Delta H^o_{diss}\right) / \Delta H^o_{298}(g) \tag{56}$$

For real metals this value is close to 1. This value shows qualitatively if the association to homo-nuclear two-atomic molecules (ΔH^o_{diss}) (non metallic character) is energetically preferred over the formation of a coordination lattice (metallic character) and vice versa. According to this relation a metallic character can also be expected for the elements 112 and 114 [28]. Element 117, for example, can be assumed to have a semi-metallic character.

The adsorption enthalpy equals the sum of desublimation enthalpy and net adsorption enthalpy. The net adsorption enthalpy is the enthalpy difference between a pure solid compound and its adsorbed state on a surface at zero surface coverage. Hence, the net adsorption enthalpy characterizes the interaction, which depends on the nature of both metals. On the other hand, the desublimation enthalpy is an exclusive property of the adsorbate.

For the calculation of the net adsorption enthalpies of transactinides on metal surfaces the partial molar enthalpies of solution and the enthalpy of displacement are required. These values can be obtained using the semi-empirical Miedema model [66-70] and the Volume-Vacancy or Surface-Vacancy model [32,70,71]. Data for these calculations are given in [34,72,73].

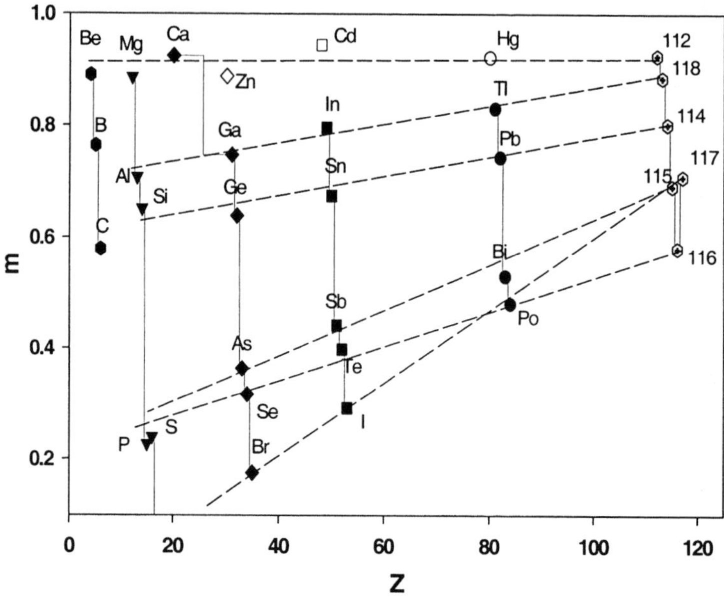

Fig. 9. The metallic character m of s- and p-elements ($m=(\Delta H^{o}_{298}(g)-0.5*\Delta H^{o}_{diss})/\Delta H^{o}_{298}(g)$) as a function of the atomic number Z and its extrapolation along the groups of the periodic table to the region of transactinides.

The net adsorption enthalpies and the predicted sublimation enthalpies (see Method ***1***) were used to calculate the adsorption enthalpies of transactinides on selected metal surfaces (Method ***11***). The metals, which are presented in Table 4, can be used as stationary phase in gas adsorption chromatographic experiments for selective gas chemical separations or, in the case of high adsorption interaction, as strong fixation materials for the sample preparation in the measurement of transactinides.

Table 4. Adsorption data of metallic elements on different stationary phases calculated with Method ***11***

/surface/ Element E	ΔH^{o}_{subl} /E/ kJ/mol	ΔH^{o}_{ads} /Quarz/ kJ/mol	ΔH^{o}_{ads} /Pd/ kJ/mol	ΔH^{o}_{ads} /Pt/ kJ/mol	ΔH^{o}_{ads} /Ni/ kJ/mol	ΔH^{o}_{ads} /Au/ kJ/mol	ΔH^{o}_{ads} /Cu/ kJ/mol
Mt	666	540 ± 20	-670	-694	-756	-651	-763
110	553	445 ± 14	-588	-596	-586	-561	-593
111	353	279 ± 6	-413	-404	-361	-390	-389
112	22	12 ± 7	-73	-23	-98	-8	-101
113	106	89 ± 4	-228	-169	-251	-115	-221
114	71	52 ± 6	-225	-169	-91	-93	-189
115	149	115 ± 3	-352	-286	-206	-179	-174
116	86	64 ± 5	-232	-167	-183	-146	-260

The experimental investigation of the metallic or non-metallic nature of the elements 112 and 114 represents an interesting and challenging scientific task.

References

1. Beyer, G.J., Novgorodov, A.F., Khalkin, A.F.: Radiokhimiya **20,** (1978) 589.
2. Eichler, B., Zimmermann, H.P., Gäggeler, H.W.: J. Phys. Chem. **A 104,** (2000) 3126.
3. Leipnitz, E., Struppe, H.G.: Handbuch der Gaschromatographie, Verlag Geest und Portig, Leipzig 1970.
4. Eichler, B., Zvara, I.: Radiochim. Acta **30,** (1982) 233.
5. Eichler, B., Zude, F., Fan, W., Trautmann, N., Herrmann, G.: Radiochim. Acta **56,** (1992) 133.
6. Eichler, B., Zude, F., Fan, W., Trautmann, N., Herrmann, G.: Radiochim. Acta **61,** (1993) 81.
7. Vahle, A., Hübener, S., Eichler, B.: Radiochim. Acta **69,** (1995) 233.
8. Eichler, B.: Radiochim. Acta **72,** (1996) 19.
9. Eichler, B., Kratz, J.V.: Radiochim. Acta **88,** (2000) 475.
10. Gäggeler, H., Dornhöfer, H., Schmidt-Ott, W.D., Greulich, N., Eichler, B.: Radiochim. Acta **38,** (1985) 103.
11. Eichler, B., Gäggeler-Koch, H., Gäggeler, H.: Radiochim. Acta **26,** (1979) 193.
12. Eichler, B.: Das Verhalten flüchtiger Radionuklide im Temperaturgradientrohr unter Vacuum. Report ZfK-346, Rossendorf 1977.
13. Gäggeler, H., Eichler, B., Greulich, N., Herrmann, G., Trautmann, N.,: Radiochim. Acta **40,** (1986) 137.
14. Eichler, B., Rhede, E.: Kernenergie **23,** (1980) 191.
15. Eichler, B., Buklanov, G.V., Timokhin, S.N.: Kernenergie **30,** (1980) 191.
16. Eichler, R., Schädel, M.: submitted to J. Phys. Chem. B 2001.
17. Zvara, I.: Radiochim. Acta **38,** (1986) 95.
18. Gilliland E.R.: Ind.Eng.Chem. **26,** (1934) 681.
19. Knudsen, M.: Ann. Phys. 4 **28,** (1909) 75.
20. Grosse, A.V.: J. Inorg. Nucl. Chem. **27,** (1965) 509.
21. Keller Jr., O.L., Burnett, J.L., Carlson, T.A., Nestor Jr., C.W.: J. Phys. Chem. **74,** (1970) 1127.
22. David, F.: Radiochem. Radioanal. Lett. **12,** (1972) 311.
23. Fricke, B., Waber, J.T.: Actinides Reviews **1,** (1971) 433.
24. Keller Jr., O.L., Nestor Jr., C.W., Carlson, T.A.: J. Phys. Chem. **77,** (1973) 1806.
25. Eichler, B.: "Voraussage des Verhaltens der Superschweren Elemente und ihrer Chloride bei der thermochromatographischen Abtrennung". In: *Joint Institute for Nuclear Research Report*, P12-7767, Dubna, (1974).
26. Keller Jr., O.L., Nestor Jr., C.W. Fricke, B.: J. Phys. Chem. **78,** (1974) 1945.
27. Fricke, B.: "Superheavy Elements". In: *Structure and Bonding*, Eds. Dunitz, J.D., Hemmerich, P., Springer-Verlag Berlin, (1975) 89-144.
28. Eichler, B., Kernenergie **19 (10),** (1976) 307.
29. Keller Jr., O.L., Seaborg, G.T.: Ann. Rev. Nucl. Sci. **27,** (1977) 139.
30. Eichler, B.: "Wechselwirkung der Transactinide um Z=114 mit Metallen". In: *Zentralinstitut für Kernforschung Report*, ZfK-374, Rossendorf, (1978).
31. Eichler, B., Reetz, T.: Kernenergie **25,** (1982) 218.
32. Eichler, B., Rossbach, H.: Radiochim. Acta **33,** (1983) 121.
33. Ionova, G.V., Pershina, V.G., Suraeva, I.I., Suraeva, N.I.: Radiochem. **37,** (1995) 282.
34. Eichler, B., "Metallchemie der Transaktinoide", In: Paul Scherrer Institut Report, 00-09, Villigen (2000).
35. Pershina, V., Fricke, B.: J. Phys. Chem. **98,** (1994) 6468.
36. Pyykkö, P.: Chem. Rev. **88,** (1988) 563.
37. Pershina, V.G.: Chem. Rev. **96,** (1996) 1977.
38. Schwerdtfeger, P., Seth, M.: "Relativistic effects of the superheavy elements". In: *Encyclopaedia of computational chemistry,* Vol. 4, Eds. von Rague-Schleyer, P.,

Allinger, N.L., Clark, T., Gasteiger, J., Kollman, P., Schaefer III, H.F., Schreiner, P.R., John Wiley and Sons, New York, (1998) 2480-2499.
39. Eley, D.D.: Ber. Bunsen-Ges. Phys. Chem. **60,** (1955) 797.
40. Ehrlich, G.J.: J. Chem. Phys. **31,** (1959) 111.
41. Pauling, L. *The Nature of Chemical Bond*, 3rd ed., Cornell University Press 1960.
42. Eichler, B., Türler, A., Gäggeler, H.W.: J. Phys. Chem. A **103 (46),** (1999) 9296.
43. Eichler, R.: "Thermochemical Predictions of Chemical Properties of Bohrium (Bh, element 107)". In: *Labor für Radio- und Umweltchemie Annual Report 2000*, Villigen (2001) 4.
44. Eichler, B., Türler, A., Gäggeler, H.W.: "Radius-Volatility Correlation of Tetrachlorides of Ti, Zr, Hf, 104, Th, and U". In: *PSI Condensed Matter Research and Material Sciences Progress Report 1994*, Annex IIIA, Ann. Rep., Villigen (1995), 77.
45. Eichler, B., Gäggeler, H.W.: "Stability and "Volatility" of Element 104 Oxychloride". In: *Labor für Radio- und Umweltchemie Annual Report 2000*, Villigen (2001) 33.
46. Eichler, B.: " On the volatility of Hahnium Pentabromide". In: *PSI Condensed Matter Research and Material Sciences Progress Report 1993*, Annex IIIA, Ann. Rep., Villigen (1994), 93.
47. Düllmann, Ch.E., Eichler, B. Türler, A., Gäggeler, H.W.: "Stability of Group 8 Tetroxides MeO_4 (Me=Ru,Os,Hs) and their Adsorption Behavior on Quartz". In: *Labor für Radio- und Umweltchemie Annual Report 2000*, Villigen (2001), 5.
48. Golutvin, J.M.: *Teploti obrazovanija i tipi khimitcheskoi svjasi w neorgenicheskikh kristallakh*, Izd. Akademii Nauk SSSR, Moskwa, (1962).
49. Bilewicz, A.: Radiochim. Acta **88,** (2000) 833.
50. Eichler, B., Kim, S.C.: Isotopenpraxis 21, (1985) 180.
51. Eichler, R., Eichler, B., Gäggeler, H.W., Jost, D.T., Dressler, R., Türler, A.: Radiochim. Acta **87,** (1999) 151.
52. Zvara, I., Chuburkov, Yu.T., Belov, V.Z., Maslov, O.D., Tsaletka, R., Shalaevskii, M.R.: Radiokhimiya **12,** (1970) 565.
53. Eichler, B. Domanov, V.P.: "Thermochromatographie trägerfreier Kernreaktionsprodukte als Chloride". In: *Joint Institute for Nuclear Research Report*, P12-7775, Dubna, (1974).
54. Eichler, B., Reetz, T., Domanov, V.P.: "Bestimmung der Adsorptionsenthalpie auf der Grundlage thermochromatographischer Daten. III. Elemente. Adsorption auf Quarz und Metallen". In: *Joint Institute for Nuclear Research Report*, P12-10047, Dubna, (1976).
55. Eichler, B., Domanov, V.P.: J. Radioanal. Chem. **28,** (1975) 143.
56. Eichler, B., Domanov, V.P., Zvara, I.: Evaluation of heat of adsorption from Thermochromatographic Data. II. Chlorides of Metals. The Adsorption on Quartz". In: *Joint Institute for Nuclear Research Report*, Dubna, (1976) GSI-tr.-4/76.
57. Eichler B.: J. Inorg. Nucl. Chem. **35,** (1973) 4001.
58. Pauling, L.: Science **134,** (1961) 3471.
59. Frederikse, H.P.R.: "Permittivity (dielectric constant) of inorganic solids". In: *Handbook of Chemistry and Physics*, 79th ed. 1998/1999, Eds. Linde, D.R., CRC-Press, Boca Raton (1998), Table 12-48.
60. Miller, T.M.: "Atomic and Molecular Polarizabilities". In: *Handbook of Chemistry and Physics*, 79th ed. 1998/1999, Eds. Linde, D.R., CRC-Press, Boca Raton (1998), Table 10-160.
61. Eichler B., Eichler, R., Gäggeler H.W.: "Van der Waals Interaction of the Elements 112, 114, and 118 with Solid Surfaces". In: *Labor für Radio- und Umweltchemie Annual Report 2000*, Villigen (2001) 10.
62. Pitzer, K.S.: J. Chem. Phys. 63, (1975) 1032.
63. Miedema, A.R., Nieuwenhuys, B.E.: Surf. Sci. **104,** (1981) 104.
64. Gurevic, L.V., *Energii razriva chimicheskikh svjazej, potenciali ionozacii i srodstvo k electronu*, Nauka Moskwa (1974).

65. Knacke, O., Kubaschewski, O., Hesselmann, K.,*Thermochemical Properties of Inorganic Substances*, 2nd ed., Springer-Verlag, Berlin, (1991).
66. Miedema, A.R.: J.Less-Common Met. **32,** (1973) 117.
67. Miedema, A.R., Boom, R., De Boer, F.R.: J. Less-Common Met. **41,** (1975) 183.
68. Boom, R., De Boer, F.R., Miedema, A.R.: J. Less-Common Met. **45,** (1976) 237.
69. Miedema, A.R.: J.Less-Common Met. **46,** (1976) 67.
70. Eichler, B.: "Bestimmung der Adsorptionswärme gasförmiger Metalle auf festen Metalloberflächen bei Nullbedeckung (Empirisches Modell)". In: *Zentralinstitut für Kernforschung Report*, ZfK-396, Rossendorf, (1979).
71. Eichler, B.: Radiochim. Acta **38,** (1985) 131.
72. Eichler, B.: "The Interaction of Element 112 with Metal Surfaces". In: *Labor für Radio- und Umweltchemie Annual Report 2000*, Villigen (2001) 7.
73. Eichler, B. : „Metallchemie der Transactinoide“. PSI-Report 00-09, Paul Scherrer Institut Villigen (2000).

Index

Chapter 7

Gas-Phase Chemistry

H.W. Gäggeler [a,b], A. Türler [c]

[a] *Labor für Radio- und Umweltchemie, Universität Bern, CH-3012 Bern, Switzerland*
[b] *Labor für Radio- und Umweltchemie, Paul Scherrer Institut, CH-5232 Villigen PSI, Switzerland*
[c] *Institut für Radiochemie, Technische Universität München, D-85748 Garching, Germany*

1. Introduction

In transactinide chemistry research, gas phase separation procedures play an important role. Already, the very first investigation of rutherfordium has been conducted in form of frontal isothermal gas chromatography in a chlorinating atmosphere [1]. The success of gas chemical separations in transactinide research is quite remarkable since gas chromatography is, in general, of minor importance in inorganic analytical chemistry.

There are several reasons for this exceptional situation. First, production of transactinides at accelerators implies a thermalization of the primary products in a gas, usually helium. It is rather straightforward to connect such a recoil chamber to a gas chromatographic system. Second, gas phase separation procedures are fast and may be performed in a continuous mode. Third, at the exit of the chromatographic column separated volatile species can be easily condensed as nearly weightless samples on thin foils. This enables detection of α decay and spontaneous fission (SF) of the separated products with supreme energy resolution.

All these advantages compensate for some disadvantages if compared to liquid-phase separations. They include modest chemical separation factors

M. Schädel (ed.), The Chemistry of Superheavy Elements, 237-289.

and a rather limited number of volatile species that are suited for gas chromatographic investigation. One should keep in mind that the retention temperature regime in quartz chromatography columns is limited to maximum temperatures of about 1000 °C. In addition, due to the short half-lives of transactinide nuclides, the kinetics of the formation of chemical compounds should be fast. So far, mostly inorganic compounds have been synthesized and separated such as halides and oxyhalides. This class of compounds, mostly in form of chlorides and bromides or oxychlorides and oxybromides, respectively, proved to be ideal for the 6d elements of groups 4 to 7.

For the group 6 and 7 elements, also the oxide/hydroxide molecules have been synthesized. For elements of group 8, the tetroxide is the species of choice, since this molecule is very volatile. For future studies with the p-elements around atomic number 114 the elements are expected to be volatile in their atomic state and should behave like noble metals or even like a noble gas.

2. Rutherfordium (Rf, Element 104)

So far, most gas chemical investigations of this element have been conducted in form of its chloride, oxychloride or bromide. Only in one experiment an attempt was made to search for a p-element behavior of Rf, based on a predicted ground state configuration of $[Rn]5f^{14}7s^27p^2$ [2] or, from a more recent calculation $[Rn]5f^{14}6d7s^27p$ [3], rather than the expected "d-like" $[Rn]5f^{14}6d^27s^2$. However, the experiment yielded no evidence for a "Pb-like" behavior of Rf [4]. This observation is not surprising, since Multi-Configuration Dirac-Fock (MCDF) calculations showed that ionization potentials, atomic and ionic radii for Rf are very similar to those of Hf [3].

2.1 VOLATILE COMPOUNDS OF GROUP-4 ELEMENTS

Due to the high sublimation enthalpies of group-4 elements, gas chromatographic separations of the atoms are not feasible in quartz columns. Under halogenating conditions, however, group-4 elements form mono-molecular pure halides such as tetra fluorides, chlorides, bromides and iodides.

A good measure of the volatility is the vapor pressure. Figure 1 depicts the vapor pressure curves for Zr and Hf halides in the gas phase over the respective solids. As can be seen, the volatility decreases according to MCl_4 > MBr_4 > MI_4 > MF_4 with M= Zr and Hf. Evidently, chlorides and bromides are clearly the best choices for gas chemical studies. Iodides have the disadvantage of a poor thermal stability and fluorides are least volatile.

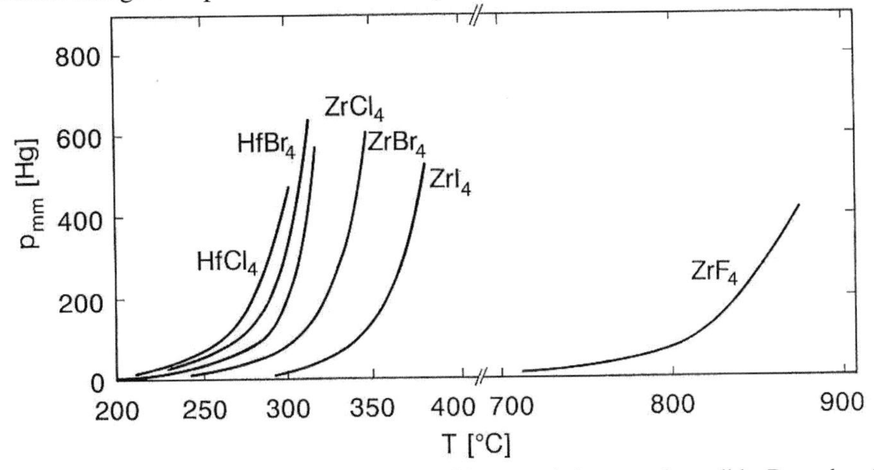

Fig. 1. Vapor pressure curves for Zr and Hf halides over their respective solids. Reproduced from [5].

In principle, in oxygen containing carrier gases also oxyhalides can be synthesized. However, for group-4 elements little is know about these compounds. It was observed that $ZrOCl_2$ and $HfOCl_2$ decompose to tetrachlorides and the oxide under elevated temperatures [6]. An alternative process is substitutive adsorption of the pure halides on the surface of the quartz chromatography column where oxychloride formation is possible in the adsorbed state only.

2.2 EARLY GAS CHEMICAL STUDIES WITH RUTHERFORDIUM

The first chemical study of Rf [1] was part of the discovery claim of this element by scientists from Dubna. For production of the isotopes $^{259,260}Rf$ the hot fusion reaction ^{22}Ne + ^{242}Pu was used. In these pioneering studies isothermal frontal gas chromatography experiments showed that in a chlorinating gas Rf forms a highly volatile molecule, see Figure 8 in Chapter 4. As a chlorinating agent 0.15 mm Hg vapor pressure of $NbCl_5/ZrCl_4$ was added to a flowing N_2 carrier gas. The gas then passed through an isothermal glass column kept at temperatures between 250 and 300 °C. From previous experiments it was known that actinides do not form sufficiently volatile chlorides that could pass the column at such moderate temperatures.

Behind the column mica solid state detectors were positioned. They were kept at lower temperatures in order to adsorb the $RfCl_4$ molecules. It was assumed that the produced isotopes of Rf decay at least partly by spontaneous fission. Mica is known to be well suited for identification of latent fission tracks. In a series of experiments that accumulated a total beam dose of $4x10^{18}$ beam particles 65 fission tracks were detected along the mica detectors. These fission events were assigned to a spontaneously-fissioning isotope of Rf, presumably ^{260}Rf. Later, this assignment was questioned since additional measurements proved that this isotope has a half-life of only 20 ms, too short for chemical study. It was therefore concluded that ^{259}Rf with a half-life of 3 s and an assumed small fission branch was the isotope that labelled the separated molecule.

After these very first experiments the Dubna group applied the thermochromatography technique, see also Chapter 4, which permits to compare the volatility of the Rf species, measured via the deposition temperature in the chromatographic tube, with those of the Hf compounds. An example of such a study is depicted in Figure 2. From the observed very similar deposition temperature of the Rf and Hf chlorides it was concluded that both elements behave very similarly, therefore convincingly proving that Rf is a d element [7]. Recently, the chromatographic peaks shown in Figure 2 have been analyzed applying a Monte-Carlo model. Based on some assumptions on the adsorption process and assuming that indeed the decay of ^{259}Rf with a half-life of 3 s was detected, the standard enthalpies of adsorption, ΔH_a^0, of -110 kJ/mol and –146 kJ/mol for Rf and Hf on the quartz chromatography column surface were deduced [8].

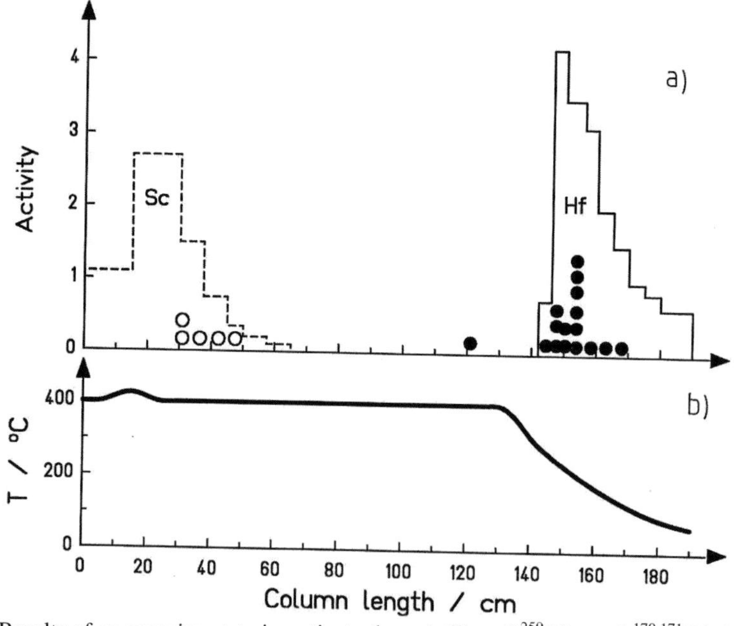

Fig. 2. Results of an experiment to investigate the volatility of ^{259}Rf – and ^{170,171}Hf chlorides; a) distribution of fission tracks (open and closed circles), ^{44m}Sc (a representative of an actinide-like element), and of ^{170,171}Hf; b) temperature profile along the column. Reproduced from [7].

Thermochromatography was also applied to investigate the volatility of Rf and Hf bromides [9]. These experiments yielded evidence that Rf bromide is more volatile than Hf bromide, and also more volatile than Rf chloride.

2.3 ON-LINE ISOTHERMAL GAS CHEMICAL INVESTIGATIONS OF RUTHERFORDIUM

2.3.1 *General remarks*

In recent years predominantly continuous isothermal chromatography has been applied in gas chemical studies of transactinides. This technique offers the possibility to combine a continuous separation of volatile species with an *in-situ* detection of the products on the basis of single atom counting. To reach this ambitious goal, novel devices have been developed such as the On-Line Gas chemistry Apparatus (OLGA) [10] or, in a modified version, the Heavy Element Volatility Instrument (HEVI) [11]; see also Chapter 4.

On-line isothermal gas chemistry has originally been developed to search for superheavy elements with atomic numbers between 112 and 118. OLGA I was restricted to an operation with inert gases. Its application concentrated on separations of volatile atoms. As "reactive" gas, traces of hydrogen gas could be added to a helium carrier gas in order to stabilize the elemental state. Model studies with the p-elements Po, Pb, Bi and At showed that at temperatures of up to 1000 °C excellent separations of these elements from d elements and from f elements could be achieved.

Improved versions of OLGA (versions II and III) enabled the applications of corrosive gases such as hydrogen chloride or hydrogen bromide, chlorine, thionyl chloride or boron tribromide vapor etc. This made it possible to synthesize volatile halides and measure their retention times in isothermal quartz columns.

2.3.2 *Isothermal gas chromatography studies of $RfCl_4$ and $RfBr_4$*

For investigations of the chlorides and bromides commonly the isotope ^{261}Rf was used as a tracer. It has a half-life of 78 s and can be produced in the fusion reaction ^{18}O + ^{248}Cm at a bombarding energy of about 100 MeV. ^{261}Rf decays via emission of two sequential α-particles via ^{257}No to ^{253}Fm, a long-lived product. Hence, identification of ^{261}Rf after chemical separation bases on the measurement of the two lifetimes and two α-decay energies of the mother and its daughter nuclide, respectively. From the four signals an unequivocal identification of every decaying atom of rutherfordium may be achieved. Simultaneous formation of Hf isotopes may be obtained by covering the ^{248}Cm target by a thin layer of Gd.

Figure 3 depicts the result from such a study with the OLGA III device [12]. Helium saturated with carbon particles served as carrier aerosol to transport the products from the collection chamber to the chemistry device. The chemical reagent was HCl gas, purified from traces of oxygen, added to the carrier gas at the entrance of the oven system.

Rf passes through the quartz column at a lower retention temperature compared to Hf. This observation received considerable attention and was interpreted as evidence for relativistic effects, since the higher volatility of $RfCl_4$ compared to that of $HfCl_4$ is unexpected on the basis of classical extrapolations.

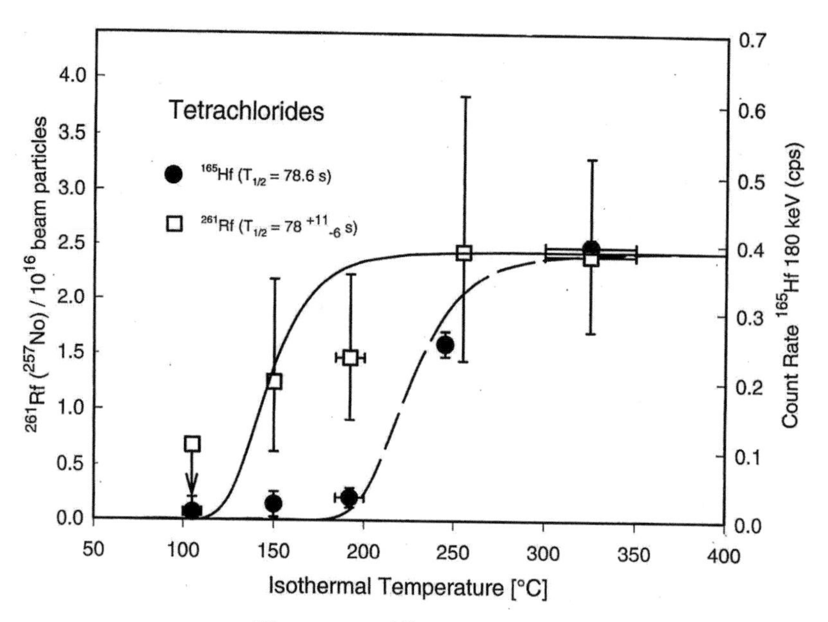

Fig. 3. Relative yields for $^{261}RfCl_4$ and $^{165}HfCl_4$ from in the same $^{18}O + ^{248}Cm/^{152}Gd$ experiment behind the isothermal quartz chromatography column as a function of the temperature. The solid lines represent Monte Carlo simulations adapted to the experimental data. Reproduced from [12].

In an earlier study with the device HEVI chlorides of Zr, Hf, and Rf were investigated employing the reactive gases HCl/CCl_4 (Zr, Hf) and HCl (Rf). MoO_3 particles in a He carrier gas served as aerosol [13]. In contact with HCl, MoO_3 forms a very volatile molybdenum oxychloride that passes through the chromatographic column without deposition on its surface. For Rf and Hf chlorides the same sequence in volatility was found as depicted in Figure 3. In addition, the volatility of Zr chloride turned out to be very similar to that of Rf chloride, hence, different to the volatility of Hf chloride. This observation is rather unexpected, since the volatilities of macro amounts of Zr and Hf tetrachlorides are nearly identical, see Figure 1.

A first study of Rf bromide with OLGA II indicated that the Rf compound is more volatile than Hf bromide [14]. For the transport KCl was used as aerosol particles and HBr/BBr_3 served as a reactive gas. In addition, Rf bromide was found to be less volatile compared to the Rf chloride. In a more recent study with HEVI, using KBr particles and HBr as a brominating agent, these findings were essentially confirmed, see Figure 4. The behavior of Zr and Hf bromides were found to be very similar being less volatile than the Rf bromide [15].

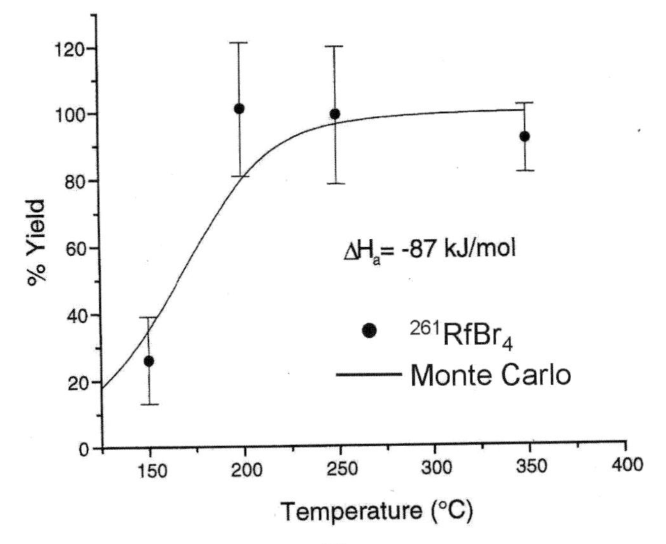

Fig. 4. Chromatographic yield curve for ^{261}Rf bromide using KBr aerosol particles for transport and HBr as reactive gas. Reproduced from [15] with the permission of Oldenbourg Verlag.

2.3.3 *Oxychlorides of Rf*

The oxychlorides of group-4 elements are expected to be less stable than the pure chlorides. $ZrOCl_2$ and $HfOCl_2$ were found to decompose to the tetrachlorides at elevated temperatures [6]. It is therefore not clear, whether $ZrOCl_2$ and $HfOCl_2$ exist in the gas phase.

In thermochromatography experiments an increase of the deposition temperature of Zr and Hf was observed as a function of the partial pressure of oxygen in a chlorinating reactive gas mixture [16]. An OLGA III study with oxygen containing chlorinating reactive gas confirmed this observation: Rf and Hf compounds were considerably less volatile compared to oxygen-free conditions, see Figure 5. It was speculated that the oxychlorides do not exist in the gas phase but only in the adsorbed state. The following transport mechanism was proposed:

$$MCl_{4(g)} + \frac{1}{2} O_2 \Leftrightarrow MOCl_{2(ads)} + Cl_{2(g)}$$

It is interesting to note that, as seen in Figure 5, $RfOCl_2$ and $HfOCl_2$ behave much more similar compared to the pure chlorides $RfCl_4$ and $HfCl_4$.

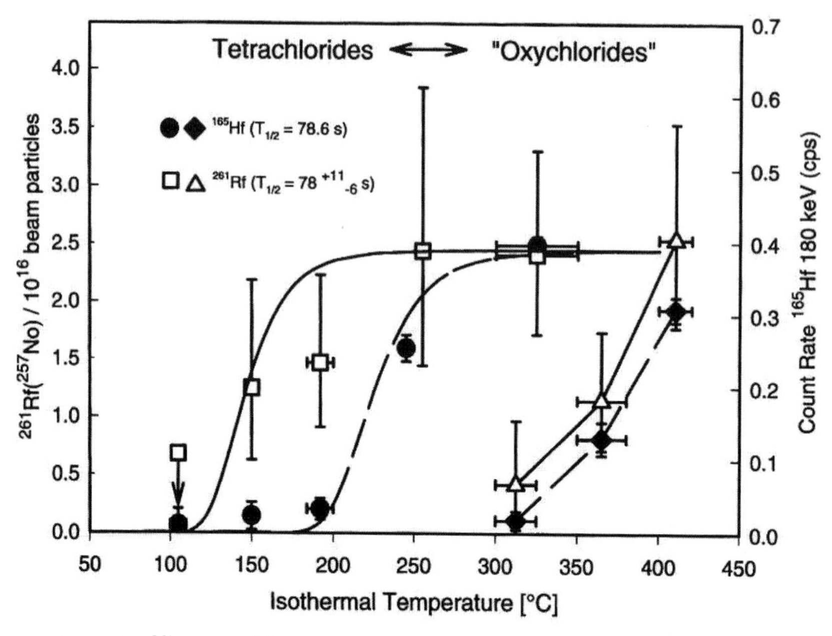

Fig. 5. Yields for ^{261}Rf and ^{165}Hf tetrachlorides (in the left part) obtained with oxygen-free HCl and (in the right part) oxychlorides from $SOCl_2$ vapor and O_2 added as a reactive gas. Lines are results from Monte Carlo simulations. Reproduced from [12].

2.3.4 *Adsorption enthalpies of Zr, Hf, and Rf chlorides and bromides on quartz*

From the measured chromatographic retention temperatures adsorption enthalpies (ΔH_a^0) of single molecules on the surface of the quartz chromatography column can be deduced. This analysis is based on certain thermodynamic assumptions of the adsorption process of single molecules with the surface of the chromatographic column [17]. In addition, a Monte Carlo model enables to describe the migration path of each single molecule along the chromatographic column under real experimental conditions [18].

Figure 6 summarizes resulting ΔH_a^0 values from isothermal gas chromatographic investigations of the pure chlorides and bromides of Zr, Hf, and Rf, respectively, using the HEVI and OLGA II devices [19]. A smooth (classical) extrapolation of the ΔH_a^0 values from Zr through Hf shows that one would expect the ΔH_a^0 values for $RfCl_4$ or $RfBr_4$ to be more negative than those of the respective Hf compounds. The experimental values for Rf show a striking reversal of this expected trend. In addition, the bromides have more negative values, hence being less volatile than the corresponding chlorides.

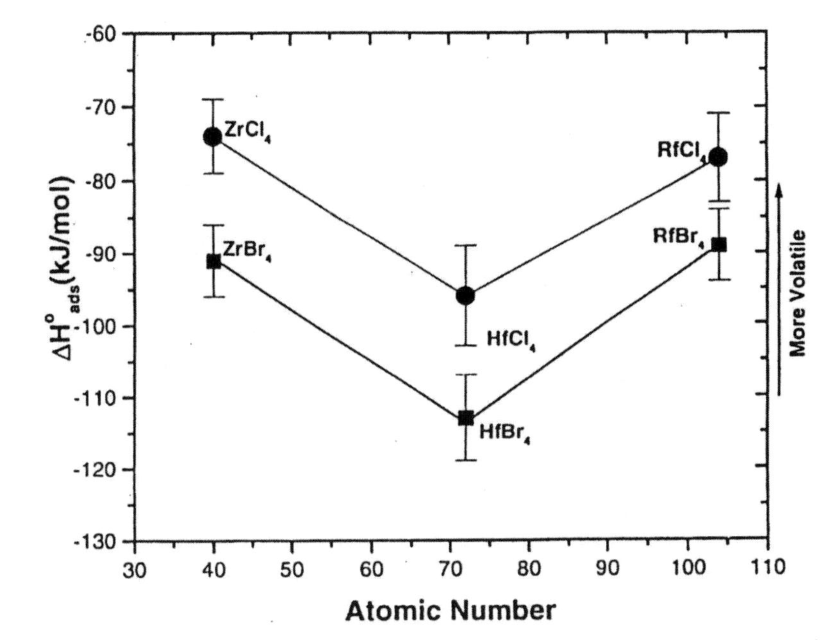

Fig. 6. Adsorption enthalpies of chlorides and bromides of Zr, Hf, and Rf on quartz surfaces, deduced from OLGA and HEVI experiments. Reproduced from [19].

Relativistic calculations of the chemical properties of these compounds predict trends that are in agreement with experimental observations, see Chapter 2. Therefore, it was argued that this "reversal" in the trend of ΔH_a^0 for chlorides and bromides when going from Zr via Hf to Rf is evidence for "relativistic effects" in the chemistry of Rf [12].

3. Dubnium (Db, Element 105)

Dubnium is expected to have a $[Rn]5f^{14}6d^{3}7s^{2}$ electronic ground state configuration. This makes dubnium a firm member of group 5 of the Periodic Table, positioned below tantalum.

All gas chemical investigations of dubnium have been performed with Db^{5+} in form of the pentahalides (chlorides and bromides) and oxyhalides. As a general rule, all these studies were extremely difficult due to the high tendency of group-5 elements to react with trace amounts of oxygen or water vapor. Hence, gas chemical investigations were only successful if the quartz chromatography columns were very carefully preconditioned with the halogenating reactive gas prior to each experiment and, in addition, applying extensive cleaning procedures to the carrier gas to remove trace amounts of oxygen and water vapor.

3.1 VOLATILE COMPOUNDS OF GROUP-5 ELEMENTS

Group-5 elements are most stable in their maximum oxidation state +5 and therefore form pentahalides, see Figure 7. Most volatile are the pentafluorides, followed by the pentachlorides and the pentabromides. Besides the pure halides, also the oxyhalides (MOX_3) are stable in the gas phase. They should be less volatile compared to the pure halides. This was confirmed experimentally for niobium, see Figure 8.

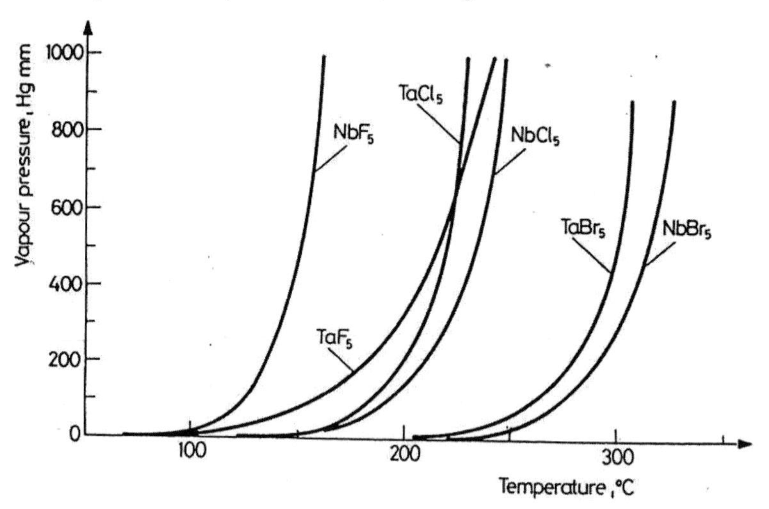

Fig. 7. Vapor pressure curves for Nb and Ta halides over the respective solids. Reproduced from [5].

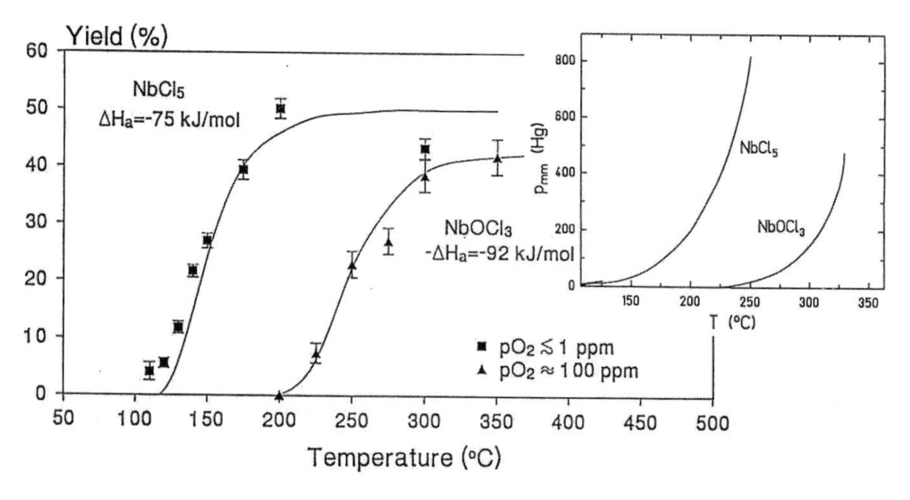

Fig. 8. Vapor pressure curves for $NbCl_5$ and $NbOCl_3$ (upper right corner) and the relative yields of $^{99g}NbCl_5$ and $^{99g}NbOCl_3$ molecules passing through an isothermal quartz column. Reproduced from [5].

3.2 EARLY GAS CHEMICAL STUDIES WITH DUBNIUM

As early as 1970 gas chemical experiments with Db were performed in a chlorinating atmosphere [20]. These studies applied the gas thermochromatography technique, see also Chapter 4, and they indicated that the deposition temperature of Db as a chloride (or oxychloride) is rather similar to that of Hf under identical conditions and significantly higher compared to the deposition temperature of Nb. Later studies were conducted in a brominating gas medium and again yielded evidence that Db bromide is less volatile compared to the homologues compound with Nb [21], see Figure 9. These investigations were performed with ^{261}Db ($T_{1/2}$ = 1.8 s) which has a small fission branch. This nuclide was produced in the ^{243}Am(^{22}Ne,4n)^{261}Db reaction.

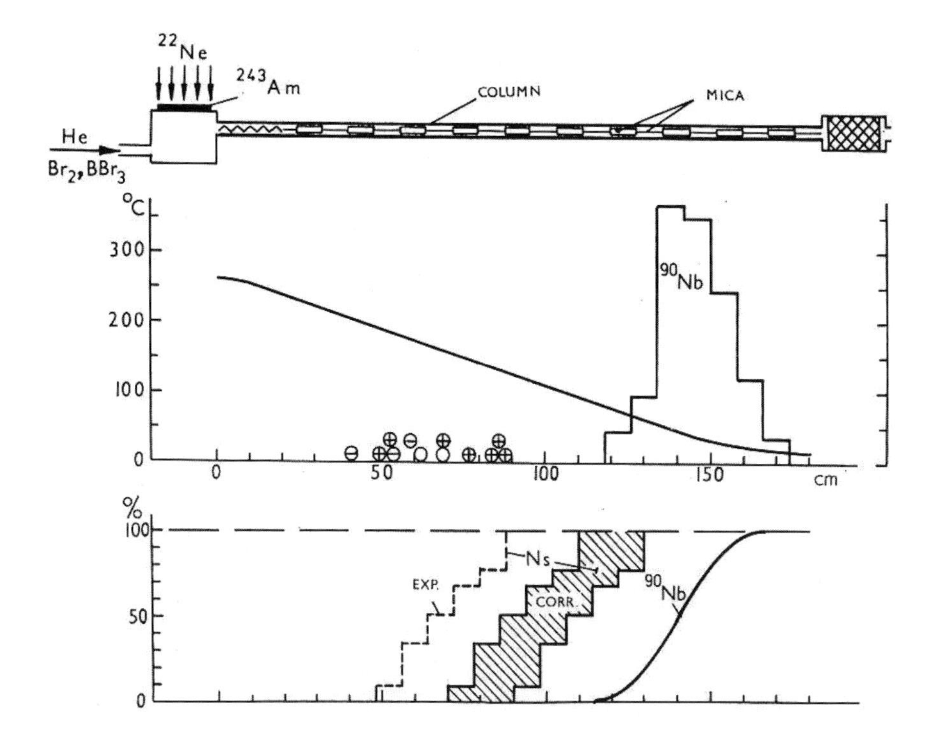

Fig. 9.
Top: Schematic of the early thermochromatography experiment with Db in a brominating atmosphere (Br_2+BBr_3).
Middle: Temperature profile along the column and measured distributions of ^{90}Nb and ^{261}Db.
Bottom: Integral distribution of ^{90}Nb (solid line) and of Db (named Ns by the authors at that time; shaded area) after corrections for the much shorter half-life of ^{261}Db compared to that of ^{90}Nb. Reproduced from [21].

From the result of this experiment it was concluded that the boiling point of $DbBr_5$ may exceed the boiling point of $NbBr_5$ by 80-100 °C and may be close to the boiling point of $PaBr_5$. The ionic radius of Db^{5+} was estimated to be close to the radius of Pa^{5+}, which is ≈ 0.9 Å, whereas the radii of Nb^{5+} and Ta^{5+} are both ≈ 0.7 Å.

3.3 ON-LINE ISOTHERMAL GAS CHEMICAL INVESTIGATIONS OF DUBNIUM

3.3.1 *Production of dubnium isotopes*

On-line gas chemical studies of dubnium have been mostly performed with ^{262}Db. This nuclide can be produced in the reaction $^{249}Bk(^{18}O,5n)$ at a beam energy of about 100 MeV. It has a half-life of 34±5 s and decays with 67 % by emission of two sequential α particles via ^{258}Lr ($T_{1/2}$=4.4 s) to the long-lived ^{254}Md ($T_{1/2}$=28m). In addition, ^{262}Db has a spontaneous fission decay branch of 33%. Hence, identification of each separated labeled molecule is based on either detection of two characteristic α-particles and their lifetimes or on the detection of a spontaneous fission decay.

3.3.2 *Chlorides and oxychlorides*

Several attempts failed to form the pure pentachloride of Db in on-line isothermal gas chromatographic investigations. Obviously, despite very thorough cleaning procedures, minute amounts of oxygen and/or water vapor in the system were still sufficient to form at least partly dubnium oxychloride, most likely $DbOCl_3$. Figure 10 depicts a measured chromatographic curve in conjunction with the data for Nb from Figure 8 [22].

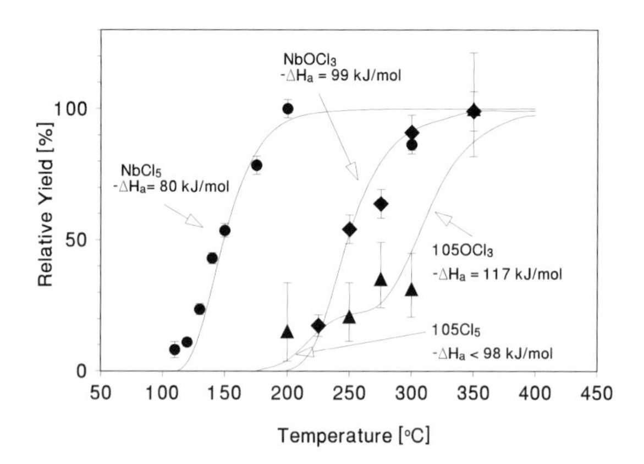

Fig. 10. Relative yield of Db (triangles) measured in an isothermal gas chromatographic experiment with purified HCl as reactive gas. Reproduced from [22] with the permission of Oldenbourg Verlag. For comparison, the data for Nb measured under identical gas chemical conditions from Figure 8 are also shown.

As chlorinating agent HCl gas was used, purified with activated charcoal at 900 °C. The shape of the yield curves suggests two components, a species with a lower volatility passing through the column above 350 °C and one with a higher volatility that is retained in the column only below 200 °C. The two species are tentatively assigned to $DbOCl_3$ and $DbCl_5$, respectively. From this follows, that $DbCl_5$ must be equal or more volatile than $RfCl_4$, see Figure 3.

It is interesting to note that none of the gas chemical investigations succeeded to investigate the behavior of tantalum chloride. Obviously, the tendency to react with oxygen is even higher for Ta compared to Nb and Db.

3.3.3 *Bromides and oxybromides*

The volatility of Dubnium bromide was studied with HBr as a reactive gas using the isotope ^{262}Db formed in the ^{249}Bk(^{18}O,5n) reaction [23]. In this experiment the retention behaviors of niobium and tantalum bromides were investigated as well. Interestingly, the volatile tantalum bromide was formed only when HBr was saturated with BBr_3 vapor. The data are shown in Figure 11. A trend in volatility of Nb≈Ta>Db was deduced. This sequence is very surprising since $DbBr_5$ is expected to be more volatile compared to $NbBr_5$ and $TaBr_5$, respectively. Evidence for a lower volatility of dubnium bromide relative to that of niobium bromide has already been found in previous thermochromatographic studies, see Figure 9.

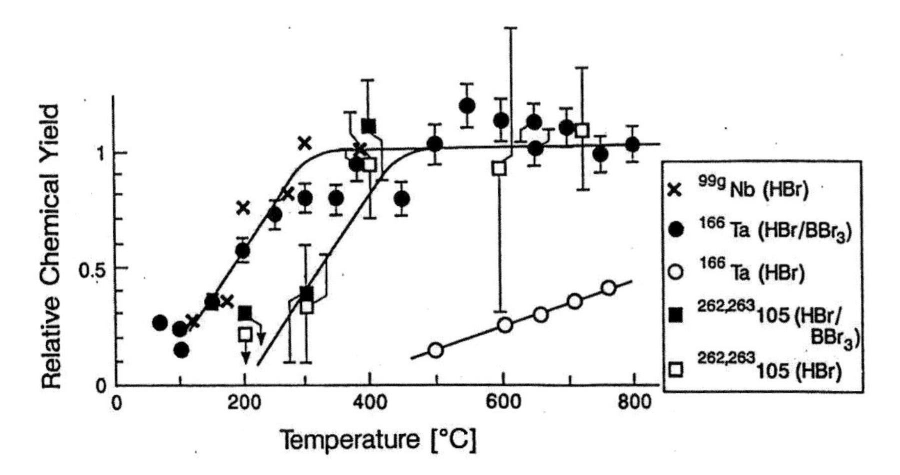

Fig. 11. Yields of Nb, Ta and Db in a gas chromatographic experiment with HBr (Nb, Db) and HBr/BBr_3 (Db, Ta) as reactive gas. Reproduced from [23] with the permission of Oldenbourg Verlag.

3.3.4 *Adsorption enthalpies of Nb, Ta and Db chlorides, oxychlorides, bromides and oxybromides on quartz*

Figure 12 depicts ΔH_a^0 values of group-5 chlorides and bromides measured with the OLGA technique (for an overview of several experimental investigations and a re-analysis of the data, see [24]). It is rather surprising to observe a different trend of the ΔH_a^0 values when going from Nb via Ta to Db, if compared to the situation in group 4 depicted in Figure 6. It was speculated [24], that e.g. in case of the bromides [23] the oxybromide of dubnium has been formed rather than the pure pentabromide. However, this would mean that under identical gas chemical conditions Nb and Ta form pure bromides while Db does not. If true, Db has a much higher tendency to react with oxygen compared to the lighter homologues Nb and Ta. So, presently it remains open whether the ΔH_a^0 values for Db shown in Figure 12 represent the behavior of the pure halides or of the oxyhalides.

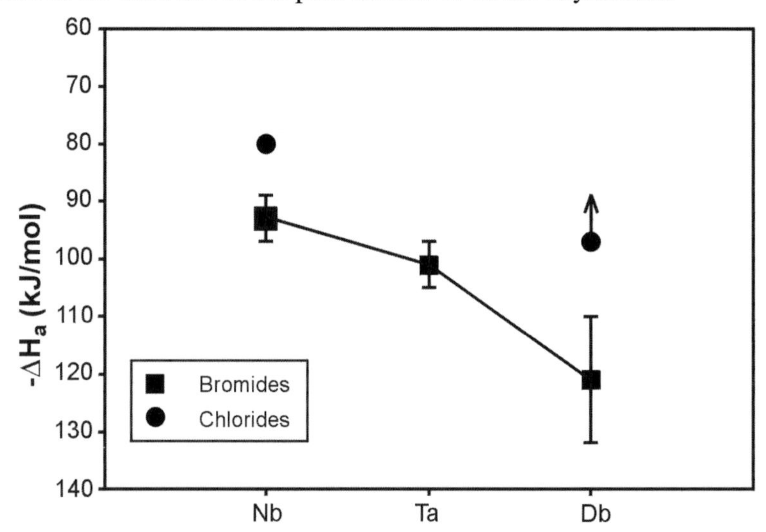

Fig. 12. Adsorption enthalpies of chlorides and bromides of Nb, Ta and Db on quartz surfaces, deduced from OLGA experiments. Data from Ref. [24].

4. Seaborgium (Sg, Element 106)

For over 20 years, ^{263}Sg with a half-life of 0.9 s was the longest-lived known Sg isotope. In addition to the minute production rates, this short half-life effectively prevented a chemical identification of Sg. In 1992 S. Timokhin et al. from Dubna studied the chemical identification of Sg as a volatile oxychloride making use of an on-line thermochromatography method [25]. This claim was substantiated by ancillary experiments [26, 27] and further studies of the behavior of homologues elements Mo and W [28]. Shortly thereafter, an international collaboration of chemists conducted on-line isothermal chromatography experiments with Sg oxychlorides [29]. The presence of Sg after chemical isolation in the gas-phase was established by directly identifying the nuclides ^{265}Sg and ^{266}Sg via the observation of their characteristic, genetically linked, nuclear decay chains [29]. Also, a first thermochemical property of a Sg compound, namely the adsorption enthalpy of Sg oxychloride on the chromatographic surface was measured in these experiments [30]. Recently, Sg was also characterized as volatile oxide hydroxide in on-line isothermal chromatography experiments [31].

4.1 VOLATILE COMPOUNDS OF GROUP-6 ELEMENTS

Seaborgium is expected to be a member of group 6 of the Periodic Table and thus homologues to Cr, Mo, and W. In the elemental state all group-6 elements are extremely refractory. The melting and boiling points are strongly increasing down the group (W has the highest melting point of all metals). While both Mo and W are chemically very similar, there is not much similarity with Cr. Both Mo and W have a wide variety of oxidation states and their chemistry is among the most complex of the transition elements. There exist a number of volatile inorganic Mo and W compounds that are suitable for gas chromatographic investigations. Mo and W form volatile halides, oxyhalides, oxide hydroxides, and also carbonyls.

4.1.1 *Halides and oxyhalides*

Among all hexahalides only the compounds MF_6 (M = Mo, W), WCl_6 and WBr_6 are known. $MoCl_6$ is not stable and exists probably only in a chlorine atmosphere in the gas phase. WCl_6 can be volatilized as a monomeric vapor while WBr_6 is decomposing to WBr_5 on moderate heating. Of the pentahalides the pentafluorides and the pentachlorides are known, W also forms the pentabromide. While MoF_5 and WF_5 have the typical tetrameric structure of the pentafluorides, $MoCl_5$ and WCl_5 form dimeric species in the solid. $MoCl_5$ is monomeric in the gas phase.

In contrast to the pure halides, the oxyhalides of group-6 elements are more stable and show a similarly high volatility. For the 6+ oxidation state the two stoichiometric types MOX_4 and MO_2X_2 (M = Mo, W; X = F, Cl) exist. The Mo compounds are less stable than those of W. Of the oxyfluorides $MoOF_4$, MoO_2F_2, and WOF_4 are known, whereas the existence of WO_2F_2 is doubtful. Of the oxychlorides all four varieties exist, however, $MoOCl_4$ decomposes to $MoOCl_3$ already at room temperature. WO_2Cl_2 disproportionates at temperatures above 200°C to form WO_3 and $WOCl_4$. However, there is no indication that single molecules of WO_2Cl_2 are unstable even at elevated temperatures.

In the 5+ oxidation state four principal compounds of the type MOX_3 (M=Mo, W; X = Cl, Br) are known. The vapor pressure of Mo and W chlorides and oxychlorides over their respective solids as a function of temperature is shown in Figure 13. The vapor pressure of the pure chlorides $MoCl_5$, WCl_5, and WCl_6 all are very similar. The volatility of MoO_2Cl_2 is higher than that of $MoOCl_4$, whereas the situation is reversed for W, where $WOCl_4$ is more volatile than WO_2Cl_2. According to tabulated enthalpies of sublimation, $MoOCl_4$ should be more volatile than MoO_2Cl_2 [32]. This change in the sequence of vapor pressures can be explained by the solid phase entropies.

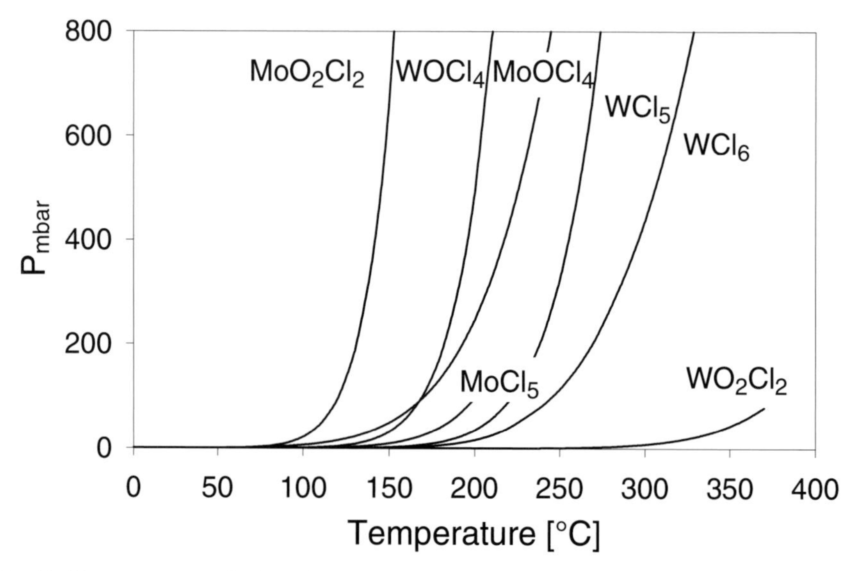

Fig. 13. Vapor pressure of Mo and W chlorides and oxychlorides over their respective solids as a function of temperature. $MoCl_5$ is melting at 197°C whereas WO_2Cl_2 disproportionates at temperatures above 200°C. Data from [32].

4.1.2 *Oxides and oxide hydroxides*

By analogy to Mo and W the oxides and oxide hydroxides of Sg are expected to be moderately volatile, whereas the heavy actinides and the transactinides Rf and Db do not form volatile oxides and oxide hydroxides. For this reason this class of compounds should be very selective with regard to a gas chromatographic isolation of Sg from the plethora of by-products of the nuclear formation reaction. Mo and W form many stable oxides, but in excess of oxygen the trioxides MO_3 (M = Mo, W) should be the main component. Macroscopic amounts of MoO_3 and WO_3 sublimate preferentially as polymers of the type $(MO_3)_n$. However, carrier-free amounts can be volatilized in dry oxygen only as monomers. In moist oxygen the more volatile oxide hydroxides $MO_2(OH)_2$ (M=Mo, W) can be formed. Extensive studies using thermochromatography and on-line isothermal chromatography [33-35] in dry and moist oxygen have revealed that the transport of Mo and W in moist oxygen is not governed by simple reversible adsorption reactions of $MO_2(OH)_2$, but by a dissociative adsorption according to the reaction

$$MO_2(OH)_2 \leftrightarrows MO_{3(ads)} + H_2O_{(g)} \qquad M = Mo, W$$

4.1.3 *Carbonyls*

A characteristic feature of d-group elements is their ability to form complexes with π-acceptor type ligands such as CO. All group-6 elements Cr, Mo, and W form very volatile and stable hexacarbonyls and constitute the only complete family of carbonyls. However, direct production of carbonyls from the elements and CO is only — if at all — accomplished at high pressures and temperatures.

4.2 GAS CHEMICAL STUDIES WITH SEABORGIUM

4.2.1 *Thermochromatography of oxychlorides*

Early on, separation procedures to chemically isolate Sg concentrated on inorganic gas chromatography of chlorides and/or oxychlorides [36]. In a number of studies the gas chromatographic behavior of halide and oxyhalide species of Mo and W were investigated with respect to a physico-chemical characterization of Sg [37-45].

In experiments by the Dubna group [25-28], the reaction $^{249}Cf(^{18}O, 4n)$ was employed to produce 0.9 s ^{263}Sg. A very similar set-up as in experiments to chemically identify Rf and Db was used. Reaction products were thermalized behind the target in a rapidly flowing stream of Ar gas and flushed to the adjoining thermochromatography column. Volatile oxychlorides were synthesized by adding air saturated with $SOCl_2$ as a reactive agent. The formed oxychloride species migrated downstream the fused silica chromatography column, to which a longitudinal, negative temperature gradient was applied, and finally deposited according to their volatility. In contrast to earlier experiments, no mica plates were inserted, but the fused silica column itself served as SF track detector. The deposition of Sg was registered after completion of the experiment by searching for latent SF tracks left by the SF decay of ^{263}Sg. Indeed, in several experiments a number of SF tracks were found in the column in the temperature region 150 - 250 ºC. They were attributed to the decay of Sg nuclides. Therefore, like its lighter homologues Mo and W, Sg must form volatile oxychloride compounds. The SF tracks were only found, when the quartz wool plug, which was inserted as a filter for aerosols, was absent. This was attributed to the increased surface and thus a much longer retention time. In Figure 14 the location of those 41 registered SF events are shown that were observed in the course of three experiments corresponding to a total beam dose of 6.1×10^{17} ^{18}O beam particles. The dotted histogram shows the data corrected for the relative detection efficiency due to the annealing of fission tracks at elevated temperatures. The solid lines denoted with "[106]" and with ^{176}W show the deposition peak for 2.5 h ^{176}W and the expected shape of the "[106]" deposition peak fitting the SF data. Based on the results of ancillary experiments with short-lived W nuclides, it was concluded that in a first, fast step volatile MO_2Cl_2 (M = W, Sg) molecules are formed and in a second, slower step the deposited MO_2Cl_2 is converted to more volatile $MOCl_4$. Therefore the Sg deposition peak was attributed to the compound SgO_2Cl_2, whereas the ^{176}W deposition peak was attributed to $WOCl_4$. Due to the occurrence of two different species as well as due to the large differences in half-life no information about the relative volatility of MO_2Cl_2 (M = Mo, W, Sg) or $MOCl_4$ (M = Mo, W, Sg) within group 6 was obtained.

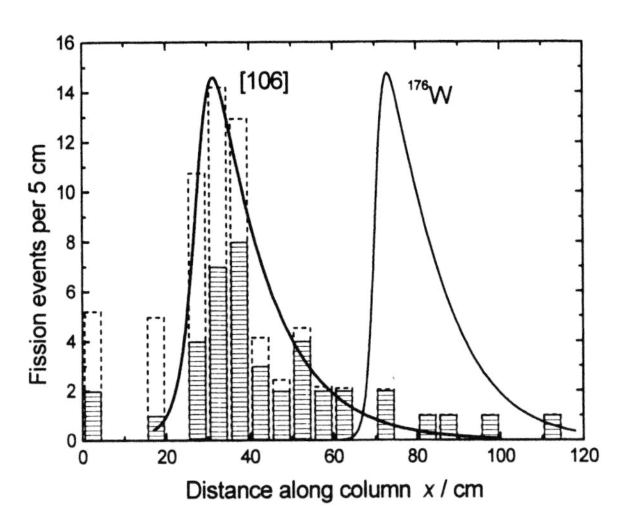

Fig. 14. Measured distribution of spontaneous fission events attributed to the decay of an isotope of element 106 (Sg). The dotted histogram shows the data corrected for the relative detection efficiency due to annealing of fission tracks. The thick solid curves show the smoothed corrected thermochromatograms for Sg, denoted "[106]", and for ^{176}W. Figure reproduced from [28] with the permission of Oldenbourg Verlag.

4.2.2 *Isothermal chromatography of oxychlorides*

In 1995 and 1996 an international collaboration of radiochemists conducted on-line isothermal chromatography experiments with Sg oxychlorides using the OLGA technique, see Chapter 4, Section 4.3, at the Gesellschaft für Schwerionenforschung (GSI) in Darmstadt [29, 30]. In this work the longer-lived Sg-isotopes ^{265}Sg and ^{266}Sg were synthesized in the reaction $^{248}Cm(^{22}Ne; 4,5n)$. Nuclear reaction products, recoiling from the target, were stopped in He gas loaded with carbon aerosols, and — adsorbed to their surface — continuously transported through a thin capillary to the OLGA set-up. The aerosols carrying the reaction products were collected on quartz wool inside the reaction oven kept at 1000°C. Reactive gases — Cl_2 saturated with $SOCl_2$ and traces of O_2 — were introduced in order to form volatile oxychlorides (thermodynamic calculations [44] indicate that Mo and W most probably form the dioxide dichloride MO_2Cl_2, M = Mo, W). Simultaneously, the carbon aerosols were converted to CO_2. The chromatographic separation takes place downstream in the adjoining isothermal section of the column. At temperatures of 300°C and above group-6 oxychloride molecules travel through the column essentially without delay. In order to increase the sensitivity of the experiment the mother-daughter recoil counting modus was implemented at the rotating wheel system ROMA, see Chapter 4, Section 4.3. In a first experiment conducted at isothermal temperatures of the chromatography column of 300°C and of 400°C the nuclide ^{265}Sg was unambiguously identified after chemical isolation by the observation of its α-decay chains [46].

In a second experiment at 350°C isothermal temperature, the results of the first experiment were confirmed by observing further ^{265}Sg α-decay chains [46]. Without changing any of the other experimental parameters, the isothermal temperature was then lowered to 250°C and the yield of ^{265}Sg was measured with a comparable sensitivity as at higher isothermal temperatures. In order to assure that the experimental set-up performed as expected, the nuclide ^{168}W was simultaneously produced from a small ^{152}Gd admixture to the ^{248}Cm target material and its yield was monitored. In Figure 15, the relative yields measured for oxychlorides of short-lived Mo, W, and Sg nuclides are shown as a function of isothermal temperature (the Sg data points measured at 300°C, 350°C and 400°C are summarized in one data point). The yield curve for ^{168}W was measured with the same chromatography column and under the same experimental conditions as they were then used for the isolation of Sg, whereas the yield curve for ^{104}Mo was determined in an earlier measurement. The solid lines show the results of a Monte Carlo simulation procedure where the migration of a molecule through the chromatography column has been modeled [47].

From the measured Sg data a first thermochemical property of a Sg compound could be deduced, namely $-\Delta H_a^0(SgO_2Cl_2)= 98^{+2}_{-5}$ kJ/mol (68% error interval). For WO_2Cl_2 $-\Delta H_a^0(WO_2Cl_2)=96\pm1$ kJ/mol was deduced, whereas for MoO_2Cl_2 $-\Delta H_a^0(MoO_2Cl_2)=90\pm3$ kJ/mol resulted. The sequence in volatility of MO_2Cl_2 (M = Mo, W, Sg) on the stationary phase is $MoO_2Cl_2 > WO_2Cl_2 \approx SgO_2Cl_2$. The probability that SgO_2Cl_2 is equally volatile or even more volatile than MoO_2Cl_2 was estimated to be less than 15%.

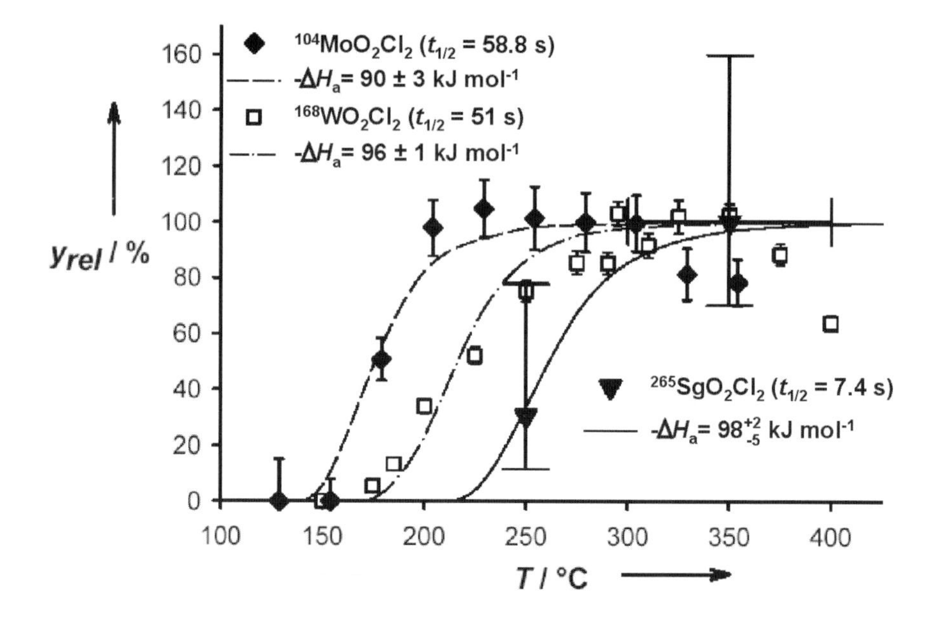

Fig. 15. Relative yield of MO_2Cl_2 (M = Mo, W, Sg) as a function of isothermal temperature in the chromatography column [30].

The experimentally determined ΔH_a^0-values, measured with trace amounts (at zero surface coverage), were directly correlated with their macroscopic sublimation enthalpies (ΔH_s^0), see Chapter 6, Part II, Section 2.3. It was therefore possible to directly estimate $\Delta H_s^0(SgO_2Cl_2)=127^{+10}_{-21}$ kJ/mol from only a few investigated molecules. $\Delta H_s^0(SgO_2Cl_2)$ is a very important quantity in order to estimate e.g. ΔH_s^0(Sg). Seaborgium is expected to have an equally or even higher ΔH_s^0 than W, the least volatile element in the Periodic Table.

4.2.3 *Isothermal chromatography of Sg oxides/oxide hydroxides*

The transport of group-6 elements Mo, W, and presumably also Sg, in moist oxygen containing gases occurs via a dissociative adsorption reaction and not via a simple reversible adsorption. Retention times for these dissociative processes in an isothermal chromatography column are generally longer, even at very high temperatures. With the High Temperature On-Line Gas Chromatography Apparatus (HITGAS) [48] retention times of about 8 to 9 s were determined from measurements with short-lived Mo and W nuclides at isothermal temperatures above 1270 K. Therefore, in experiments to characterize Sg as an oxide hydroxide, the longer-lived isotope ^{266}Sg ($T_{1/2}$≈21 s), produced in the reaction ^{248}Cm(^{22}Ne, 4n), was used despite its lower production cross section of only about 60 pb, compared to the shorter-lived ^{265}Sg ($T_{1/2}$=7.4 s) produced in the 5n evaporation channel with a maximum cross section of about 240 pb [46,49]. By condensing the separated volatile species directly on metal foils mounted on the circumference of the rotating wheel of the ROMA detection system, see Chapter 4, Section 4.3, the time-consuming reclustering step could be avoided. However this reduced the detection efficiency, since, due to the thickness of the metal foils, final samples could be assayed only in a 2π geometry. Furthermore, contaminants like various Po isotopes cannot be removed in the oxide/hydroxide chemical system, which makes detection of genetically linked α-decay chains difficult. Fortunately, ^{266}Sg decays by α-particle emission to the relatively short-lived ^{262}Rf ($T_{1/2}$=2.1 s) which decays by spontaneous fission (100%).

In an experiment conducted at GSI a ^{248}Cm target was bombarded with 119 MeV ^{22}Ne ions. Reaction products recoiling from the target were stopped in He gas loaded with MoO_3 aerosol particles and swept to the HITGAS set-up. At the entrance to the chromatography column moist O_2 was added to the gas-jet. The temperature of the quartz chromatography column was 1325 K in the reaction– and 1300 K in the isothermal zone. Loosely packed quartz wool in the reaction zone served as filter for aerosol particles. A total beam dose of $6.3 \cdot 10^{17}$ ^{22}Ne ions was accumulated. The search for genetically linked decay chains $^{266}Sg \xrightarrow{\alpha} {}^{262}Rf \xrightarrow{sf}$ revealed two candidate events. The probability that both of these events were entirely random was only 2%. Therefore, as expected, Sg appeared to be volatile under the conditions of the experiment, presumably as Sg oxide hydroxide. In the O_2-$H_2O_{(g)}$/$SiO_{2(s)}$-system Sg showed typical group-6 element properties. Under the given conditions this coincides also with a U(VI)-like behavior. U is also known to form a volatile oxide hydroxide.

In Figure 16, the relative yields of Mo and W oxide hydroxides in open quartz columns using humid O_2 as reactive carrier gas component are shown as a function of isothermal temperature. The solid lines are the result of a Monte Carlo model based on a microscopic description of the dissociative adsorption process [34] with $\Delta H^0_{diss.ads}$ ($MoO_2(OH)_2$) = -54 kJ/mol and $\Delta H^0_{diss.ads}$ ($WO_2(OH)_2$) = -56 kJ/mol. The dashed line represents a hypothetical yield curve assuming that group-6 oxide hydroxides are transported by simple reversible adsorption with ΔH_a^0 = -220 kJ/mol [31].

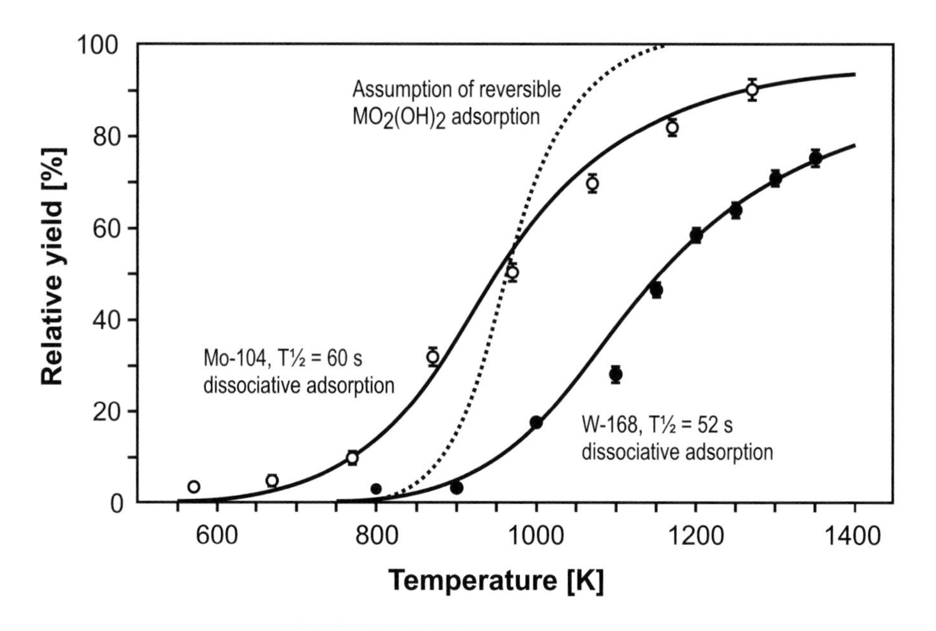

Fig. 16. Relative yields in isothermal gas chromatography of ^{104}Mo (O) and ^{168}W (●) oxide hydroxides in quartz columns using humid oxygen as reactive carrier gas component. Sg was observed at an isothermal temperature of 1300 K.

In order to answer the question about the sequence of volatility of oxide hydroxides within group 6, further experiments have to be conducted at lower isothermal temperatures.

5. Bohrium (Bh, Element 107)

The fourth transactinide element bohrium is expected to be a homologue of Mn, Tc, and Re and thus to belong to group 7 of the Periodic Table. Two early attempts to chemically identify Bh as volatile oxides or oxide hydroxides failed [50,51]. For the synthesis of Bh nuclides the reactions $^{249}Bk(^{22}Ne; 4,5n)^{267,266}Bh$ and $^{254}Es(^{16}O; 4,5n)^{265,266}Bh$ were employed. The decay properties of the nuclides $^{265\text{-}267}Bh$ were entirely unknown at the time. With the recent identification of the nuclides ^{266}Bh ($T_{1/2} \approx 1$ s) and ^{267}Bh ($T_{1/2} = 17^{+14}_{-6}$ s) [52] in bombardments of ^{249}Bk with ^{22}Ne ions and the recognition that the rapid formation of volatile oxide hydroxides is apparently hindered [53], R. Eichler et al. paved the way to the first successful chemical identification of Bh as oxychloride compound [54]. However, due to the very low formation cross sections of only about 70 pb for ^{267}Bh (produced in the reaction $^{249}Bk(^{22}Ne, 4n)$) [52], any experiment aiming at a chemical identification of Bh was predestined to be a "tour de force". Nevertheless, in a one month long experiment conducted at the Paul Scherrer Institute (PSI), Switzerland, an international collaboration of radiochemists observed a total of 6 α-decay chains originating from ^{267}Bh after chemical isolation and established the sequence in volatility $TcO_3Cl > ReO_3Cl > BhO_3Cl$ [55].

5.1 VOLATILE COMPOUNDS OF GROUP-7 ELEMENTS

In contrast to elements in groups 4 and 5, but similar to group 6, the 7 valence electrons of group 7 elements allow for a large number of stable oxidation states and thus a wide variety of inorganic compounds. An increased stabilization is observed for the half-filled d-shells, which is especially evident for the 3d shell of Mn, which is considerably more volatile than its neighbors Cr and Fe in the same period. However, Mn behaves chemically markedly different from its homologues Tc and Re. Compounds of Mn are chemically most stable in the oxidation state +2 whereas compounds in the oxidation states +4 and +7 are strong oxidizing agents. Compounds of Tc and Re in high oxidation states are much more stable towards reduction and the oxidation state +2 is of minor importance. Due to the lanthanide contraction, atomic and ionic radii of Tc and Re are very similar and thus these elements are chemically very much alike. Some typical mononuclear compounds of group-7 elements are listed in Table 1. Of all the compounds listed in Table 1 the oxides, oxide hydroxides, and the oxychlorides turned out to be the most promising candidates for a chemical separation and identification of bohrium.

Table 1. Typical mononuclear compounds of group 7 elements

Compound	Mn	Tc	Re
Oxides	MnO, MnO_2	TcO_2, TcO_3	ReO_2, ReO_3
Hydoxides	$MnOH$, $Mn(OH)_2$		
Oxide hydroxides		$HTcO_4$	$HReO_4$
Sulfides	MnS, MnS_2	TcS_3	ReS_3
Halides X = F, Cl, Br, I	MnX_2, MnX_3, MnX_4	TcX_3, TcX_4, TcX_5, TcX_6	ReX_3, ReX_4, ReX_5, ReX_6
Oxyhalides X = F, Cl, Br, I		$TcOX_3$, $TcOX_4$, TcO_3X	$ReOX_3$, $ReOX_4$, ReO_3X

5.1.1 *Oxides and oxide hydroxides*

Oxides and oxide hydroxides of Tc and Re are typically formed in an O_2/H_2O containing gas phase. They were extensively studied, mostly using the method of thermochromatography [56-67]. The technique has also been applied to develop Tc and Re generator systems for nuclear medical applications [68,69]. In their works, M. Schädel et al. [70] and R. Eichler et al. [53] studied the oxide and the oxide hydroxide chemistry of trace amounts of Re in an O_2/H_2O-containing system with respect to its suitability for a first gas chemical identification of Bh. They investigated the behavior of long-lived Re nuclides in thermochromatographic systems as well as the one of short-lived Re nuclides in on-line isothermal chromatography. The results of these studies [53] are summarized in Figure 17 and can be described as follows:

Thermochromatography of oxides and oxide hydroxides. In thermochromatography experiments three different processes can be distinguished, reflected in the deposition peaks B, C, and D in Figure 17, depending on the pretreatment of the column surface and the oxidation potential of the carrier gas. These are:

1. The rapid formation of the perrhenic acid ($HReO_4$) and a gas chromatographic transport of the rather volatile $HReO_4$ governed by mobile adsorption processes to relatively low deposition temperatures of less than 100 °C (deposition peak D in Figure 17). This behavior is observed if the employed quartz columns are pretreated in excess of 1000 °C with H_2 and with O_2/H_2O or H_2O_2 as reactive component of the carrier gas.

2. The rapid formation of the rhenium trioxide (ReO_3) and a gas chromatographic transport of ReO_3 governed by mobile adsorption processes to deposition temperatures of about 500 °C (deposition peak B in Figure 17). This behavior is observed if the employed quartz columns are pretreated in excess of 1000 °C with O_2 and with O_2, O_2/H_2O, or H_2O_2 as reactive component of the carrier gas.

3. The formation of the rhenium trioxide (ReO_3) and a consecutive transport of ReO_3 by mobile adsorption <u>and</u> a superimposed transport reaction. The latter leads to a reversible formation of more volatile $HReO_4$ via a surface catalyzed reaction and thus, through the intermediate $HReO_4$, to a transport of ReO_3 to lower adsorption temperatures (deposition peak C in Figure 17). This behavior is observed if the employed quartz columns are pretreated in excess of 1000 °C with O_2 and with O_2 as reactive component of the carrier gas.

Also, a small fraction of Re remains as non-volatile compound at the starting position (peak A in Figure 17). Due to their high volatility the oxide hydroxides appear to be especially interesting for an on-line gas chromatographic study of Bh. A high volatility of the investigated compound gives rise to high separation factors from less volatile by-products, such as heavy actinides, but also from Po, Pb and Bi nuclides. Due their very similar α-decay energies, they usually hamper a sensitive detection of trans-actinides.

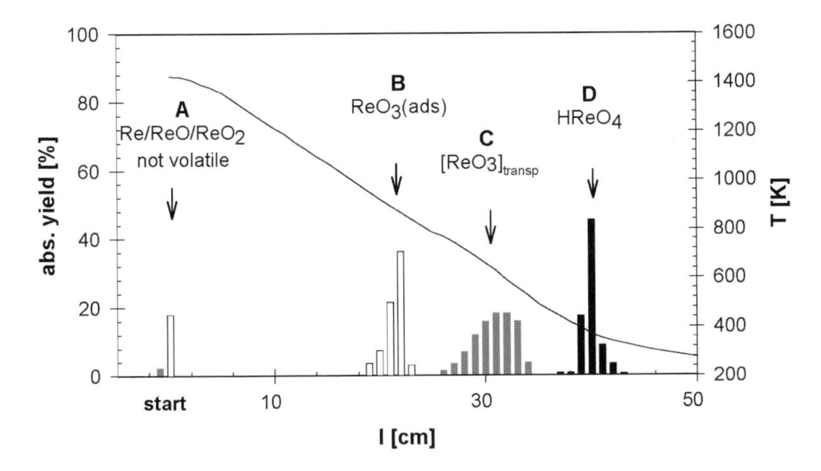

Fig. 17. Merged thermochromatograms of Re in the system He, O_2, H_2O. Figure from [53] with the permission of Oldenbourg Verlag.

Isothermal chromatography of oxides and oxide hydroxides. Based on thermochromatographic studies on-line methods for the gas chromatographic isolation of volatile group-7 oxides or oxide hydroxides were investigated using the OLGA technique [53], see Chapter 4, Section 4.3. The nuclide ^{169m}Re ($T_{1/2}$=16 s) with its α-decay branch (E_α=5.0 MeV) is ideally suited to model the behavior of its heavier group-7 homologue Bh and was produced in the fusion reaction ^{156}Dy(^{19}F, 6n). Transfer reaction products such as $^{152\text{-}155}$Er and $^{151\text{-}154}$Ho served as model elements for the behavior of heavy actinides.

Reaction products were transported attached to carbon aerosol particles in He from the target site to the OLGA set-up. In order to obtain the volatile $HReO_4$ 100 ml/min O_2 (containing 500 ppm O_3) saturated with H_2O_2 at room temperature were added as reactive components. The carbon aerosols were stopped on a quartz wool plug in the reaction oven at 1373 K, where the reaction products were oxidized and the aerosols converted to CO_2. The yield of volatile Re oxides was measured as a function of the temperature in the adjoining isothermal section of the column. The resulting temperature vs. yield curve is shown in Figure 18. Unfortunately, high yields of a volatile Re compound were only observed at temperatures of 900 K and above. The deduced $\Delta H_a^0(ReO_3)$ = -176±10 kJ/mol was in good agreement with $\Delta H_a^0(ReO_3)$ = –190±10 kJ/mol (evaluated from peak B shown in Figure 17) measured in thermochromatography experiments and this value identified the volatile species as ReO_3 [53]. The much more volatile $HReO_4$ was not observed at the given experimental conditions. A kinetic hindrance of the formation of $HReO_4$ was excluded, since also much longer-lived Re nuclides were not observed after chemical separation.

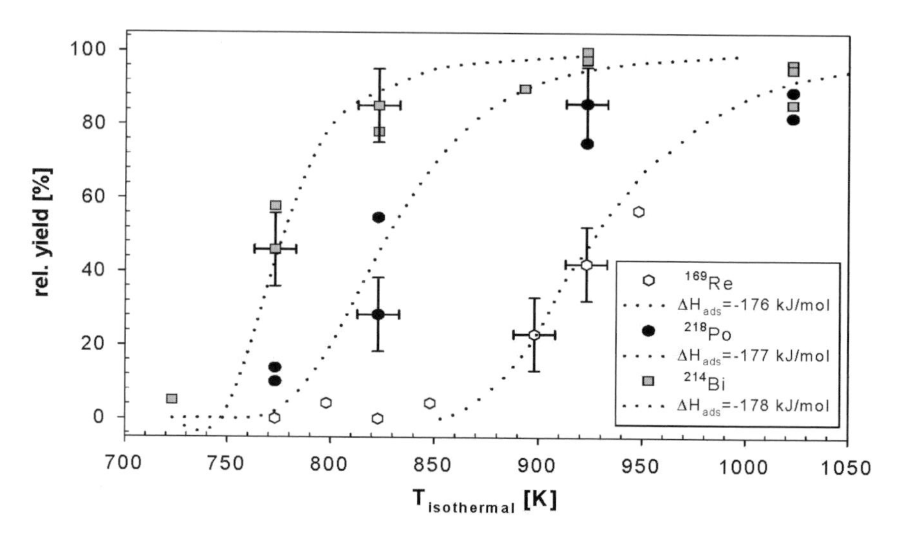

Fig. 18. Temperature vs. yield curve from isothermal chromatography of $^{169m}ReO_3$, ^{218}Po ($T_{1/2}$=3.05 m, presumably as $^{218}PoO_2$), and ^{214}Bi ($T_{1/2}$=19.9 m, presumably as BiOOH). Figure from [53] with the permission of Oldenbourg Verlag.

Po and Bi as possibly interfering contaminants were also investigated under the same experimental conditions and, as shown in Figure 18, were found to be similarly volatile as ReO_3. In conclusion, an on-line isolation of very volatile group-7 oxide hydroxides was not accomplished. The isolation of less volatile trioxides appeared not to be promising due to the interference of Po and Bi by-products hampering the unambiguous identification of Bh nuclides after chemical isolation. Nevertheless, the oxide system provided an excellent separation from lanthanides [53] and actinides [50, 51], separation factors of $\geq 10^3$ were deduced.

5.1.2 *Chlorides and oxychlorides*

Since the oxide and the oxide hydroxide systems are not well suited to rapidly isolate single atoms of group-7 elements [53], chlorides and oxychlorides were investigated as potential candidate compounds for an on-line gas chemical isolation of Bh [54]. This approach had already been successful in studies of volatile Db and Sg oxychlorides, see Sections 3 and 4 of this chapter. However, a number of different chloride/oxychloride species exist within group 7 and thus the chemical speciation of the formed compounds appears to be complicated. For Re the pure chlorides $ReCl_3$, $ReCl_4$, $ReCl_5$, and $ReCl_6$ are known, as well as the oxychlorides $ReOCl_3$, $ReOCl_4$, and ReO_3Cl.

Thermochromatography of chlorides and of oxychlorides. Only few thermochromatographic studies of chloride and oxychloride compounds of group-7 elements Tc and Re were known [71-73]. Therefore, R. Eichler et al. [54] reinvestigated the thermochromatographic behavior of Tc and Re in quartz columns in the gas chemical system $He_{(g)}/O_{2(g)}/HCl_{(g)}$ using trace amounts of the nuclides ^{101}Tc and ^{104}Tc obtained from a thin ^{252}Cf fission source and $^{183,184}Re$ produced from proton irradiations of $^{nat.}W$. Due to the large variety of chloride and oxychloride compounds of group-7 elements Tc and Re many different deposition zones were expected in thermochromatography experiments. Surprisingly, only one single deposition zone for Tc and Re was observed at rather low deposition temperatures, see Figure 19, indicating the formation of a very volatile compound. Variation of the carrier gas mixture He (Vol.% 0-60), O_2 (Vol.% 0-80%), and HCl (Vol.% 10-100) did not yield any other volatile compound. The deposition zone of Tc was observed at a lower temperature as the one for Re and coincided with the condensation zone of H_2O which was formed in the reaction of HCl and O_2 at 1400 K in the reaction oven. For this reason, only an upper limit of the adsorption enthalpy of the Tc compound was established. By using an empirical correlation of the measured adsorption enthalpies of trace amounts of a number of chloride and oxychloride species with their macroscopic boiling point identified the formed volatile species quite clearly as the trioxychlorides (MO_3Cl, M=Tc, Re). The observed properties of group-7 oxychlorides in thermochromatography experiments appeared quite promising for a first chemical identification of Bh. Therefore, the gas chemical system $He_{(g)}/O_{2(g)}/HCl_{(g)}$ was further investigated in on-line isothermal chromatography using short-lived Tc and Re nuclides.

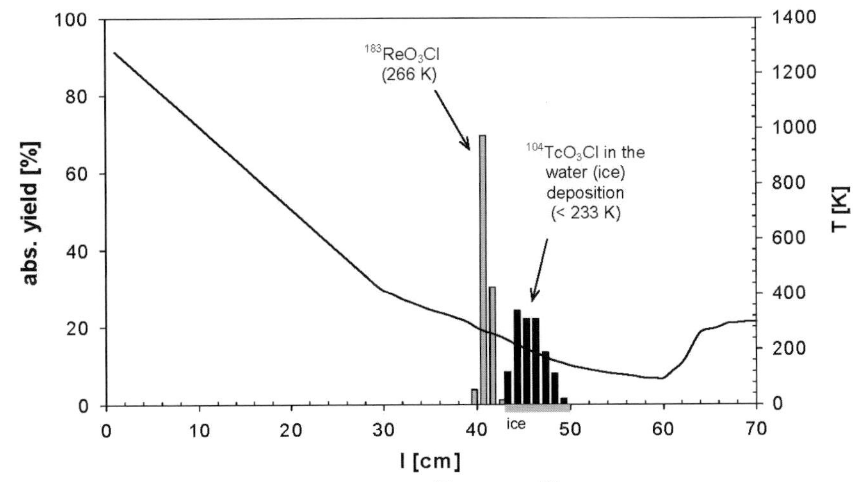

Fig. 19. Merged thermochromatograms of ^{104}Tc and ^{183}Re in the gas chemical system $He_{(g)}/O_{2(g)}/HCl_{(g)}/SiO_{2(s)}$. Figure reproduced from [54] with the permission of Oldenbourg Verlag.

Isothermal chromatography of chlorides and oxychloride. As in studies of group-7 oxides and oxide hydroxides short-lived ^{169m}Re were used. Short-lived $^{106-108}$Tc, $^{98-101}$Nb, and $^{99-102}$Zr were obtained from a ^{252}Cf fission source. As reactive gases HCl and O_2 were added to the He/C-aerosol gas-jet shortly before the reaction oven. Nuclear reaction products were oxidized and chlorinated together with the carbon aerosols. At the column exit the separated volatile molecules were adsorbed on the surface of CsCl aerosol particles of a second gas-jet ("reclustering") and were rapidly transported to a detection system. Yields of 80% were observed compared to the activity of ^{169m}Re entering the OLGA III set-up. However, this approach did not work for the more volatile TcO_3Cl. Obviously, the adsorption enthalpy of TcO_3Cl on a CsCl surface was too low to allow an efficient reclustering. Using aerosols with a reducing surface, such as $FeCl_2$, significantly improved the yield. This interesting property allowed a distinction between a "Tc-like" and a "Re-like" behavior in a first experiment with Bh; see below. Ancillary experiments with the ^{218}Po and ^{218}Bi were conducted to investigate the separation of volatile Tc and Re trioxychlorides from Po and Bi contaminants. As shown in Figure 20, the separation of Tc and Re trioxychlorides from the less volatile BiOCl or $PoOCl_2$ is excellent. Separation factors from lanthanides, serving as models for heavy actinides, were $>10^3$. $^{99-102}$Zr and $^{98-101}$Nb, serving as model elements for the lighter transactinides Rf and Db, were separated at isothermal temperatures up to 470 K with separation factors of $>10^2$. An average overall process time of about 3 s was evaluated for Re. Thus, the gas chemical system $He_{(g)}/O_{2(g)}/HCl_{(g)}/SiO_{2(s)}$ was shown to fulfill all requirements for a first successful chemical identification of Bh [54].

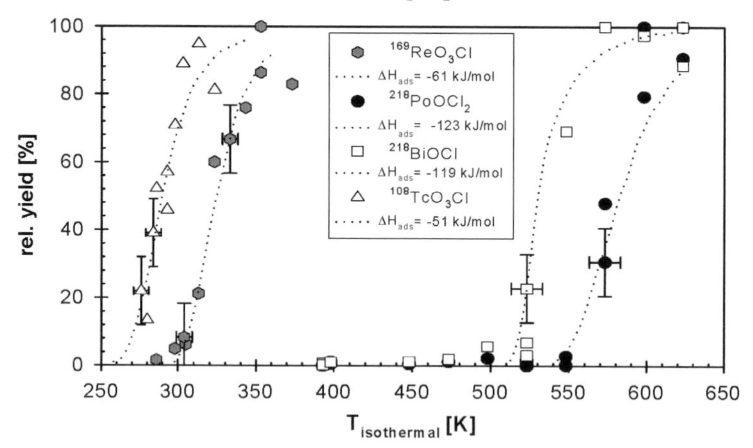

Fig. 20. Yield curves vs. isothermal temperature measured for oxychloride compounds of the nuclides ^{108}Tc (△), ^{169}Re (⬢), ^{218}Po (●), and ^{218}Bi (□) in the chemical system $He_{(g)}/O_{2(g)}/HCl_{(g)}/SiO_{2(s)}$. The dotted lines indicate the results of simulations with the microscopic model of Zvara [47] with the adsorption enthalpies indicated. Figure reproduced from [54] with the permission of Oldenbourg Verlag.

5.2 GAS CHEMICAL STUDIES OF BOHRIUM

5.2.1 *Thermochromatography of oxides and/or oxide hydroxides*

A first attempt to chemically identify element 107 as eka-rhenium was conducted by I. Zvara and co-workers already in 1984 [50]. As in earlier experiments with Rf, Db and Sg, see above Sections 2.2, 3.2 and 4.2.1, respectively, they searched for latent tracks in thermochromatographic columns imprinted by the stopping of a fission fragment from the decay of a spontaneously fissioning isotope of Bh. Moist air with a water vapor pressure of 600 Pa at a flow rate of 0.75 l/min passed behind a 150 $\mu g/cm^2$ thick ^{249}Bk target that was irradiated with ^{22}Ne ions. Eight experiments were conducted with varying conditions concerning the operation of the thermochromatographic column. In the last 3 experiments an optimum purification from actinides was achieved. No SF tracks were observed in a temperature range from 800 °C down 20 °C, while the nuclide ^{177}Re (produced from an admixture of ^{159}Tb to the target material) was adsorbed at around 200 °C. This negative result was interpreted that either the half-lives of the produced Bh nuclides were shorter than 2 s or that the production cross sections were lower than 100 pb [50]. Even though the reached cross section limits are very close to the cross sections measured later by P.A. Wilk et al. [52], the studies by R. Eichler et al. [53] showed that the rapid formation of a volatile oxide hydroxide is hindered.

5.2.2 *On-line gas chromatography of oxides*

Later, in a different attempt M. Schädel et al. [51] bombarded a ^{254}Es target with ^{16}O ions to produce the isotopes ^{266}Bh and ^{265}Bh at the 88-Inch Cyclotron of the Lawrence Berkeley National Laboratory (LBNL). Reaction products recoiling from the target were thermalized in He containing 20% O_2 and, attached to the surface of KCl aerosols, were transported to the on-line chromatography set-up OLGA. In the reaction oven of OLGA, kept at 1050 °C, the KCl aerosols were stopped and were destroyed on a quartz wool plug. The water content of the gas mixture was kept below 100 ppm in order to form only the trioxide species, as determined in test experiments with Re. Volatile oxides, which passed through the second part of the column with a negative temperature gradient ranging from 1050 °C down to 500 °C at the exit of the column, were deposited on thin Ni catcher foils (0.67 mg/cm^2) that were coated with 50 $\mu g/cm^2$ Ta. The catcher foils were mounted on the circumference of a rotating wheel and stepped periodically between pairs of surface barrier detectors to register α decays and SF decays. A spectrum of all α-events from all runs (93 MeV and 96 MeV bombarding energy) and all detectors revealed that a small portion of heavy actinide isotopes had passed through the gas chromatographic column, but that the decontamination factor was better than 10^3. Even though a couple of α-decays were registered with energies between 8.4 and 9.2 MeV none of these could be conclusively attributed to the decay of a Bh isotope. No genetically linked decay chains were observed. Assuming a transport time of about 1 s and a half-life of the produced Bh isotopes of more than 2 s, cross section limits of about 3 to 10 nb were reached (95% confidence level). These upper limits were larger than the calculated production cross sections by more than one order of magnitude. As outlined by the authors, the experiment clearly failed to chemically identify Bh.

5.2.3 *Isothermal chromatography of oxychlorides*

In an experiment at the PSI Philips cyclotron, the first successful chemical isolation and identification of Bh was accomplished [55]. A target of 670 $\mu g/cm^2$ ^{249}Bk covered with a 100 $\mu g/cm^2$ layer of ^{159}Tb was prepared at LBNL on a thin 2.77 mg/cm^2 Be foil. The target was irradiated for about four weeks with typically 1.6×10^{12} particles of ^{22}Ne per second at a beam energy in the middle of the target of 119±1 MeV, producing 17-s ^{267}Bh in the reaction $^{249}Bk(^{22}Ne, 4n)$. ^{176}Re was simultaneously produced in the reaction $^{159}Tb(^{22}Ne, 5n)$ and served as a yield monitor for the chemical separation process. Nuclear reaction products recoiling from the target were attached to carbon aerosol clusters and were transported with the carrier gas flow through a capillary to the modified OLGA III set-up. As reactive gases a mixture of HCl and O_2 was added. After chemical separation final products were attached to CsCl aerosols and were transported to the rotating wheel detection system ROMA where α-particle and SF decays were registered event by event in almost 4π geometry. Measurements were performed at isothermal temperatures of 180°C, 150°C, and 75°C. At each isothermal temperature a beam integral of 1×10^{18} ^{22}Ne particles was accumulated.

Throughout the experiment close to 180.000 samples were measured. A total of 6 genetically linked decay chains attributed to the decay of ^{267}Bh were observed; four at an isothermal temperature of 180°C, two at 150°C and none at 75°C. Due to a small contamination with Po and Bi nuclides, and a statistical treatment of this background, 1.3 of the 4 decay chains observed at 180°C had to be attributed to accidental correlations unrelated to the decay of ^{267}Bh. At 150°C this correction was only 0.1 out of 2 observed decay chains. The properties of the observed decay chains are shown in Figure 21.

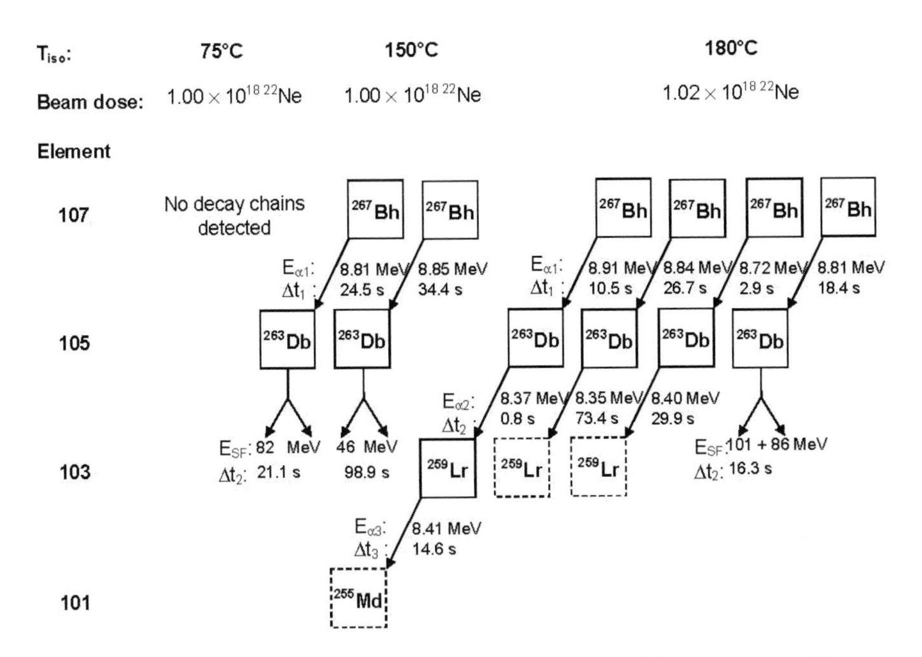

Fig. 21. The six nuclear decay chains attributed to the decay of ^{267}Bh leading to ^{263}Db and ^{259}Lr. Given are the observed decay energies and the lifetimes between end of sample collection (Δt_1) and after the previous α-decay (Δt_2, Δt_3). These decays were observed at 180°C and 150°C, which allowed the unambiguous identification of Bh after chemical separation, presumably as volatile BhO_3Cl. No ^{267}Bh was detected at 75°C isothermal temperature. Figure reproduced from [55].

Interestingly, for $^{169}ReO_3Cl$ a relatively high yield of 80% was observed at 75°C as compared with the yield at 180°C isothermal temperature. This indicates that BhO_3Cl, which was not observed at 75°C, is less volatile than ReO_3Cl. The fact that ^{267}Bh was identified after chemical separation already excludes a "Tc-like" behavior of Bh, since CsCl was used as the recluster aerosol material, which was not suitable to recluster the very volatile TcO_3Cl [54]. The relative yields of the compounds $^{108}TcO_3Cl$, $^{169}ReO_3Cl$, and (most likely) $^{267}BhO_3Cl$ as a function of isothermal temperature are shown in Figure 22. The deduced enthalpies of adsorption on the column surface were $-\Delta H_a^0(TcO_3Cl)$ = 51±3 kJ/mol, $-\Delta H_a^0(ReO_3Cl)$ = 61±3 kJ/mol, and $-\Delta H_a^0(BhO_3Cl) = 75^{+6}_{-9}$ kJ/mol (68% confidence interval). Therefore, the sequence in volatility is $TcO_3Cl > ReO_3Cl > BhO_3Cl$. The probability that BhO_3Cl is equally or more volatile than ReO_3Cl is less than 10%.

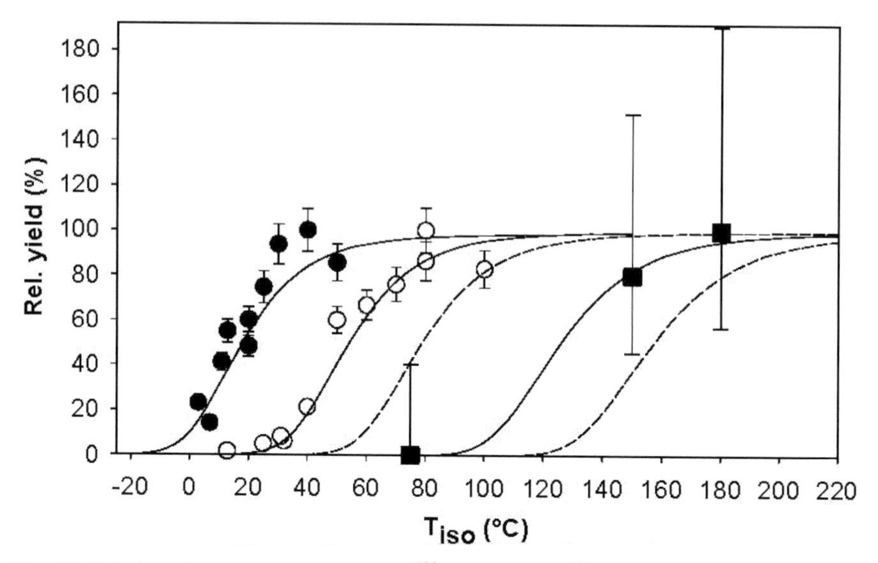

Fig. 22. Relative yields of the compounds $^{108}TcO_3Cl$ (○), $^{169}ReO_3Cl$ (●), and (most likely) $^{267}BhO_3Cl$ (■) as a function of isothermal temperature. The error bars indicate a 68% confidence interval. The solid lines indicate the results of simulations with the microscopic model of I. Zvara [18] with the adsorption enthalpies given in the text. The dashed lines represent the calculated relative yield concerning the 68% confidence interval of the standard adsorption enthalpy of BhO_3Cl from -66 to -81 kJ/mol. Figure reproduced from [55].

This sequence in volatility agrees well with predictions from fully relativistic density-functional calculations for group-7 oxychlorides that have been performed by V. Pershina et al. [74]. The results of these calculations showed that the electronic structure of BhO_3Cl is very similar to that of TcO_3Cl or ReO_3Cl. Increasing dipole moments and electric dipole polarizabilities in the group suggest a decreasing volatility in the sequence $TcO_3Cl > ReO_3Cl > BhO_3Cl$. However, also classical extrapolations down the groups of the Periodic Table making use of empirical correlations of thermochemical properties predict BhO_3Cl to be more stable and less volatile than ReO_3Cl or TcO_3Cl [75]. As in the case of Sg oxychlorides the experimentally determined ΔH_a^0-value can be used to estimate a macroscopic sublimation enthalpy (ΔH_s^0) of BhO_3Cl, see Chapter 6, Part II, Section 2.3, using an empirical linear correlation function. It was therefore possible to directly estimate $\Delta H_s^0(BhO_3Cl) = 89^{+21}_{-18}$ kJ/mol from only a few investigated molecules.

6. Hassium (Hs, Element 108)

The experimental chemical investigation and characterization of the next heavier transactinide element Hassium (Hs, element 108) has, for some years, constituted a daunting task even though from the very beginning the selection of a volatile compound was absolutely clear. Hassium, as a presumed member of group 8 of the Periodic Table and thus a homologue of Fe, Ru, and Os, should form stable and at the same time very volatile HsO_4 molecules, very similar to OsO_4.

The discovery of Hs was reported in 1984 [76] with the identification of the nuclide ^{265}Hs with a half-life of only 1.5 ms, far too short for all of the currently available chemical separator systems. Only in 1996, the much longer-lived isotope ^{269}Hs with a half-life of the order of about 10 s was observed in the α-decay chain of the nuclide ^{277}112 [77]. However, the production cross section of only about 1 pb for the reaction ^{208}Pb(^{70}Zn, 1n)^{277}112 was discouragingly small. A somewhat larger production cross section of about 7 pb could be expected for the direct production of ^{269}Hs in the reaction ^{248}Cm(^{26}Mg, 5n) [78].

Thus, the state of the art techniques, that have successfully been applied to chemically identify Bh had to be improved by at least one order of magnitude! This goal was indeed accomplished by introducing novel techniques for irradiation, separation and detection. The successful Hs chemistry experiment was conducted in the spring of 2001 again in the framework of an international collaboration at the Gesellschaft für Schwerionenforschung, Darmstadt, and proved that Hs behaves like a typical member of group 8 and forms volatile Hs-oxide molecules, very likely HsO_4 [79].

6.1 VOLATILE COMPOUNDS OF GROUP-8 ELEMENTS

Group-8 elements Fe, Ru and Os are known to exist in a large number of oxidation states: Fe is known in all states from -2 through +6, Ru in the states -2 through +8 (with the exception of +1) and Os in all states from -2 through +8 explaining the large variety of compounds. Ru and Os are the elements with the highest maximum valency within their periods and the only elements which can form an 8+ oxidation state (with the exception of Xe, which is known to form tetrahedral XeO_4 [80]). While the chemistry of Ru and Os is quite similar, Fe behaves differently. The reason is the existence of the lanthanide series which is inserted in the sixth period of the Periodic Table. Therefore, investigations of the chemical properties for a future Hs chemistry experiment concentrated on Ru and Os. The most important volatile compounds of Ru and Os are the tetroxides MO_4 (M=Ru,

Os). There also exist a number of volatile Ru and Os halides and oxyhalides. The fluorides and oxyfluorides are of importance, but experimentally difficult to handle. Quite naturally, early considerations [81,82] and experimental developments [83-96] for a first Hs chemistry exclusively concentrated on the tetroxides. This strategy is justified, since classical extrapolations [97] as well as fully relativistic density functional theory calculations on the group-8 tetroxides [98] predict the existence of a volatile and very stable HsO_4.

6.1.1 *Thermochromatography of oxides*

The volatilization and deposition of carrier-free radionuclides of the elements Re, Os, Ir, Mo, Tc, and Ru in a thermochromatography column were studied using air as a carrier gas [83]. The columns were filled with quartz powder (200 μm). Os was completely volatilized and adsorbed at -40 °C. The deduced enthalpy of adsorption on the quartz surface was $-\Delta H_a^0(OsO_4)=50\pm5$ kJ/mol. Ru was deposited at much higher temperatures around 400 °C and identified as RuO_3. Later, in different on-line TC experiments consistently values for $-\Delta H_a^0(OsO_4)$ between 39 and 41 kJ/mol were determined [79, 85, 96].

The transport of Ru oxides in a temperature gradient tube appears to be more complicated. First indications that also Ru can be volatilized in the form of RuO_4 were obtained in [85], but not observed in [86]. The transport of Ru appears to occur by chemical transport reactions where the chemical species change during the chromatographic process. In experiments by Ch. Düllmann et al. [92] using O_2 as carrier gas, two deposition peaks were observed, see Figure 23. The location of the deposition peak at higher temperatures varied over a wide range in different experiments and disappeared completely in experiments lasting more than one hour. This peak was therefore attributed to Ru transported by a transport reaction of the type

$$RuO_{3(ads)} \leftrightarrow RuO_{4(g)} + \frac{1}{2} O_{2(g)}.$$

However, RuO_3 does not exist in macroscopic quantities. The second peak at lower deposition temperatures was attributed to RuO_4 transported by mobile adsorption, with an adsorption enthalpy of $-\Delta H_a^0(RuO_4)=55\pm4$ kJ/mol.

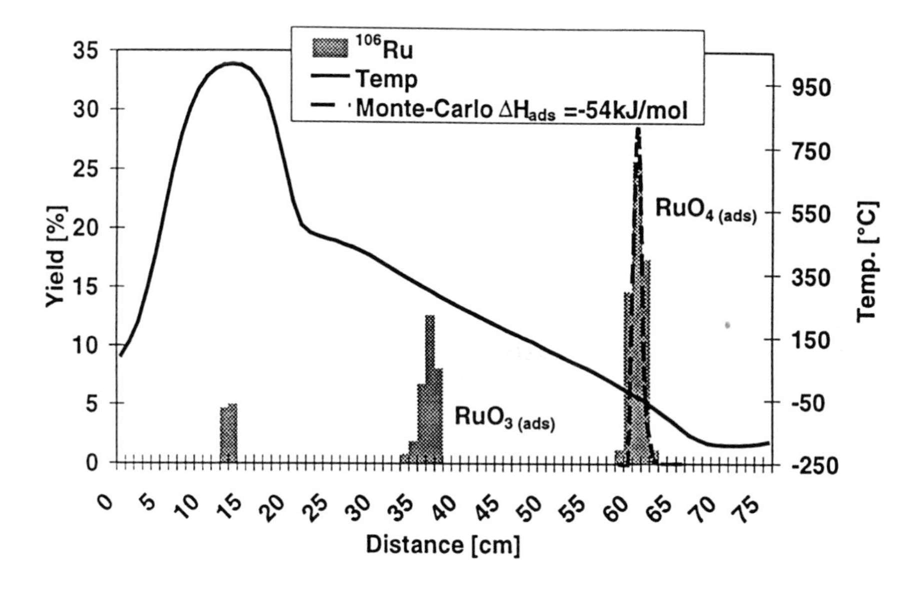

Fig. 23. Thermochromatography of ^{106}Ru in O_2 gas (20 ml/min) in an empty quartz column. The solid line represents the temperature profile in the column. Two different Ru zones were observed after completion of the experiment (for details see text). Some of the Ru was not volatilized at the starting position. The dashed lines indicates the modeled deposition zone of a species transported by mobile adsorption with $-\Delta H_a^0(RuO_4)$=54 kJ/mol. Figure reproduced from [92].

6.1.2 *Isothermal chromatography of oxides*

In order to directly detect the nuclear decay of Hs isotopes in a future experiment, test experiments with short-lived, α-particle emitting Os isotopes were conducted. The experiments by A. Yakushev et al. [94] with $^{171-174}$Os ($T_{1/2}$ = 8.3 - 45 s) demonstrated several important aspects of a future Hs experiment. First, by installing an oven as closely as possible to the recoil chamber high yields of OsO_4 were obtained simply by using a mixture of Ar/O_2 as carrier gas. The addition of aerosol particles was not necessary. Indeed, OsO_4 was already formed "in-situ" in the recoil chamber [93], but the chemical yield of short-lived Os nuclides could substantially be improved by heating the attached oven [95]. The OsO_4 molecules were transported with minimal losses by the carrier gas at room temperature through an open quartz chromatography column. At the exit, the Ar/O_2 stream was mixed with a stream of Ar loaded with Pb aerosols on which the OsO_4 was adsorbed and reduced to non-volatile OsO_2. The Pb aerosols were collected on a stepwise moving tape by impaction and transported in front of 6 PIPS detectors to register the α-particle decay of the isolated Os nuclides. The overall yield (not including the detection efficiency) was measured to be 50 - 60%.

Experiments by A. von Zweidorf et al. [93] also employed the *in-situ* production of Ru- and Os tetroxides in oxygen containing carrier gases, but were aimed at studies of the adsorption properties of volatile tetroxides on different surfaces at room temperature. Glass fiber filters soaked with 1 M NaOH yielded the best results, but also freshly prepared Na surfaces provided good adsorption of OsO_4.

Using a similar set-up as in [94] Ch. Düllmann et al. [97] measured a yield vs. isothermal temperature breakthrough curve of $^{173}OsO_4$, see Figure 24. This set-up was named In-situ Volatilization and On-line Detection (IVO). Volatile OsO_4 were synthesized in-situ in the recoil chamber and transported by the He/O_2 carrier gas to a quartz chromatography column that could be operated between ambient temperature and -80 °C. At the exit of the column volatile molecules were adsorbed on the surface of Pb aerosols and transported to the ROMA counting system. In order to prevent the build-up of ice in the column, all gases had to be carefully dried. In agreement with thermochromatography experiments [79,85,96] $-\Delta H_a^0(OsO_4)$ = 38.0±1.5 kJ/mol was determined. The decontamination from interfering elements i.e. Po was determined to be $>10^4$. The yield of the IVO technique was of the order of 50% and therefore about a factor of 3 more efficient than the OLGA system used in experiments with Bh. But the gain of at least one order of magnitude was not yet accomplished.

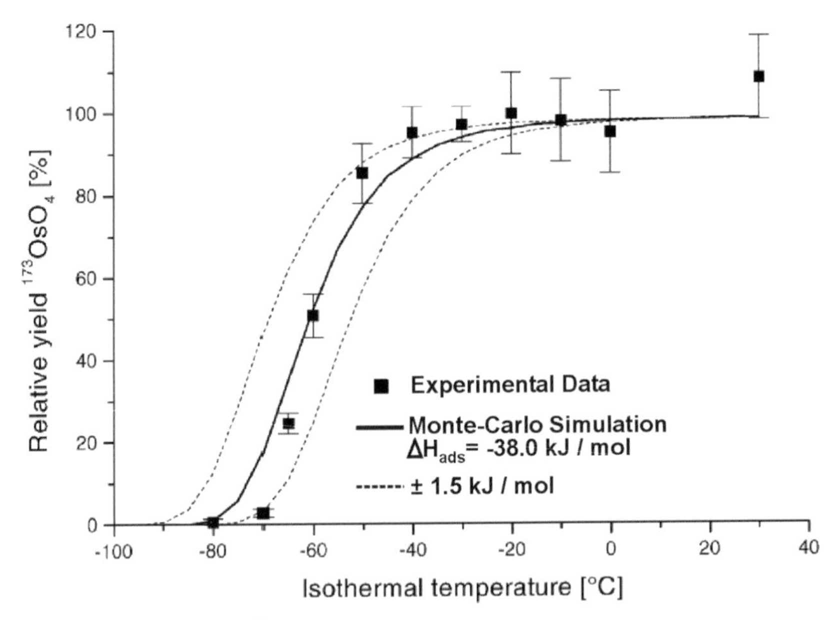

Fig. 24. Relative yields of $^{173}OsO_4$ ($T_{1/2}$ = 22.4 s) as a function of isothermal temperature. Figure reproduced from [95].

6.2 EARLY ATTEMPTS TO CHEMICALLY IDENTIFY HASSIUM

A first unsuccessful attempt to chemically identify Hs as volatile HsO_4 was reported by B. Zhuikov et al. [88] from Dubna. The reaction $^{40}Ar+^{235}U$ was employed to produce short-lived α-decaying isotopes of element 110 and their Hs daughter nuclides. These were rapidly isolated as volatile tetroxides to detect their SF decay. Atoms recoiling from the target were stopped in air and transported to a thermochromatography column, where the purification from actinides took place on a hot quartz wool filter. OsO_4 was adsorbed quantitatively on Lavsan (polyethylenphtalat) fission track detectors covered with 50 $\mu g/cm^2$ of Pb. No SF decays were registered resulting in a production cross section limit of 10 pb for nuclides with half-lives longer than 150 ms.

In a second experiment, using the reaction $^{249}Cf(^{22}Ne, 4n)^{267}Hs$, B. Zhuikov et al. [88] searched also for short-lived α-particle emitting isotopes of Hs. Recoiling atoms were thermalized in a mixture of Ar + 2% O_2 and were continuously swept from the target chamber through a Teflon capillary to a quartz column kept at 1000-1100 °C and filled with CaO to retain non-volatile nuclear reaction products; i.e. actinides, Ra, Fr, and Po. Volatile species were then transported through a Teflon capillary and blown onto the surface of a Si detector covered with 50 $\mu g/cm^2$ of Pb. At the opposite side an annular Lavsan track detector (also coated with 50 $\mu g/cm^2$ of Pb) was located for registering fission fragments. The whole counting device was placed inside a shielding of Cd and paraffin in order to decrease the background. In model experiments with Os, OsO_4 was efficiently absorbed on the Pb surfaces. The decontamination from actinides was excellent (separation factor $>10^6$) as well as that from Po ($>10^3$). Nevertheless, no α-particles in the energy range above 8.5 MeV and no SF events were registered and an upper limit of 100 pb for the production cross sections of α-decaying nuclides with half-lives in the range between 50 ms and 12 h and of 50 pb for spontaneously fissioning nuclides was established.

A similar experiment was reported by R.J. Dougan et al. [90]. A set-up called On-line Separation and Condensation AppaRatus (OSCAR) was installed at the LBNL 88-Inch Cyclotron. Nuclear reaction products were collected with a KCl aerosol gas-jet and were transported from the target chamber to the OSCAR set-up where O_2 was added. The aerosol particles were destroyed on a hot quartz wool plug and the formation of tetroxides occurred at a temperature of 650°C. Non volatile reaction products were retained on the quartz wool plug whereas the volatile tetroxides were swept by the carrier gas flow to a condensation chamber, where they were deposited on a Ag disk, which was cooled with liquid N_2. An annular Si surface barrier detector registered α-particle and SF decays of nuclides adsorbed on the disk surface. The OSCAR set-up was used to search for α-decaying ^{272}Hs, the expected EC decay daughter of ^{272}Mt (estimated EC-decay half-life: 25 m), produced in the ^{254}Es(^{22}Ne, 4n) reaction. However, no α-decays between 8.7 and 11 MeV were observed and an upper limit for the production cross section of 1 nb was derived.

The experiments by B. Zhuikov et al. [88] and R.J. Dougan et al. [90] clearly demonstrated that the chemical selectivity would have been sufficient for a first chemical identification of Hs. However, the overall efficiency and the longtime stability of the experiments to reach the required sensitivity were not sufficient.

6.3 ON-LINE THERMOCHROMATOGRAPHY OF HASSIUM

In order to reach the required sensitivity and to obtain at the same time meaningful chemical information about Hs and its compounds different experimental developments had to be combined. First, the most promising approach to synthesize relatively long-lived Hs isotopes appeared to be the reaction ^{248}Cm(^{26}Mg; 4,5n)^{270,269}Hs. Second, the rate of production could be increased by using an intense ^{26}Mg beam impinging on rotating ^{248}Cm targets. Such a technically very challenging irradiation set-up was constructed and put into operation by M. Schädel et al., see Chapter 4, Section 2.1.3.. Third, the "in-situ" production [95] allowed for highest possible chemical yields of group-8 tetroxides. Fourth, in order to compare the volatility of HsO_4 with those of its lighter homologues of group 8, thermochromatography is the method of choice since the position of every detected atom contributes chemical information. The only problem of the thermochromatographic technique so far was the unambiguous identification of the decaying nuclide. This problem has been solved by U. Kirbach et al. [96] who have built a rectangular thermochromatography column consisting of PIN diodes. In the actual Hs experiment an improved version namely the Cryo On-Line Detector (COLD) was used, see Chapter 4, Section 4.4..

The required gain in sensitivity of one order of magnitude compared to the OLGA set-up used in experiments with Bh was thus accomplished. With the rotating target wheel, synthesis of about 3 atoms of ^{269}Hs per day could be expected, assuming a production cross section of 7 pb. The overall efficiency of the set-up (including detection of a 3 member α-decay chain) amounted to 30 - 50%, resulting in the expected detection of about one decay chain per day of experiment.

In an experiment to produce Hs isotopes conducted in May of 2001 at the GSI, valid data was collected during 64.2 h. During this time 1.0×10^{18} ^{26}Mg beam particles passed through the ^{248}Cm target. Only α-lines originating from ^{211}At, ^{219,220}Rn and their decay products were identified. While ^{211}At and its decay daughter ^{211}Po were deposited mainly in the first two detectors, ^{219,220}Rn and their decay products accumulated in the last three detectors, where the temperature was sufficiently low to partly adsorb Rn. During the experiment, seven correlated decay chains were detected, see Figure 25.

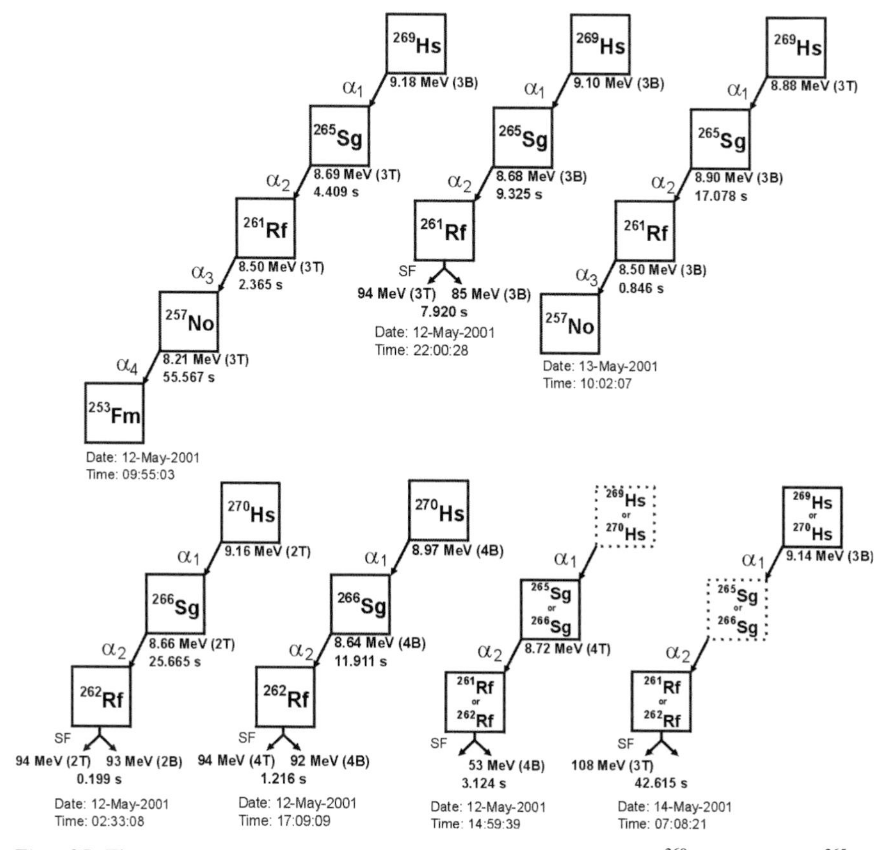

Fig. 25. The seven nuclear decay chains attributed to the decay of ^{269}Hs leading to ^{265}Sg, ^{261}Rf and ^{257}Lr or ^{270}Hs leading to ^{266}Sg and ^{262}Rf registered in the COLD detector after chemical isolation of volatile tetroxides. Figure reproduced form [79].

All decay chains were observed in detectors 2 through 4 and were assigned to the decay of either ^{269}Hs or the yet unknown ^{270}Hs. The characteristics of three decay chains agreed well with literature data on ^{269}Hs and its daughter nuclides [77, 99], while two other decay chains were attributed to the decay of ^{270}Hs. The last two decay chains were incomplete and a definite assignment to ^{269}Hs or ^{270}Hs could not be made. No additional three-member decay chains with a total length of ≤300 s were registered in detectors 2 to 10. The background count-rate of α-particles with energies between 8.0 and 9.5MeV was about 0.6 h^{-1} per detector, leading to very low probabilities of $\leq 7\times10^{-5}$ and $\leq 2\times10^{-3}$ for any of the first five chains and any of the last two chains, respectively, being of random origin. In addition, four fission fragments with energies >50 MeV that were not correlated with a preceding α-particle were registered in detectors 2 through 4.

The longitudinal distribution of the 7 decay chains originating from Hs is depicted in Figure 26. The maximum of the Hs distribution was found at a temperature of -44 ± 6 °C. The distribution of $^{172}OsO_4$ ($T_{1/2}$ = 19.2 s) measured before and after the experiment showed a maximum in detector 6 at a deposition temperature of -82 ± 7 °C. As in experiments with lighter transactinide elements the Monte Carlo model of I. Zvara [18], that describes the microscopic migration of a molecule in a gas chromatographic column, was used to evaluate the adsorption enthalpy of HsO_4 and OsO_4 on the silicon nitride detector surface. The modeled distributions with $-\Delta H_a^0(HsO_4)$ = 46 ± 2 kJ/mol (68% confidence interval) and $-\Delta H_a^0(OsO_4)$ = 39 ± 1 kJ/mol are shown as solid lines in Figure 26.

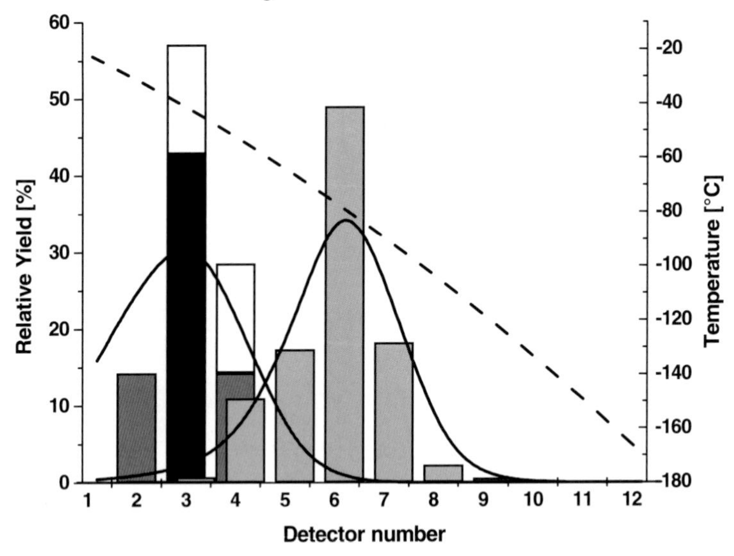

Fig. 26. Relative yields of HsO_4 and OsO_4 for each of the 12 detector pairs. Measured values are represented by bars: $^{269}HsO_4$: black; $^{270}HsO_4$: dark grey; $^{269 \text{ or } 270}HsO_4$: white; $^{172}OsO_4$: light grey. The dashed line indicates the temperature profile (right-hand scale). The maxima of the deposition distributions were evaluated as -44 ± 6 °C for HsO_4 and -82 ± 7 °C for OsO_4. Solid lines represent results of a simulation of the adsorption process with standard adsorption enthalpies of -46.0 kJ/mol for $^{269}HsO_4$ and -39.0 kJ/mol for $^{172}OsO_4$. Figure reproduced from [100].

The higher deposition temperature of about 40 °C and the thus about 7 kJ/mol higher adsorption enthalpy seems to indicate a slightly lower volatility of HsO_4 compared to its lighter homologue OsO_4. This experimental result was somewhat unexpected since according to both, classical extrapolations and relativistic molecular calculations, HsO_4 was predicted to be about as volatile as OsO_4. Nevertheless, the high volatility of the Hs oxide species clearly suggests that it is HsO_4 since, by analogy with the known properties of the Os oxides, all other Hs oxides are expected to be much less volatile and unable to reach the detector system. The observed formation of a volatile Hs oxide (very likely HsO_4) provides strong experimental evidence that Hs behaves chemically as an ordinary member of group 8 of the Periodic Table.

7. Future studies with spherical superheavy elements

With the reported synthesis of long-lived isotopes of elements 112 and 114 in ^{48}Ca induced reactions on ^{238}U and $^{242,244}Pu$ targets [101-103], the focus of chemists has shifted to the chemical exploration of superheavy elements, rather than continuing with Mt, element 110 and so on. Keeping in mind that there are currently no long-lived isotopes known of elements 109 through 111 (except for $^{280,281}110$ the daughter nuclei of $^{288,289}114$) and that there are also no obvious choices for their chemical investigation, the jump to elements 112 and 114 appears quite natural. While early attempts to chemically identify superheavy elements did not reach the required sensitivity to investigate elements that can be produced only at the picobarn level, see Chapter 8, two recent attempts to characterize element 112 have provided very interesting results.

7.1 VOLATILITY OF GROUP-12 TO GROUP-18 ELEMENTS AND COMPOUNDS

Due to the expected high volatility of elements with atomic numbers 112 to 118 in the elemental state [104], see also Chapters 2 and 6, gas phase chemical studies will play an important role in investigating the chemical properties of the newly discovered superheavy elements. An interesting question is, if e.g. elements 112 and 114 are indeed relatively inert gases (similar to a noble gas) [105] due to closed s^2 and $p_{1/2}{}^2$ shells, respectively, or if they retain some metallic character and are thus adsorbed quite well on certain metal surfaces, see Chapter 6, Part II, Section 3.2. Extrapolations by B. Eichler et al. [106] point to Pd or Cu as ideal surfaces for the adsorption of superheavy elements.

Besides the elemental state, also volatile compounds of superheavy elements have been considered. K. Bächmann et al. [107] extrapolated the boiling points of hydrides, methyl- and ethyl compounds of elements 113 through 117. N. Trautmann et al. [108] showed, that short-lived ^{216}Po ($T_{1/2}$ = 0.15 s) was volatilized using ethyl radicals, probably as diethylpolonium.

Since isotopes of the group 14 to 16 elements Pb (^{212}Pb, ^{213}Pb), Bi (^{212}Bi, ^{213}Bi) and Po (^{211m}Po, ^{212m}Po) severely interfered with the detection of transactinide halides and oxyhalides in the case of Rf, Db, and Sg, the formation of volatile halides (or oxyhalides) of elements 114, 115, and 116 can be expected. However, due to the high α-decay energies of the above mentioned isotopes a very clean separation between Pb, Bi, and Po and the transactinide elements must be accomplished in order to study heavy transactinide halides (or oxyhalides).

7.2 ISOTHERMAL CHROMATOGRAPHY OF ELEMENT 112

A first attempt to chemically identify one of the recently found long-lived isotope of element 112, namely $^{283}112$ (SF, $T_{1/2} \approx 3$ m), in the elemental state was made by A. Yakushev et al. [109] in Dubna. The nuclide was produced by bombarding a $^{nat.}U$ target with ^{48}Ca ions. Simultaneously, short-lived Hg isotopes were produced from a small admixture of Nd to the target material. In test experiments short-lived Hg isotopes could be isolated in the elemental form from other reaction products and were transported in He quantitatively through a 30 m long Teflon™ capillary at room temperature.

Adsorption of Hg nuclides on silicon detectors, as in the successful experiment with HsO_4, proved experimentally not feasible, since Hg was adsorbed on quartz surfaces only at temperatures of -150 °C and below. However, Hg adsorbed quantitatively on Au, Pt, and Pd surfaces at room temperature. As little as 1 cm^2 of Au or Pd surface was sufficient to adsorb Hg atoms nearly quantitatively from a stream of 1 l/min He. Therefore, detector chambers containing a pair of Au or Pd coated PIPS detectors were constructed. Eight detector chambers (6 Au and 2 Pd) were connected in series by Teflon™ tubing. The detector chambers were positioned inside an assembly of 84 ^3He filled neutron detectors (in a polyethylen moderator) in order to simultaneously detect neutrons accompanying spontaneous fission events, see Figure 27.

In an experiment conducted in January of 2000 a total beam dose of 6.85×10^{17} ^{48}Ca ions was accumulated. The chemical yield for the simultaneously produced ^{185}Hg ($T_{1/2}$ = 49 s) was 80%. If element 112 behaved chemically like Hg and all efficiencies measured for Hg were also valid for element 112, detection of $3.4^{+4.3}_{-2.2}$ SF events could be expected assuming the cross section value for the production of $^{283}112$ measured in [101]. However, no SF events were observed. Therefore, no unambiguous answer as to the chemical and physical properties of element 112 was obtained.

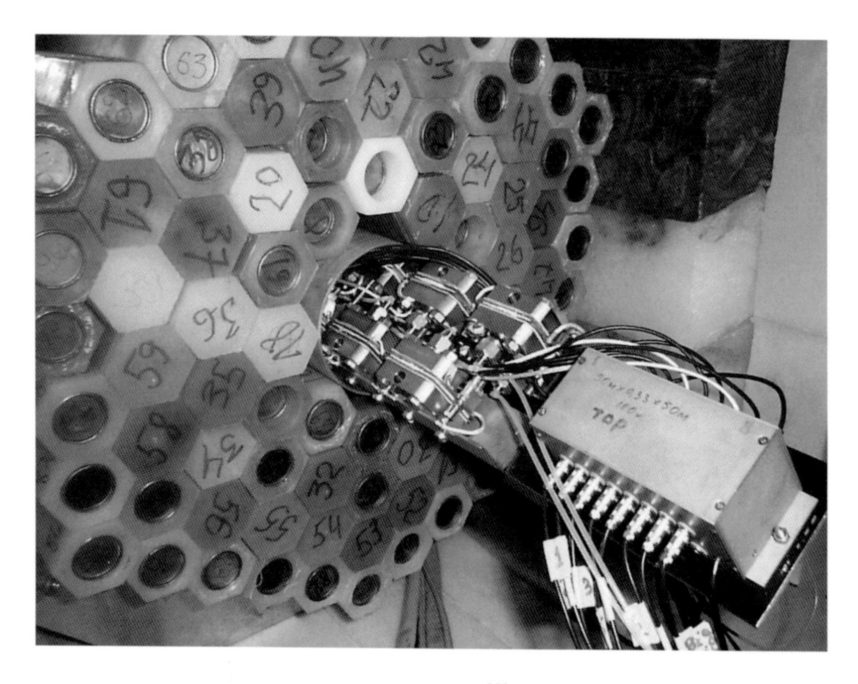

Fig. 27. Detector arrangement for the detection of $^{283}112$ consisting of pairs of Au and Pd coated PIPS detectors inside an assembly of 84 ^{3}He filled neutron detectors. Photograph reproduced from [109].

In a next experiment, the question whether element 112 remained in the gas phase and passed over the Au and Pd surfaces was addressed [110]. Therefore, a special ionization chamber to measure α-particle and SF fragments of nuclei remaining in the gas was added at the exit of the Au or Pd coated PIPS detector array.

A total beam dose of 2.8×10^{18} ^{48}Ca ions was accumulated. Again zero SF events were registered on the Au and Pd coated PIPS detectors, confirming the result of the first experiment. However, 8 SF events accompanied by neutrons were registered in the ionization chamber, while only one background count was expected. Therefore, the majority of the SF events were attributed to the decay of a nuclide of element 112, since there are no other known volatile nuclides decaying by SF.

From this experiment it appears that the interaction of element 112 with an Au or Pd surface is much weaker than for Hg. The obtained enthalpies of adsorption were $-\Delta H_a^0(\mathrm{Hg}) \geq 75$ kJ/mol and $-\Delta H_a^0(\text{element } 112) \leq 55$ kJ/mol. Such a vastly different chemical behavior as in the present case of element 112 compared to its lighter homologue Hg has not been observed for any of the lighter transactinides so far and might reflect the predicted inertness and enhanced volatility due to relativistic effects. Obviously, in a next step, the enthalpy of adsorption of element 112 on Au surfaces has to be measured experimentally. This can be done with a cryo thermochromatography detector containing Au coated detectors. The temperature in the gradient should start at room temperature and reach down to the adsorption temperature of Rn, which should be still above the temperature of liquid N_2.

References

1. Zvara, I., Chuburkov, Yu.T., Caltetka, R., Shalaevski, M.R., Shilov, B.V.: At. Energ. **21,** 1966) 8.
2. Keller, O.L.: Radiochim. Acta **37,** (1984) 169.
3. Johnson, E., Fricke, B., Keller, O.L., Nestor, C.W., Tucker, T.C.: J.Chem. Phys. **93,** (1990) 8041.
4. Zhuikov, B.L., Chuburkov, Yu.T., Timokhin, S.N., Kim U. Jin, Zvara, I.: Radiochim. Acta **46,** (1989) 113.
5. Gäggeler, H.W. : J. Radioanal. Nucl. Chem. **183,** (1994) 261.
6. Morozove, A.I., Karlova, E.V.: Russ J. Inorg. Chem. **16,** (1971) 12.
7. Zvara, I., Belov, V.Z., Chelnikov, L.P., Domanov, V.P., Hussonois, M., Korotkin, Yu.S., Shegolev, V.A., Shalaevski, M.R.: Inorg. Nucl. Chem. Lett. **7,** (1971) 1109.
8. Türler, A.: "Gas Phase Chemistry of the Transactinide Elements Rutherfordium, Dubnium, and Seaborgium", *in*: Habilitation Thesis, Bern University (1999).
9. Zvara, I., Timokhin, S.N., Chuburkov, Yu.T., Domanov, V., Gorski, B.: Joint Institute for Nuclear Research, Laboratory for Nuclear Reactions, Scientific Report 1989-1990, E7-91-75, Dubna, p.34.
10. Gäggeler, H.W., Jost, D.T., Baltensperger, U., Weber, A., Kovacs, A., Vermeulen, D., Türler, A.: Nucl. Instrum. Meth. in Phys. Res. **A309,** (1991) 201.
11. Kadkhodayan, B., Türler, A., Gregorich, K.E., Nurmia, M.J., Lee, D., Hoffman, D.C.,: Nucl. Instrum. Meth. in Phys. Res. **A317,** (1992) 254.
12. Türler, A., Buklanov, G.V., Eichler, B., Gäggeler, H.W., Grantz, M., Hübener, S., Jost, D.T., Lebedev, V.Ya., Piguet, D., Timokhin, S.N., Yakushev, A.B., Zvara, I.: J. Alloys. Comp. **271-273,** (1998) 287.
13. Kadkhodayan, B. Türler, A., Gregorich, K.E., Baisden, P.A., Czerwinski, K.R., Eichler, B., Gäggeler, H.W., Hamilton, T.M., Jost, D.T., Kacher, C.D., Kovacs, A., Kreek, S.A., Lane, M.R., Mohar, M.F., Neu, M.P., Stoyer, N.J., Sylwester, E.R., Lee, D.M., Nurmia, M.J., Seaborg, G.T., Hoffman, D.C.: Radiochim. Acta **72,** (1996) 169.
14. Türler, A., Gäggeler, H.W., Gregorich, K.E., Barth, H., Brüchle, W., Czerwinski, K.R., Gober, M.K., Hannink, N.J., Henderson, R.A., Hoffman, D.C., Jost., D.T., Kacher, C.D., Kadkhodayan, B., Kovacs, A., Kratz. J.V., Kreek, S.A., Lee, D.M., Leyba, J.D., Nurmia, M.J., Schädel, M., Scherer, U.W., Schimpf, E., Vermeulen, D., Weber, A., Zimmermann, H.P., Zvara, I.: J. Radioanal. Nucl. Chem. **160,** (1992) 327.
15. Sylwester, E., Gregorich, K.E., Lee, D.M., Kadkhodayan, B., Türler, A., Adams, J.L., Kacher, C.D., Lane, M.R., Laue, C.A., McGrath, C.A., Shaughnessy, D.A., Strellis, D.A., Wilk, P.A., Hoffman, D.C.: Radiochim. Acta **88,** (2000) 837.
16. Domanov, V.P., Kim U. Zin,: Radiokhimiya **31,** (1989) 12.
17. Eichler, B. Zvara, I.: Radiochim. Acta **30,** (1982) 233.
18. Zvara, I.: Radiochim. Acta **38,** (1985) 95.
19. Gregorich, K.E., in: Radiochemistry of Rutherfordium and Hahnium, Proc. "The Robert A. Welch Foundation, 41st Conference on Chemical Research – The Transactinide Elements", Houston, Texas, Oct. 27-28 (1997), p.95.
20. Zvara, I., Belov, V.Z., Korotkin, Yu.S., Shalaevski, M.R., Shchegolev, V.A., Hussonois, M., Zager, B.A.: Joint Institute for Nuclear Research, Dubna, Report P12-5120, May 15 (1970).
21. Zvara, I., Belov, V., Domanov, V.P., Shalaevski, M.R., Sov. Radiochem. **18,** (1976) 371.
22. Türler, A., Eichler, B., Jost, D.T., Piguet, D., Gäggeler, H.W., Gregorich, K.E., Kadkhodayan, B., Kreek, S.A., Lee, D.M., Mohar, M., Sylwester, E., Hoffman, D.C., Hübener, S.: Radiochim. Acta **73,** (1996) 55.

23. Gäggeler, H.W., Jost, D.T., Kovacs, J., Scherer, U., Weber, A., Vermeulen, D., Türler, A., Gregorich, K.E., Henderson, R., Czerwinski, K., Kadkhodayan, B., Lee, D.M., Nurmia, M., Hofman, D.C., Kratz, J.V., Gober, M., Zimmermann, H.P., Schädel, M., Brüchle,W., Schimpf, E., Zvara, I.: Radiochim. Acta **57**, (1992) 93.
24. Türler, A.: Radiochim. Acta **72,** (1996) 7.
25. Timokhin, S.N., Yakushev, A.B., Perelygin, V.P., Zvara, I., "Chemical Identification of Element 106 by the Thermochromatographic Method". In: Proceedings of the "International School Seminar on Heavy Ion Physics", Dubna, 10-15 May 1993, pp. 204-206.
26. Timokhin, S.N., Yakushev, A.B., Honggui, Xu, Perelygin, V.P., Zvara, I.: J. Radioanal. Nucl. Chem., Letters **212**, (1996) 31.
27. Yakushev, A.B., Timokhin, S.N., Vedeneev, M.V., Honggui, Xu, Zvara, I.: J. Radioanal. Nucl. Chem. **205** (1996) 63.
28. Zvara, I., Yakushev, A.B., Timokhin, S.N., Honggui, Xu, Perelygin, V.P., Chuburkov Yu.T.: Radiochim. Acta **81** (1998) 179.
29. Schädel, M., Brüchle, W., Dressler, R., Eichler, B., Gäggeler, H.W., Günther, R., Gregorich, K.E., Hoffman, D.C., Hübener, S., Jost, D.T., Kratz, J.V., Paulus, W., Schumann, D., Timokhin, S., Trautmann, N., Türler, A., Wirth, G., Yakushev, A.: Nature **388**, (1997) 55.
30. Türler, A. Brüchle, W., Dressler, R., Eichler, B., Eichler, R., Gäggeler, H.W., Gärtner, M., Gregorich, K.E., Hübener, S., Jost, D.T., Lebedev, V.Y., Pershina, V.G., Schädel, M., Taut, S., Timokhin, S.N., Trautmann, N., Vahle, A., Yakushev, A.B.: Angew. Chem. Int. Ed. **38**, (1999) 2212.
31. Hübener, S., Taut, S., Vahle, A., Dressler, R., Eichler, B., Gäggeler, H.W., Jost, D.T., Piguet, D., Türler, A., Brüchle, W., Jäger, E., Schädel, M., Schimpf, E., Kirbach, U., Trautmann, N., Yakushev, A.B.: Radiochim. Acta **89**, (2001) 737.
32. Knacke, O., Kubaschewski, O., Hesselmann K. (Eds.), *Thermochemical Properties of Inorganic Substances II*, Springer-Verlag, Berlin, 1991.
33. Vahle, A., Hübener, S., Eichler, B.: Radiochim. Acta **69**, (1995) 233.
34. Vahle, A., Hübener, S., Dressler, R., Eichler, B., Türler, A.: Radiochim. Acta **78**, (1997) 53.
35. Vahle, A., Hübener, S., Funke, H., Eichler, B., Jost, D.T., Türler, A., Brüchle, W., Jäger E.: Radiochim. Acta **84**, (1999) 43.
36. Zvara, I., Eichler, B., Belov, V.Z., Zvarova, T.S., Korotkin, Yu.S., Shalayevski, M.R., Shchegolev, V.A., Hussonnois, M.: Sov. Radiochemistry **16**, (1974) 709.
37. Helas, G., Hoffmann, P., Bächmann, K.: Radiochem. Radioanal. Letters **30**, (1977) 371.
38. Bayar, B., Votsilka, I., Zaitseva, N.G., Novgorodov, A.F.: Sov. Radiochemistry **20**, (1978) 64.
39. Tsalas, S., Bächmann, K.: Anal. Chim. Acta **98**, (1978) 17.
40. Rudolph, J., Bächmann, K., Steffen, A., Tsalas, S.: Microchim. Acta **I** (1978) 471.
41. Rudolph, J., Bächmann, K.: J. Radioanal. Chem. **43**, (1978) 113.
42. Rudolph, J., Bächmann, K.: Microchim. Acta **I**, (1979) 477.
43. Tsalas, S., Bächmann, K., Heinlein, G.: Radiochim. Acta **29**, (1981) 217.
44. Gärtner, M., Boettger, M., Eichler, B., Gäggeler, H.W., Grantz, M., Hübener, S., Jost, D.T., Piguet, D., Dressler, R., Türler, A., Yakushev, A.B.: Radiochim. Acta **78**, (1997) 59.
45. Lebedev, V.Ya., Yakushev, A.B., Timokhin, S.N., Vedeneev, M.B., Zvara, I.: Czech. J. Phys. 49/S1, (1999) 589.
46. Türler, A., Dressler, R., Eichler, B., Gäggeler, H.W., Jost, D.T., Schädel, M., Brüchle, W., Gregorich, K.E., Trautmann, N., Taut, S., Phys. Rev. C **57**, (1998) 1648.
47. Zvara, I.: Radiochim. Acta 38, (1985) 95.
48. Vahle, A., Hübener, S., Dressler, R., Grantz, M.: Nucl. Instr. Meth. A **481**, (2002) 637.

49. Lazarev, Y.A., Lobanov, Y.V., Oganessian, Y.T., Utyonkov, V.K., Abdullin, F.S., Buklanov, G.V., Gikal, B.N., Iliev, S., Mezentsev, A.N., Polyakov, A.N., Sedykh, I.M. Shirokovsky, I.V., Subbotin, V.G., Sukhov, A.M., Tsyganov, Y.S., Zhuchko, V.E., Lougheed, R.W., Moody, K.J., Wild, J.F., Hulet, E.K., McQuaid, J.H.: Phys. Rev. Lett. **73**, (1994) 624.
50. Zvara, I., Domanov, V.P., Hübener, S., Shalaevskii, M.R., Timokhin, S.N., Zhuikov, B.L., Eichler, B., Buklanov, G.V.: Sov. Radiochem. **26**, (1984) 72.
51. Schädel, M., Jäger, E., Brüchle, W., Sümmerer, K., Hulet, E.K., Wild, J.F., Lougheed, R.W., Dougan, R.J., Moody, K.J.: Radiochim. Acta **68**, (1995) 7.
52. Wilk, P. A., Gregorich, K. E., Türler, A., Laue, C. A., Eichler, R., Ninov, V., Adams, J. L., Kirbach, U.W.,. Lane, M.R., Lee, D.M., Patin, J.B., Shaughnessy, D.A., Strellis, D.A., Nitsche, H., Hoffman, D.C.: Phys. Rev. Lett. **85**, (2000) 2697.
53. Eichler, R., Eichler, B., Gäggeler, H.W., Jost, D.T., Dressler, R., Türler, A.: Radiochim. Acta **87**, (1999) 151.
54. Eichler, R., Eichler, B., Gäggeler, H.W., Jost, D.T., Piguet, D., Türler, A.: Radiochim. Acta **88**, (2000) 87.
55. Eichler, R., Brüchle, W., Dressler, R., Düllmann, Ch.E., Eichler, B., Gäggeler, H.W., Gregorich, K.E., Hoffman, D.C., Hübener, S., Jost, D.T., Kirbach, U.W., Laue, C.A., Lavanchy, V.M., Nitsche, H., Patin, J.B., Piguet, D., Schädel, M., Shaughnessy, D.A., Strellis, D.A., Taut, S., Tobler, L., Tsyganov, Y.S., Türler, A., Vahle, A., Wilk, P.A., Yakushev, A.B.: Nature **407**, (2000) 63.
56. Merinis, J., Bouissieres, G.: Anal. Chim. Acta **25**, (1961) 498.
57. Schäfer, H.: *Chemische Transportreaktionen*, Verlag Chemie, Weinheim (1962).
58. Bayar, B., Novgorodov, A.F., Zaitseva, N.G.: Radiochem. Radioanal. Lett. **15**, (1973) 231.
59. Bayar, B., Vocilka, I., Zaitseva, N.G., Novgorodov, A.F.: Sov. Radiochem. **16** (1974) 333.
60. Bayar, B., Novgorodov, A.F., Vocilka, I., Zaitseva, N.G.: Radiochem. Radioanal. Lett. **19**, (1974) 45.
61. Eichler, B., Domanov, V.P.: J. Radioanal. Chem. **28**, (1975) 143.
62. Bayar, B., Vocilka, I., Zaitseva, N.G., Novgorodov, A.F.: Radiochem. Radioanal. Lett. **34**, (1978) 75.
63. Adilbish, M., Zaitseva, N.G., Kovach, Z., Novgorodov, A.F., Sergeev, Yu.Ya., Tikhonov, V.I.,: Sov. Radiochem. **20**, (1978) 652.
64. Steffen, A., Bächmann, K.: Talanta **25**, (1978) 677.
65. Novgorodov, A.F., Adilibish, M., Zaitseva, N.G., Kovalev, A.S., Kovach, Z.: Sov. Radiochem. **22**, (1980) 590.
66. Domanov, V.P., Hübener, S., Shalaevskii, M.R., Timokhin, S.N., Petrov, D.V., Zvara, I.: Sov. Radiochem. **25**, (1983) 23.
67. Eichler, B.: Radiochim. Acta **72**, (1996) 19.
68. Rösch, F., Novgorodov, A.F., Qaim, S.M.: Radiochim. Acta **64**, (1994) 113.
69. Novgorodov, A.F., Bruchertseifer, F., Brockmann, J., Lebedev, N.A., Rösch, F.: Radiochim. Acta **88**, (2000) 163.
70. Schädel. M., Jäger, E., Schimpf, E., Brüchle, W.: Radiochim. Acta **68**, (1995) 1.
71. Merinis, J., Bouissieres, G.: Radiochim. Acta **12**, (1969) 140.
72. Neidhart, B., Bächmann, K., Krämer, S., Link, I: Radiochem. Radioanal. Lett. **12**, (1972) 59.
73. Tsalas, S., Bächmann, K.: Analytica Chimica Acta **98**, (1978) 17.
74. Pershina, V, Bastug, T.: J. Chem. Phys. **113**, (2000) 1441.
75. Eichler, R.: PSI internal report TM-18-00-04 (in german) (2000).
76. Münzenberg, G., Armbruster, P., Folger, H., Hessberger, F.P., Hofmann, S., Keller, J., Poppensieker, K.E., Reisdorf, W., Schmidt, K.-H., Schött, H.-J., Leino, M.E., Hingmann, R.: Z. Phys. A **317**, (1984) 235.

77. Hofmann, S., Ninov, V., Hessberger, F.P., Armbruster, P., Folger, H., Münzenberg, G., Schött, H.J., Popeko, A.G., Yeremin, A.V., Saro, S., Janik, R., Leino, M.: Z. Phys. A **354**, (1996) 229.
78. Kratz, J.V.: "Chemical properties of the Transactinide Elements". In: Heavy Elements and Related New Phenomena. Eds. Greiner, W., Gupta, R.K., World Scientific, Singapore, (1999) 129-193.
79. Düllmann, Ch.E., Brüchle, W., Dressler, R., Eberhardt, K., Eichler, B., Eichler, R., Gäggeler, H.W., Ginter, T.N., Glaus, F., Gregorich, K.E., Hoffman, D.C., Jäger, E., Jost, D.T., Kirbach, U.W., Lee, D.M., Nitsche, H., Patin, J.B., Pershina, V., Piguet, D., Qin, Z., Schädel, M., Schausten, B., Schimpf, E., Schött, H.-J., Soverna, S., Sudowe, R., Thörle, P., Timokhin, S.N., Trautmann, N., Türler, A., Vahle, A., Wirth, G., Yakushev, A.B., Zielinski, P.M.: Nature **418**, (2002) 859.
80. Gundersen, G., Hedberg, K., Huston, J.L.: J. Chem. Phys. **52**, (1979) 812.
81. Bächmann, K., Hoffmann, P.: Radiochim. Acta **15**, (1971) 153.
82. Fricke, B.: Structure and bonding **21**, (1975) 90.
83. Eichler, B., Domanov, V.P.: J. Radioanal. Chem. **28**, (1975) 143.
84. Eichler, B.: Radiochem. Radioanal. Letters **22**, (1975) 147.
85. Domanov, V.P. Zvara, I.: Radiokhimiya **26**, (1984) 770.
86. Eichler, B., Zude, F., Fan, W., Trautmann, N., Herrmann, G.: Radiochim. Acta **56**, (1992) 133.
87. Zude, F., Fan, W., Trautmann, N., Herrmann, G., Eichler, B.: Radiochim. Acta **62**, (1993) 61.
88. Zhuikov, B.L., Chepigin, V.I., Kruz, H., Ter-Akopian, G.M., Zvara, I.: unpublished report 1985, see also Chepigin, V.I., Zhuikov, B.L., Ter-Akopian, G.M., Zvara, I.: Fizika tiazhelykh ionov - 1985. Sbornik annotacii. Report JINR **P7-86-322**, Dubna, (1986) 15.
89. Zhuikov, B.L., Kruz, H., Zvara, I.: Fizika tiazhelykh ionov - 1985. Sbornik annotacii. Report JINR **P7-86-322**, Dubna, (1986) 26.
90. Dougan, R.J., Moody, K.J., Hulet, E.K., Bethune, G.R.: Lawrence Livermore National Laboratory Annual Report FY 87, UCAR **10062/87**, (1987) 4-17.
91. Hulet, E.K., Moody, K.J., Lougheed, R.W., Wild, F., Dougan, R.J., Bethune, G.R.: Lawrence Livermore National Laboratory Annual Report FY 87, UCAR **10062/87**, (1987) 4-9.
92. Düllmann, Ch.E., Türler, A., Eichler, B., Gäggeler, H.W.: "Thermochromatographic Investigation of Ruthenium with Oxygen as Carrier Gas" In: Extended Abstracts of "1st International Conference on Chemistry and Physics of the Transactinide Elements", Seeheim, Germany, 26-30 September 1999, P-M-13.
93. von Zweidorf, A., Kratz, J.V., Trautmann, N., Schädel, M., Nähler, A., Jäger, E., Schausten B., Brüchle W., Schimpf, E., Angert, R., Li, Z., Wirth, G.: "The Synthesis of Volatile Tetroxides of Osmium and Ruthenium" In: Extended Abstracts of "1st International Conference on Chemistry and Physics of the Transactinide Elements", Seeheim, Germany, 26-30 September 1999, P-M-15.
94. Yakushev, A.B., Vakatov, V.I., Vasko, V., Lebedev, V.Ya., Timokhin, S.N., Tsyganov, Yu.S., Zvara, I.: "On-line Experiments with Short-lived Osmium Isotopes as a Test of the Chemical Identification of the Element 108 - Hassium" In: Extended Abstracts of "1st International Conference on Chemistry and Physics of the Transactinide Elements", Seeheim, Germany, 26-30 September 1999, P-M-17.
95. Düllmann, Ch.E., Eichler, B., Eichler, R., Gäggeler, H.W., Jost, D.T., Piguet, D., Türler, A.: Nucl. Instr. Meth. A **479**, (2002) 631.
96. Kirbach, U.W., Folden III, C.M., Ginter, T.N., Gregorich, K.E., Lee, D.M., Ninov, V., Omtvedt, J.P., Patin, J.B., Seward, N.K., Strellis, D.A., Sudowe, R., Türler, A., Wilk, P.A., Zielinski, P.M., Hoffman, D.C., Nitsche, H.: Nucl. Instr. Meth. A **484**, (2002) 587.
97. Düllmann, Ch.E., Eichler, B., Eichler, R., Gäggeler, H.W., Türler, A.: J. Phys. Chem. B **106**, (2002) 6679.
98. Pershina, V., Bastug, T., Fricke, B., Varga, S.: J. Chem. Phys. **115**, (2001) 792.

99. Hofmann, S., Heßberger, F.P., Ackermann, D., Münzenberg, G., Antalic, S., Cagarda, P., Kindler, B., Kojouharova, J., Leino, M., Lommel, B., Mann, R., Popeko, A.G., Reshitko, S., Saro, S., Uusitalo, J., Yeremin, A.V.: Eur. Phys. J. A **10**, (2001) 5.
100. Düllmann, Ch.E.: "Chemical Investigation of Hassium (Hs, Z=108)", PhD thesis, Bern University (2002).
101. Oganessian, Yu.Ts., Yeremin, A.V., Gulbekian, G.G., Bogomolov, S.L., Chepigin, V.I., Gikal, B.N., Gorshkov, V.A., Itkis, M.G., Kabachenko, A.P., Kutner, V.B., Lavrentev, A.Yu., Malyshev, O.N., Popeko, A.G., Roháč, J., Sagaidak, R.N., Hofmann, S., Münzenberg, G., Veselsky, M., Saro, S., Iwasa, N., Morita, K.: Eur. Phys. J. A **5**, (1999) 63.
102. Oganessian, Yu.Ts., Yeremin; A.V., Popeko A. G., Bogomolov, S.L., Buklanov, G.V., Chelnokov, M.L., Chepigin, V.I., Gikal, B.N., Gorshkov, V.A., Gulbekian, G.G., Itkis, M.G., Kabachenko, A.P., Lavrentev, A.Yu., Malyshev, O.N., Roháč, J., Sagaidak, R.N., Hofmann, S., Saro, S., Giardina, G., Morita, K.: Nature **400**, (1999) 242.
103. Oganessian, Yu.Ts., Utyonkov, V.K., Lobanov, Yu.V., Abdullin, F.Sh., Polyakov, A.N., Shirokovsky, I.V., Tsyganov, Yu.S., Gulekian, G.G., Bogomolov, S.L., Gikal, B.N., Mezentsev, A.N., Iliev, S., Subbotin, V.G., Sukhov, A.M., Ivanov, O.V., Buklanov, G.V., Subotiv, K., Itkis, M.G., Moody, K.J., Wild, J.F., Stoyer, N.J., Stoyer, M.A., Lougheed, R.W.: Phys. Rev. C **62**, (2000) 041604(R).
104. Eichler, B.: Kernenergie **19**, (1976) 307.
105. Pitzer, K.S.: J. Chem. Phys. 63, (1975) 1032.
106. Eichler, B., Rossbach, H.: Radiochim. Acta **33**, (1983) 121.
107. Bächmann, K., Hoffmann, P.: Radiochim. Acta **15**, (1971)153.
108. Trautmann, W., Hoffmann, P., Bächmann, K.: J. Organometallic Chem. **92**, (1975) 191.
109. Yakushev, A.B., Buklanov, G.V., Chelnokov, M.L., Chepigin, V.I., Dmitriev, S.N., Gorshkov, V.A., Hübener, S., Lebedev, V.Ya., Malyshev, O.N., Oganessian, Yu.Ts., Popeko, A.G., Sokol, E.A., Timokhin, S.N., Türler, A., Vasko, V.M., Yeremin, A.V., Zvara, I.: Radiochim. Acta **89**, (2001) 743.
110. Yakushev, A.B., Belozerov, A.V., Buklanov, G.V., Chelnokov, M.L., Chepigin, V.I., Dmitriev, S.N., Eichler, B., Gorshkov, V.A., Gulyev, A.V., Hübener, S., Itkis, M.G., Lebedev, V.Ya., Malyshev, O.N., Oganessian, Yu.Ts., Popeko, A.G., Sokol, E.A., Soverna, S. Szeglowski, Z., Timokhin, S.N., Türler, A., Vasko, V.M., Yeremin, A.V., Zvara, I.: submitted to Radiochim. Acta (2002).

Index

Chapter 8

Historical Reminiscences

G. Herrmann
Institut für Kernchemie, Universität Mainz, Fritz-Straßmann-Weg 2, D-55128 Mainz, Germany

1. Introduction

In 1955, J.A. Wheeler [1] concluded from a courageous extrapolation of nuclear masses and decay half-lives the existence of nuclei twice as heavy as the heaviest known nuclei; he called them *superheavy nuclei.* Two years later, G. Scharff-Goldhaber [2] mentioned in a discussion of the nuclear shell model, that beyond the well established proton shell at Z=82, lead, the next proton shell should be completed at Z=126 in analogy to the known N=126 neutron shell. Together with a new N=184 shell, this shell closure should lead to local region of relative stability. These early speculations remained without impact on contemporary research, however.

The situation changed in 1966 due to three theoretical papers. In a study of nuclear masses and deformations, W.D. Myers and W.J. Swiatecki [3] emphasized the enormous stabilization against fission gained by shell closures. Fission barriers even higher than that of uranium could thus occur for nuclei at the next proton shell closure beyond the existing nuclei, making them quite stable against spontaneous fission. This was in sharp contrast to the liquid-drop nuclear model which predicts in the same region vanishing barriers and, hence, prompt disruption by fission. Remarkably, although the discussion in Ref. [3] focused on Z=126 as the next proton shell closure,

M. Schädel (ed.), The Chemistry of Superheavy Elements, 291-318.

Z=114 was mentioned as an alternative, with reference to unpublished calculations by H. Meldner, who presented his results [4] later at the "Why and How ..." symposium 1966 [5], the seminal event for superheavy element research. Simultaneously, A. Sobiczewski et al. [6] also derived, that 114 should be the next magic proton number. Other groups using different theoretical approaches soon agreed. A fantastic perspective was, thus, opened: an island of superheavy elements located not too far from the then heaviest known element, 103, and, hence, perhaps within reach.

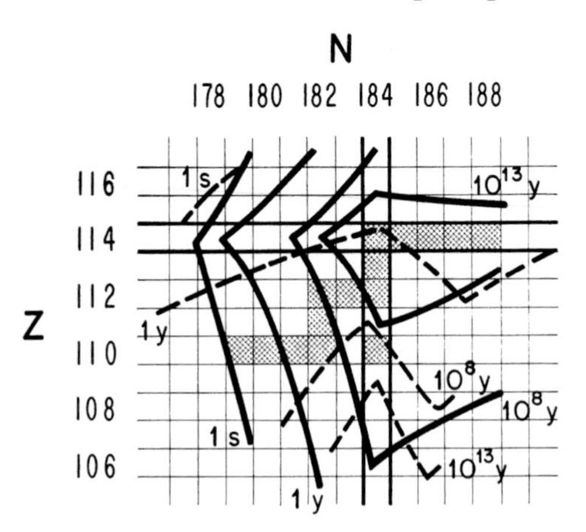

Fig. 1. Topology of the island of superheavy nuclei around the shell closures at proton number Z=114 and neutron number N=184 as predicted in 1969. Thick solid lines are contours of spontaneous fission half-lives, broken lines refer to α-decay half-lives. Shaded nuclei are stable against β-decay. Reproduced from S.G. Nilsson, S.G. Thompson and C.F. Tsang [10], Copyright (2002), with permission from Elsevier Science.

First theoretical estimates [7-10] of decay half-lives around the doubly magic nucleus *Z*=114, *N*=184 revealed an island-like topology as is depicted in Figure 1 [10]. Considered were the three major decay modes: spontaneous fission, α-decay, and β^--decay or electron capture. Spontaneous fission half-lives were calculated to peak sharply at the doubly magic nucleus, descending by orders of magnitude within short distances in the *Z-N* plane, thus causing the island-like shape. In contrast, half-lives for α-decay should decrease rather uniformly with increasing proton number, with some zigzag at the nuclear shell closures. The β-stable nuclei would cross the plane as a diagonal belt. For *Z*=114, *N*=184 an enormous spontaneous fission half-life of 2×10^{19} y was estimated, but only 10 y resulted for α-decay. In a limited region, long half-lives for both decay modes were expected to overlap, most spectacular at *Z*=110, *N*=184 for which an overall half-life of 2×10^{8} y should result – sufficiently long for the occurrence of superheavy elements in Nature! Additional stability was expected for odd elements such as 111 or 113 [8,11] due to the well-known hindrance of spontaneous fission and α-decay for odd proton numbers.

Such predictions of very long overall half-lives immediately stirred up a gold-rush period of hunting for superheavy elements in natural samples. Everybody could feel encouraged to participate. Nearly nothing would be needed: little money, nearly no equipment, no research group, no permission by the laboratory director, no accelerator beam time, no proposal to funding agencies – not even a garage. Just an intelligent choice of a natural sample and a corner in the kitchen at home could be sufficient to make an outstanding discovery: new and superheavy elements in Nature. The detector would be a simple microscope from school days showing fission fragment tracks accumulated in the sample during geological times. Such tracks (Figure 2) are caused by radiation damage of the solid when the energetic fragments are slowed down, and they can be made visible by chemical etching.

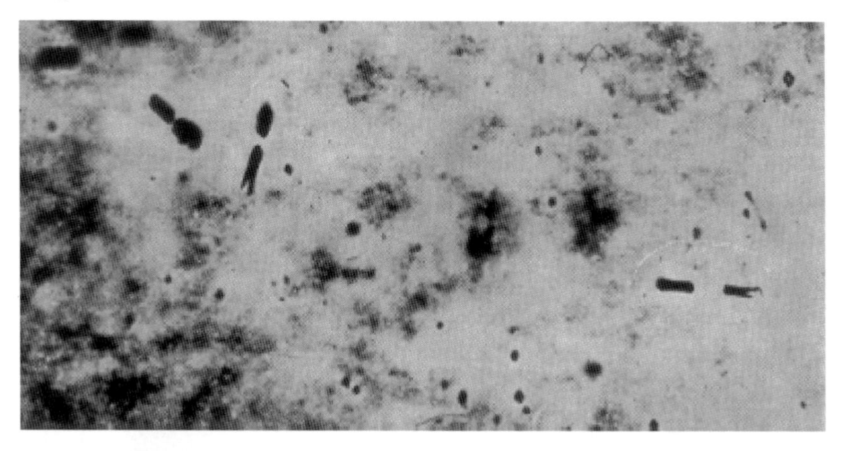

Fig. 2. Tracks of fission fragments in mica showing the characteristic forward-backward orientation of the two fragments emerging from the same fission event. By courtesy of R. Brandt (1974); reproduced from G. Herrmann, Phys. Scr. **10A**, (1974) 71.

Due to the topology of the island, superheavy nuclei should decay by spontaneous fission, either immediately or after a sequence of other decay steps. In a detailed theoretical exploration [12] of the *Z*-*N* plane around the island, the longest-lived nuclide again turned out to be Z=110, N=184, decaying with 3×10^9 y half-life by α-particle emission to ^{290}108. From there, two subsequent β^--transitions should lead via ^{290}109 to ^{290}110, where the chain should terminate by spontaneous fission with 140 d half-life. The doubly magic ^{298}114, half-life 790 y, should also decay into ^{290}110 by two α-particle emissions via ^{294}112 as the intermediate.

Since spontaneous fission is extremely rare in Nature, detection of fission events in natural samples would give a strong hint. Alpha-particle spectra would be less specific, because the energies predicted for superheavy nuclei fell into the range covered by the natural decay series deriving from uranium and thorium, and elaborate chemical treatment would, hence, be required to identify the searched components.

First attempts to produce superheavy nuclei in the laboratory were already undertaken in the late sixtieth. Complete fusion of heavy projectiles with heavy targets – so far very successful in the extension of the Periodic Table by the heaviest elements – was considered the only practical way. However, a large gap had to be bridged between conceivable targets, such as uranium or curium, and the island. The then existing heavy-ion accelerators could not provide the required medium-element projectiles in adequate intensities or not at all. These demands had a strong impact on accelerator technology in order to upgrade existing and build novel facilities.

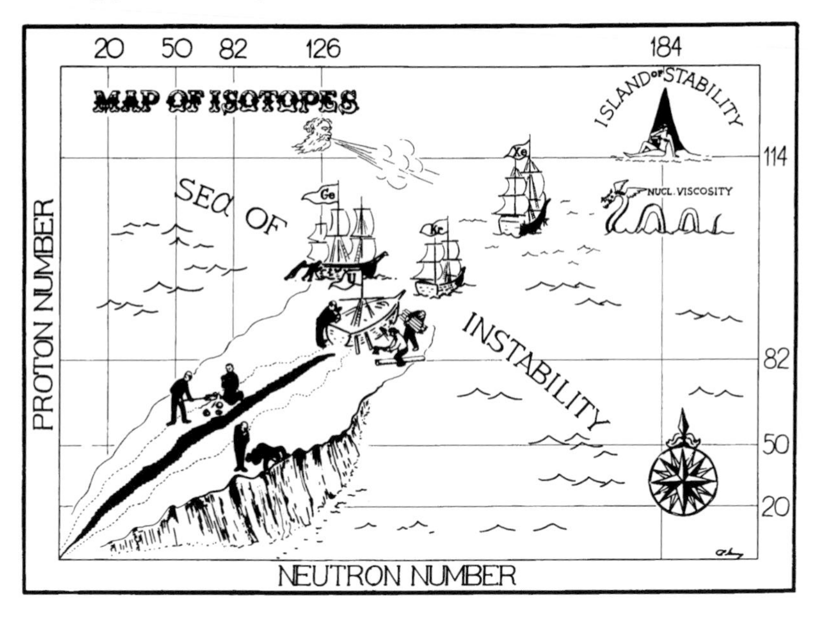

Fig. 3. Allegorical view of heavy-ion accelerator projects launched in the early seventieths for a journey to the island of superheavy elements. The flags indicate characteristic projectile beams offered by the facilities, see text. Cartoon provided by G.N. Flerov [13].

The situation is illustrated in a cartoon, Figure 3, which enjoyed the audiences of pertinent conferences in the early seventieth. Several sailors are shown in attempts of crossing the sea of instability, fighting against hostile forces. Already on the way were the crews of the JINR (Joint Institute for Nuclear Research) at Dubna/Soviet Union (now Russia) with the Heavy-ion U-300 Cyclotron (Xe) and of the IPN (Institut de Physique Nucléaire) at Orsay/France with the ALICE cyclotron (Kr), those of the LBL (Lawrence Berkeley Laboratory) at Berkeley/USA were just launching the SuperHILAC linear accelerator (Ge), and at Darmstadt/Germany, the UNILAC (U) of the GSI (Gesellschaft für Schwerionenforschung) was still under construction.

Turning to chemistry, the crucial question was: where are the superheavy elements located in the Periodic Table and how well do they fit its architecture? The answer had immediate implications for the ongoing "search for" campaigns, either for the selection of natural samples, or the design of chemical identification procedures in synthesis experiments. In a naive continuation of the Table, element 110 is located below platinum, 112 below mercury, 114 below lead, and 118 becomes the next noble gas below radon. Quantum-mechanical calculations of ground-state electronic configurations [14,15] confirmed this view. The electron configurations should indeed be analogous to the closest homologues; e.g., two *7s* valence electrons were predicted for element 112, as there are two *6s* electrons in mercury.

Extrapolations within the respective groups of the Periodic Table should thus be an appropriate approach to predict the chemical behavior of superheavy elements [14]. Examples are the detailed treatments [16] of chemical and physical properties of the *7p* elements 113 and 114, eka-thallium and eka-lead. Predictions of properties common to several superheavy elements were carried out for the design of group separations as a first step in chemical search experiments. Examples are the high volatility of elements 112 to 116 in the metallic state [17], or the formation of strong bromide complexes of elements 108 to 116 in solution [18].

However, deviations from straightforward extrapolations within the Periodic Table were also considered [19]. As a consequence of relativistic effects on the electronic structure, the *s*- and *p*-orbitals of heavy elements should shrink whereas higher lying orbitals should expand. Consequently, the two *s* electrons in element 112 and also the two $p_{1/2}$ electrons in element 114 could form closed electron shells, and eka-mercury and eka-lead could both be chemically inert gases like element 118, eka-radon.

Beyond element 121, eka-actinium, a series of *6f* elements may occur, in analogy to the *5f* actinide elements following actinium. But the *5g* orbital – the first *g* orbital at all – may be filled in competition. Consequently, a series of 32 superactinide elements [20] may exist in which inner electron shells are filled whereas the configuration of the valence electrons remains unchanged. The chemistry of such elements [14,15] would be most exciting.

Within a few years, many aspects of superheavy nuclei and elements were touched. A review [21] covering the literature until the end of 1973 was based on already 329 references, and status reports [22-27] published from time to time illustrate how the field developed. During this exploratory period continuing until the early eighties the optimistic perspectives of occurrence in Nature and easy synthesis at accelerators were proven in a trial-and-error approach, as is typical for new and attractive research topics. This Chapter gives an overview over this pioneering period.

2. Search for Superheavy Elements in Nature

Since the solar system and soon later the Earth's crust were formed about 4×10^9 y ago, a half-life of some 10^8 y for superheavy nuclei would be long enough for their survival until presence, provided they are produced in the nucleosynthesis. Heavy elements beyond iron are created in gigantic stellar explosions, the supernovae, which produce free neutrons in tremendous densities and initiate the so-called *r*-process, *r* stands for "rapid". Starting at seed nuclei around iron (the theoretical view at that time), several neutrons are captured to form very neutron-rich isotopes which decay quickly by β^--transitions to the next heavier element. These daughter products again capture neutrons and undergo β^--decay and so forth. In this way, the *r*-process path proceeds parallel to the belt of known nuclei to the heaviest elements, but shifted to much higher neutron numbers – as is sketched in Figure 4 – with discontinuities at magic neutron numbers.

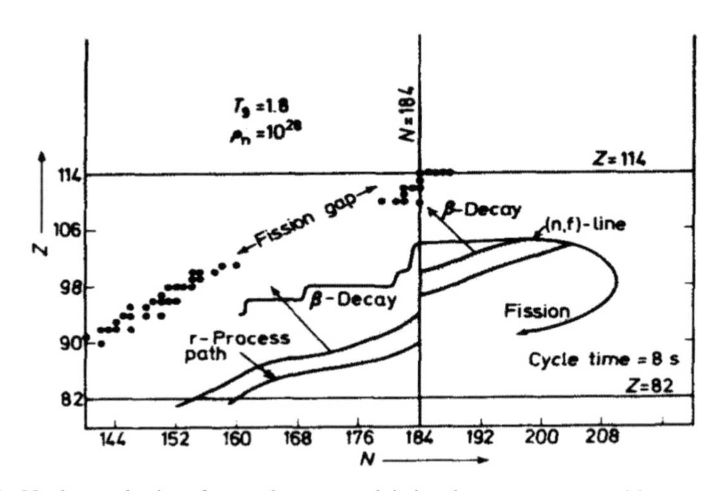

Fig. 4. Nucleosynthesis of superheavy nuclei in the r-process: rapid neutron capture alternating with ß⁻-transitions during a supernova explosion. Shown in the Z-N plane is the r-process path of very neutron-rich nuclei extending to Z≈100, from where β^--decay chains directed towards the belt of β-stable nuclei would lead to Z≈114, N≈184 nuclei (dots). From D.N. Schramm and W.A. Fowler [28], reprinted with permission from Nature, Copyright (2002) Macmillan Magazines Limited.

As soon as the stellar explosion ceases, the very short-lived nuclei decay towards the region of β-stability by a chain of fast β^--transitions, thereby further increasing the atomic number. Somewhere at very high atomic numbers, the *r*-process is terminated by neutron-induced and β^--delayed fission. Figure 4 resulted [28] for a stellar temperature of 1.8×10^9 K (T_9), a neutron density of 10^{28} cm^{-3} (ρ_n) and a cycle time of only 8 s for the whole process. This study predicted continuation of the process up to $Z\approx104$, sufficient to feed Z=114, N=184 magic nuclei. Other early treatments [29-31] denied the production of superheavy elements in the *r*-process. The question remained controversial for quite some time [32].

But even if the half-lives of superheavy nuclides would not exceed the 10^8 y level, there was hope to discover them in Nature. Although now extinct, they may have left detectable traces such as fission tracks or fission products in certain samples. Another possible source could be the cosmic radiation impinging on Earth whose heavy component may be formed by *r*-process nucleosynthesis in our galaxy not longer than 10^7 y ago [33] and may, hence, contain superheavy nuclei with half-lives down to some 10^5 years.

In this context, attention was drawn [11] to quite a number of earlier reports on natural α-particle emitters with energies which do not match with any known natural radioactive source but fall into the ranges expected for superheavy nuclides: were the superheavies already there, but unrecognized?

2.1 TERRESTRIAL SAMPLES

Any search for superheavy elements in terrestrial material begins with the choice of a sample. Relevant geochemical aspects were discussed in Refs. [34,35]. First searches were reported in 1969 by the Berkeley [10] and the Dubna groups [36]. In Berkeley, a search for element 110, eka-platinum, in natural platinum ores with standard analytical techniques remained negative at a concentration limit of 1 ppb, and low-level counting techniques did not reveal any activity above background. In Dubna, however, fission tracks discovered in lead glasses were tentatively attributed to element 114, eka-lead, present at a concentration of 10^{-11} to 10^{-12} gram per gram of sample, assuming a half-life of 1×10^9 y for the radioactive source. This latter convention is also followed here for comparison of results of such "search for" experiments.

Examinations of the same and of other lead-bearing samples for spontaneous fission events with large proportional counters in Dubna seemed to confirm these findings, but further measurements [37] of thin samples sandwiched between two plastic fission-track detectors showed that the events were background caused by cosmic-ray induced reactions of lead. Other groups [38] found no evidence for spontaneous fission activities in lead and other samples at a lower detection limit of 10^{-13} g/g achieved with the sandwich technique. Even lower limits down to 10^{-17} g/g can be reached by etching internal fission tracks in suitable minerals, where they were accumulated over millions of years. Searches [35,39] remained inconclusive in a variety of minerals containing heavy elements including lead, however.

A more generally applicable technique for spontaneous fission detection is counting of the neutrons emitted in the fission process. Although neutron detection is less efficient than fission fragment or α-particle counting, it can compete because much larger samples, up to tens of kilograms, can be inspected. With a simple arrangement of six ^{3}He-filled neutron counters, a sensitivity of 10^{-11} g/g was reached [40] in two days of counting, allowing a quick survey of a great variety of samples. Activities were indeed found with metallic platinum to bismuth, but with identical rates, an unlikely situation. Furthermore, as Figure 5 shows, the rates fell on a curve representing relative cross-sections for high-energy spallation reactions as a function of atomic number. Thus, the neutrons were due to cosmic-ray induced nuclear reactions during the counting operation.

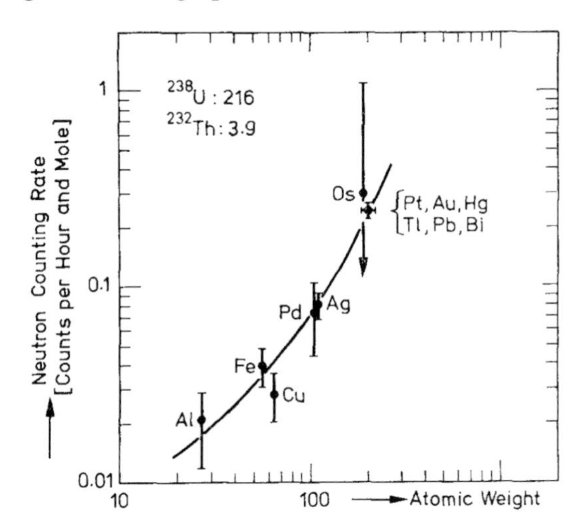

Fig. 5. Neutron counting as detection method for spontanous fission events of superheavy nuclei. The recorded neutron rates (points) were found to follow the relative cross sections of cosmic-ray induced spallation reactions (curve) and were, thus, due to background events. The numbers are rates for natural uranium and thorium. From W. Grimm, G. Herrmann and H.-D. Schüssler [40].

More advanced applications of neutron counting were based on the expectation that spontaneous fission events of superheavy nuclei should be accompanied by the emission of about ten neutrons [41,42], distinctly more than two to four observed for any other spontaneous fission decay. Such neutron bursts can be recognized by recording neutron multiplicities – events with several neutrons in coincidence – with ^{3}He-filled counting tubes [43,44] or large tanks filled with a liquid scintillator sensitive to neutrons [45].

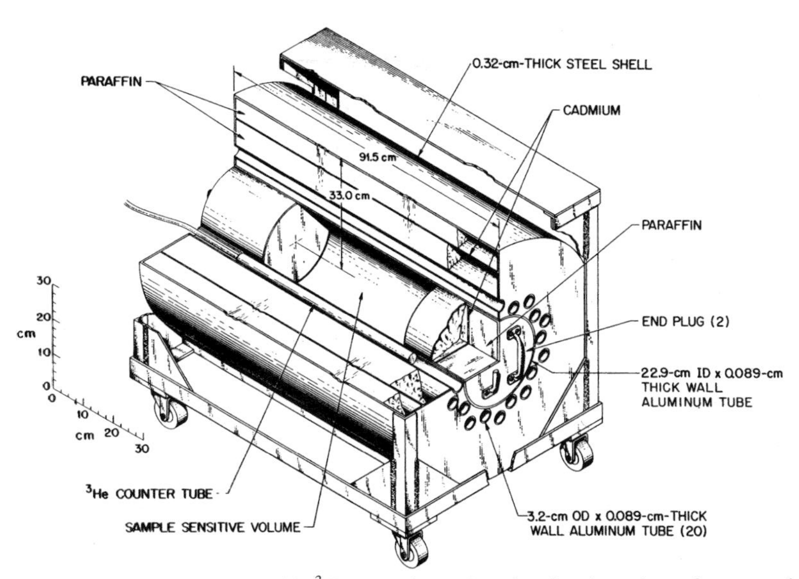

Fig. 6. Large neutron counter with ^{3}He counting tubes for the detection of neutron bursts emitted in the spontaneous fission of superheavy nuclei. Reproduced from R.L. Macklin et al. [43], Copyright (2002), with permission from Elsevier Science.

Figure 6 shows a multiplicity detector [43] with twenty ^{3}He counting tubes arranged in two rings around the central sample chamber which accommodated up to 100 kg of a sample. The tubes were embedded in paraffin for slowing down the neutrons. Bursts of ≥10 neutrons in the sample at an emitter concentration of 10^{-14} g/g should result in about one event per day with multiplicity four or larger. A similar sensitivity was reached [45] with a scintillator-based neutron detector. To suppress background by cosmic-ray induced neutron showers, such detectors were operated below ground and with an electronic anticoincidence shielding triggered by incoming high-energy particles. With the latter instrument [45] no positive results at the 10^{-14} g/g level were obtained for lead ores and samples from industrial lead processing. The publication gives an illustrative example how researchers were misled for some time by a tiny contamination of a sample by the synthetic 2.5-y ^{252}Cf, common as a source of spontaneous fission events in research laboratories.

A quite unconventional approach to internal fission-event detection was a device called the spinner. It operates on the same principle as the cloud chamber. The instrument, Figure 7, consists of a glass cylinder with glass arms filled with about one liter of the sample solution. Upon rotation, a negative pressure develops in the solution through the action of centrifugal forces. The solvent does not evaporate, however, but remains in a metastable state until a strongly ionizing event in the solution destroys this state and produces a bubble which is detected optically. The spinner can be operated at rates as low as one event per month corresponding to the detection of 10^{-13} to 10^{-14} g/g of a spontaneous fission activity. No such events were observed [46] in salts of platinum to lead and natural lead sulfide (galena).

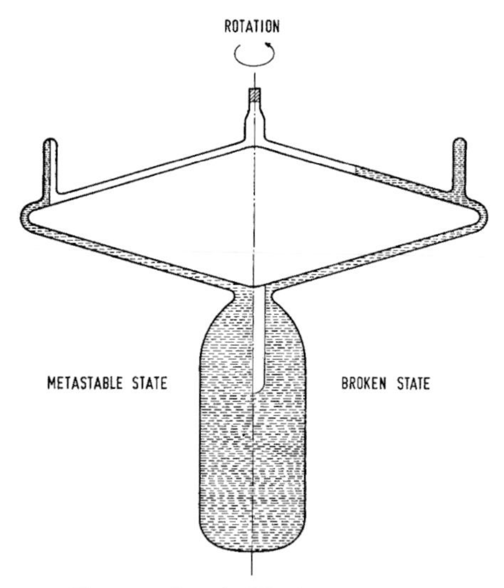

Fig. 7. The spinner detector. The container is filled with a liquid containing the sample. Upon rotation, a metastable state develops (at left) which breaks down after an ionizing event, as is indicated by the formation of a central bubble (at right). From K. Behringer et al. [46].

After the unsuccessful inspection of hundreds of samples with highly sensitive techniques attempts were made to improve the sensitivity of such searches by enrichment of superheavy elements from very large quantities. The extreme case [47] was a search for element 114 in flue dust collected during the industrial processing of 10^3 to 10^4 tons of galena. Eka-lead should be more volatile than lead and, hence, be enriched in flue dust. After further concentration by chemical and mass separations, the final samples were exposed to neutrons but no fission events were found with track detectors. The deduced concentration limit of 10^{-19} to 10^{-23} g/g is the lowest ever achieved in searches in Nature.

Among the "hottest" natural samples were brines from hot springs at the Cheleken Peninsula in the Caspian Sea which are known to be rich in volatile elements, probably due to the collection of material escaping from large depth in the Earth's mantle. They thus may carry superheavy elements deposited in deeper layers of the Globe. After processing of some 2000 m^3 of spring water through 850 kg of an anion exchange resin, a weak spontaneous fission activity appeared in the resin, and fractions eluted from the resin showed rates up to five events per day with neutron multiplicities as depicted

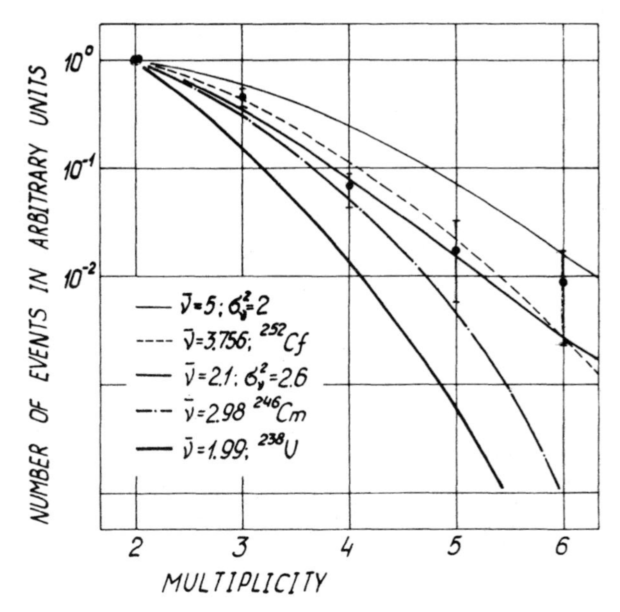

Fig. 8. Spontaneous fission activity in hot spring water at the Cheleken peninsula after concentration by ion exchange and precipitation methods. Shown is the measured neutron multiplicity distribution (dots) compared with measured distributions for ^{238}U, ^{246}Cm and ^{252}Cf spontaneous fission and calculated distributions for sets of ν and σ^2, the average number of neutrons per fission and its variance. Reproduced from G.N. Flerov et al. [48], Fig. 1, copyright (2002), with permission from Springer-Verlag.

in Figure 8 [48]. Evidently, natural ^{238}U can be ruled out as the source, but not a contamination by 2.6-y ^{252}Cf. Attempts to concentrate the activity for identification of its atomic number failed [49]. A search for such activities in similar brines, Salton Sea in California and Atlantis II at the floor of the Red Sea, gave no positive evidence [50].

Very unexpected news arrived in the summer of 1976: Evidence for element 126 and possibly 124 and 116 in Nature was reported [51] in a study of giant radioactive halos in biotite. Radioactive halos are a known phenomenon since the early days of radioactivity research. They are found in certain minerals as spherical zones of discoloration around a central mineral grain, Figure 9, and are due to radiation damage by α-particles escaping from radionuclides enclosed in the grain. Cuts through such halos reveal a ring structure with diameters corresponding to the ranges in the host mineral of α-particles from the natural decay series. There are, however, ranges which cannot be associated with known natural radionuclides. In particular, in biotite from Madagascar, giant halos were observed [52] with ranges equivalent to about 14 MeV α-particles, an energy predicted for nuclides around element 126. Such halos have as central grains relatively large crystals of monazite, a lanthanide-thorium-uranium phosphate.

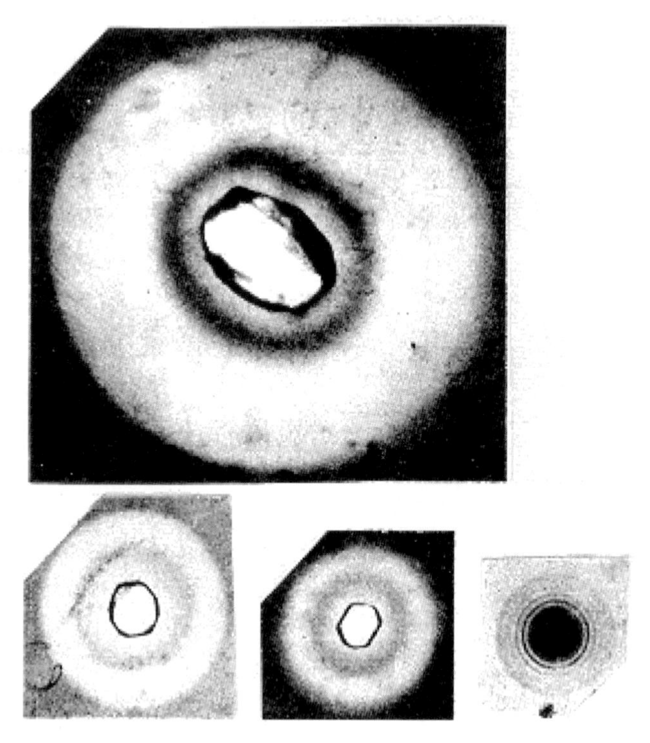

Fig. 9. Radioactive halos around large central monazite inclusions in biotite from Madagascar. Top: giant halo, bottom from left to right: thorium and uranium halos around, and a well-resolved uranium halo with a small central grain. All photographs are on the same scale; the outer diameter of the halo at top is 250 μm. From R.V. Gentry [53].

In order to verify the supposed presence of elements around $Z \approx 126$, monazite inclusions were irradiated with a sharply collimated proton beam and the proton induced x-ray spectra of the elements were recorded. As can be seen in Figure 10, two well separated groups of strong peaks appear [51], the L x-rays of uranium and thorium, and the K x-rays of the lanthanide elements. In between, from about 24 to 29 keV energy, much weaker peaks are identified and assigned to the $L_{\alpha 1}$ x-rays of elements 126, 116 and 124.

The concentrations of the superheavy elements, as estimated from peak intensities, were surprisingly high, 10 to 100 ppm. If such concentrations would also hold for bulk monazite, tons of superheavy elements would be easily accessible in some regions, e.g. at Indian beaches. In the concept of a superactinide series of elements, the occurrence of elements 126 and 124 in monazites would not be unlikely because they should be homologues of uranium and thorium [14]. Element 116 is expected to be a homologue of polonium. Since polonium is known to be strongly enriched by some marine invertebrates, it was suggested [54] to search for element 116 in crustacea

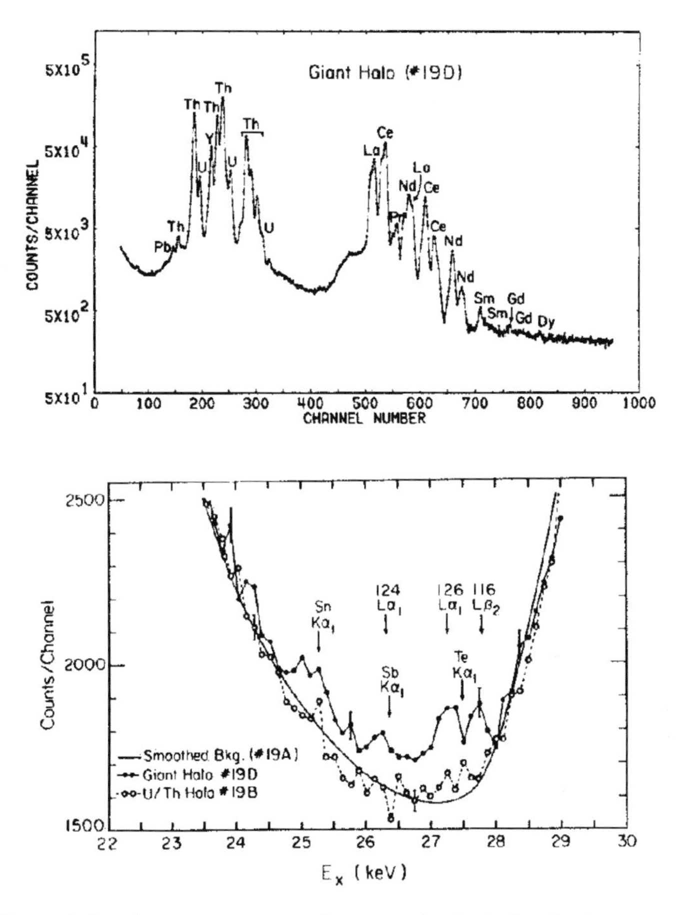

Fig. 10. Proton-induced x-ray spectrum of a monazite inclusion in the center of a giant radioactive halo (at top). The region in the gap around channel 400 is shown enlarged at the bottom (solid line) together with the spectrum of a U-Th halo (dashed line) and a smoothed background. From R.V. Gentry et al. [51].

such as lobsters, shrimp, and crabs in coastal waters at beaches rich in monazite sand – perhaps a gourmet's recommendation.

Objections against these findings were soon raised. The strongest peak attributed to element 126 could experimentally be accounted for [55] by a prompt γ-ray from the (p,n) nuclear reaction with natural ^{140}Ce, a major component of the monazite crystals. The weaker peaks were shown to stem from to *K* x-rays from traces of ordinary elements such as antimony and tellurium [56]. When a more specific technique for the excitation of x-ray spectra was applied to the inclusions, namely by monochromatic synchrotron radiation tuned to the x-ray absorption edges, the evidence for superheavy elements vanished [57,58]. Furthermore, attempts failed [59] to detect them in bulk monazites through isolation of an A>294 fraction with a mass separator. Chemical treatments [60] of large bulk samples also remained without success. The conclusion is that giant halos are not due to superheavy elements, but a generally accepted explanation what they are is still lacking.

2.2 EXTRA-TERRESTRIAL SAMPLES

No indications of spontaneous fission activities on the moon surface were obtained by neutron multiplicity counting of 3 kg of lunar rocks [45].

Much attention was paid to evidence for extinct superheavy elements in a class of primitive meteorites, the carbonaceous chondrites. These are low-temperature condensates from a solar gas that have more or less escaped reheating and other differentiating processes and may, thus, represent the stuff from which the solar system was made. They contain a surplus of the neutron-rich xenon isotopes 131 to 136 [61] first attributed to spontaneous fission of now extinct ^{244}Pu. But when this assignment became questionable after closer inspection, it was suggested [62,63] that superheavy elements might be the progenitors. This hypothesis was supported by correlations [64] between the concentrations of excess xenon and of volatile elements such as thallium, bismuth and indium in meteorites, which pointed to the homologues elements 111 to 116, a suspicion later narrowed to elements 115 or 114, 113 [65]. The strange xenon was found to be strongly enriched [66] in a host phase comprising less than 0.5 % of the meteorite, isolated after dissolution of the bulk in strong acids.

However, light xenon isotopes from 129 to 124 were also over-abundant [61,67,68] in such meteorites and enriched [66] in the tiny host phase although they are not formed in fission. Whether there are at least two anomalous xenon components of different origin, remained controversial for years [69]. Eventually, the fission origin of the anomalous xenon was ruled out [70] because in a host phase containing the excess xenon no enrichment was detected for the adjacent barium isotopes 130 to 138, which are abundant fission products.

Stimulated by these studies, samples of primitive meteorites were inspected by neutron multiplicity counting. In the Allende meteorite available in large quantities, a weak fission activity at the 10^{-14} g/g level was reported [71-73] but could not be chemically enriched [74-76].

Such "search for" experiments coincided with studies of the heavy element abundances in the cosmic radiation, carried out by exposure of particle track detectors – nuclear emulsions or plastic sheets – in balloon flights to high altitudes and analysis of the recorded tracks for atomic number and abundance. A survey [33] of all data obtained until 1970 showed one single event beyond $Z \approx 100$. With the data collected in the Skylab space station, the limit became more stringent: no superheavy nucleus in spite of recorded 204 tracks with atomic number 74 to 87 [77]. A similar limit was deduced [78] after exposure in a satellite. In a study of cosmic-ray induced tracks in olivine crystals enclosed in iron-stone meteorites and, hence, exposed in space over millions of years, unusually long tracks were found and attributed [79,80] to superheavy elements, but this conclusion could not be maintained [81,82] after calibration experiments with energetic ^{238}U beams at accelerators.

The largest collector surface for elements impinging on the Globe is the sea, of course. Heavy elements deposited in seawater are enriched in certain sediments such as manganese nodules, iron-manganese hydroxides. Fission tracks were found [83] in feldspar inclusions in such nodules, but no evidence was obtained [40,45] for spontaneous fission activities by counting nodules with neutron detectors.

3. Early Attempts to Synthesize Superheavy Elements

First attempts – in the early seventieths – to produce superheavy nuclei by nuclear synthesis in the laboratory were based on complete fusion of a projectile and a target nucleus chosen to attain by amalgamation the proton number of the desired element. In order to initiate fusion, close contact between projectile and target is required. The projectile has to carry sufficient kinetic energy to overcome the mutual electrostatic repulsion between the two partners. This energy is not completely consumed in the nuclear reaction, so that the compound nucleus retains excitation energy, which is high enough for the prompt evaporation of several neutrons. Examples for fusion-evaporation reactions are $^{249}Cf(^{12}C,4n)^{257}104$ and $^{242}Pu(^{22}Ne,5n)^{259}104$, at that time tried in Berkeley and Dubna, respectively, for the discovery of element 104. In such heavy systems, the yields of fusion products are much lower than expected according to the cross sections for compound nucleus formation. The deficit is attributed to past-fusion losses by fission of the compound nucleus and at the intermediate steps in the evaporation chain. As a precaution, the excitation energy of the compound nucleus is kept as low as possible by choosing low projectile energies.

For the synthesis of superheavy nuclei by complete fusion, larger projectiles with at least twice as many protons and neutrons are required. It soon became obvious that in such reactions the deficit in the yields is even larger than can be explained by post-fusion losses. This problem stimulated systematic studies of heavy-ion reactions all the way down from the first touch of two interacting nuclei until final fusion. New types of reactions –

called deep-inelastic reaction and quasi-fission – turned out to play a significant role. The partners first stick together in a dumbbell-shaped, dinuclear collision complex, which rotates, but the electrostatic repulsion between the partners drives the complex apart before amalgamation to a compound nucleus can occur. In this intermediate state nucleons are exchanged between the partners, so that with some probability one side can grow on the expense of the other one. The results are broad distributions of proton and neutron numbers and also excitation energies in the reaction products. This hindrance of fusion becomes significant especially for systems with a large total number of protons, as in superheavy element syntheses [25]. On the other hand, when very heavy projectiles such as uranium became available at the Darmstadt UNILAC in the middle of the seventieths, such incomplete transfer reactions offered an alternative approach to reach superheavy elements.

In the early experiments identification of synthetic superheavy nuclei was mainly attempted by radioactivity detection. With α-particle spectroscopy, signals in the energy range expected for superheavy nuclei were searched for, and with fission fragment detection half-lives not attributable to known spontaneous fission activities were looked at. Because of the complexity of the reaction product mixtures, additional selective steps were often included, such as chemical or mass separations. A principal problem was that – in the view at that time – the searched-for island would be completely isolated from known heavy nuclei by a region of instability against prompt fission. Hence, assignments of new elements by their connection, through radioactive decay, with nuclides of known elements – very successful for the identification of new actinide elements – were not possible. Thus, any positive evidence for superheavies could only be considered as a first hint and would call for further examinations, a requirement not always followed during this gold-rush period, leading to premature claims which soon disappeared again.

3.1 COMPLETE FUSION REACTIONS

As long as elements around 126 were the goal, the perspectives for superheavy element synthesis by complete fusion reactions looked very promising. Based on data for the production of heavy actinides and without knowing the fusion hindrance, cross sections as large as tens of millibarn were extrapolated [84,85], e.g. for the ^{232}Th(^{80}Kr,2n)$^{310}126_{184}$ reaction. However, with element 114 in the focus the situation became different in that the doubly magic Z=114, N=184 is extremely neutron rich. Its neutron-to-proton ratio cannot be achieved by any realistic projectile-target combination. Close approach to Z=114 means that the neutron number remains far below N=184, and this deficit would further increase by evaporation of several neutrons from the compound nucleus to products located close to the border or even outside the anticipated stability island. On the other hand, to meet N=184 would require an overshooting of Z=114 by about ten protons [84,86].

The first attempt to synthesize element 114 was made in 1969 in Berkeley [10] by bombarding ^{248}Cm – a synthetic, high-Z, neutron-rich target – with ^{40}Ar projectiles. The experiment was negative at a cross section limit of about 10 nanobarn. The compound nucleus $^{288}114$ contains only 174 neutrons, and it should evaporate four neutrons to $^{284}114_{170}$, probably located outside the island. With the larger variety of heavy-ion beams soon becoming available, attempts to reach the island by complete fusion reactions were carried out on a broad basis. Within a few years, twenty different reactions were tried [24] covering compound nuclei with proton numbers from 110 to 128 and neutron numbers from 168 to 194.

The detection techniques applied in those early attempts were often surprisingly simple searches for spontaneous fission activities. The whole product mixture was collected on a catcher foil and exposed to mica, glass or polymer sheets to produce tracks of spontaneous fission events. By quickly rotating the catcher between detector foils during bombardment, this technique allows the detection of short-lived nuclides down to millisecond half-lives [88].

A few examples may illustrate this exploratory period. Fusion of ^{76}Ge with ^{232}Th and ^{238}U aimed at the synthesis of the elements 122 and 124 was studied at Dubna [88] without positive results. High-energy α-particles, 13 to 15 MeV energy – as predicted for elements around 126 – from short-lived emitters were observed at the ALICE Orsay in bombardments of thorium with krypton. This was taken as an indication [89] for the synthesis of a compound nucleus of element 126. But attempts in the same laboratory to secure the evidence by a direct mass identification failed [90]. In these ambitious experiments, the magnetic rigidity, kinetic energy and time-of-flight of fragments from ^{84}Kr+^{232}Th, ^{208}Pb and ^{238}U interactions were measured. Whereas most of the early fusion studies employed relatively light projectiles and heavy targets, more symmetric combinations were also investigated, such as ^{136}Xe+^{181}Ta, which could fuse to element 127 [91].

In this time, very surprising news appeared: can superheavy nuclei be made by heavy-ion reactions without using a heavy-ion accelerator? From tungsten plates bombarded over long periods in the beam dump of the 24 GeV proton beam at CERN Geneva, a weak, long-lived spontaneous fission activity was chemically isolated in the mercury fraction and assigned to element 112, eka-mercury [92]. Production by a two-stage process was postulated:

generation of a broad distribution of energetic recoil atoms by proton-induced spallation of tungsten, followed by fusion of such atoms with tungsten. Attempts to confirm the results in other laboratories were unsuccessful [46,93,94], a conclusion finally shared by most of the original authors [95].

After this first round of fusion experiments, further efforts were focused on experiments with ^{48}Ca as projectile, a very neutron-rich and doubly magic isotope (Z=20, N=28), rare (0.19 % natural abundance) and expensive. At Berkeley [96,97] targets of ^{248}Cm were bombarded, at Dubna ^{246,248}Cm, ^{243}Am, ^{242}Pu, ^{233}U, ^{231}Pa, and ^{232}Th [98,99].

The ^{48}Ca+^{248}Cm reaction was generally considered to be most promising for the synthesis of superheavy elements because the compound nucleus Z=116, N=180 provides at a moderate overshooting of the proton shell a relatively close approach to the neutron shell. Also, the decay chain after evaporation of four neutrons was expected [12] to be suitable for detection. It should start at $^{292}116_{176}$ by α-decay with a few seconds half-life, followed within several minutes by two electron captures in $^{288}114_{174}$ and $^{288}113_{175}$ and ends at $^{288}112_{176}$ by spontaneous fission with a 50 min half-life.

In Figure 11 we refer to a later elaborate series of ^{48}Ca+^{248}Cm experiments performed by a large international collaboration [100] at the UNILAC and the SuperHILAC. With a variety of techniques, a half-life range of 14 orders of magnitude – from 1 μs to 10 y – was covered, but again without any evidence for superheavy nuclei. In the Figure, the resulting upper limits of the production cross sections are plotted as a function of the assumed half-life. Each technique has a region of highest sensitivity, limited by decay losses before isolation at shorter, and too low decay rates at longer half-lives. The region of very short-lived nuclei (curves 1 and 2) was inspected with two recoil-fragment separators with fragment detection in surface-barrier detectors. For intermediate half-lives (3 to 5), chemical on-line separations were applied, and off-line chemistry for long-lived products (6 to 8). The chemical procedures were based on volatilization at high temperature (3 and 8) aimed at elements 112 through 116 [17], and at room temperature (4 and 6) for 112 and 114 [19]. Anion exchange of bromide complexes (5 and 7) was applied for 108 to 116 [18]. The cross section limit achieved was about 200 picobarn with some extra sensitivity gained for long-lived products by fission fragment-fission neutron coincidence counting [101].

The ^{48}Ca+^{254}Es reaction with a Z=119 compound nucleus was also studied [102] with negative results.

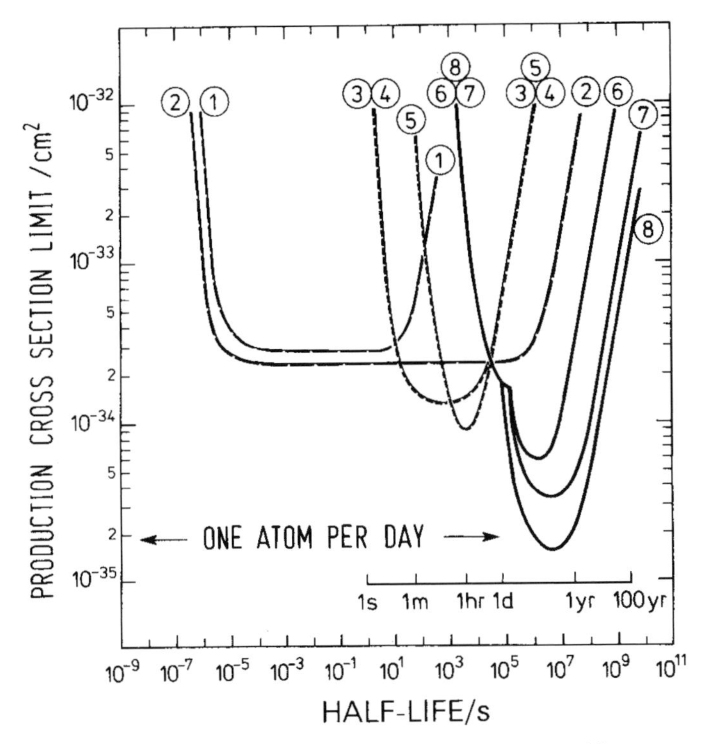

Fig. 11. Search for superheavy nuclides in the reaction of ^{48}Ca with ^{248}Cm: upper limits of production cross sections plotted versus half-life. The curves refer to different separation techniques (see text): recoil-fragment separators (curves 1, 2), fast on-line chemistry (3-5), and off-line chemistry with low-background counting (6-8). From P. Armbruster et al. [100].

The ^{48}Ca+^{248}Cm system has recently been reinvestigated at Dubna [103], now with positive evidence for the formation of ^{292}116 (see Ch. 1). The half-lives in the postulated decay chain – milliseconds to seconds – fall into the region covered in Figure 11. But the production cross section is two orders of magnitude below the level reached in the previous studies. If these results can further be substantiated, they would give a hint why so many attempts to make superheavy elements by complete fusion failed. Over the years, the question was: Is production the problem or is it survival – nuclear reaction or nuclear stability? Are the cross sections too low or the half-lives too short? These recent results would point to the production as the key problem.

3.2 INCOMPLETE TRANSFER REACTIONS

Would incomplete transfer processes open an alternative approach to the synthesis of superheavy nuclei? Would, e.g., in the collision of two $^{238}U_{146}$ nuclei one partner take up enough protons and neutrons to grow to the doubly magic $^{298}114_{184}$, whereas the complementary partner would shrink to $^{178}70_{108}$, a known neutron-rich isotope of ytterbium?

A radiochemical study [104] of the element distribution in the $^{238}U+^{238}U$ reaction at the UNILAC revealed the expected broad distribution of reaction products. Below uranium, where losses by sequential fission of transfer products are not significant, the observed yields decreased exponentially from Z=92 down to Z=73. This trend was well reproduced [105] by a theoretical model treating nucleon transfer in the intermediate collision complex as a diffusion process. By extrapolation of the model to Z=70 nuclei about 100 microbarn total production cross section resulted, associated with broad distributions of neutron numbers and excitation energies.

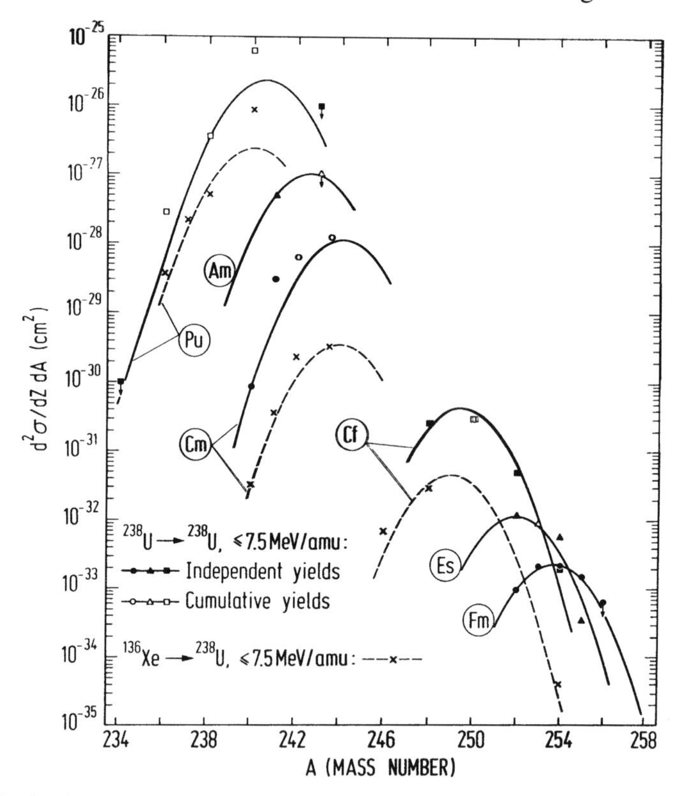

Fig. 12. Production cross sections of transuranium nuclides in the interaction of ^{238}U with ^{238}U (solid lines) plotted versus mass number. Also shown are data for the $^{136}Xe+^{238}U$ interaction (dashed lines). From M. Schädel et al. [104].

The Z<92 distributions should be reflected in complementary Z>92 distributions, including the Z=114 isotopes. However, beyond uranium, fission of the freshly formed transfer products becomes very significant and leads to severe losses. As is shown in Figure 12, the production cross sections of surviving transuranium nuclides decreased from plutonium to fermium by eight orders of magnitude. Nonetheless, an extrapolation for surviving Z=114 fragments gave about 10 picobarn cross section [25], a not completely hopeless situation.

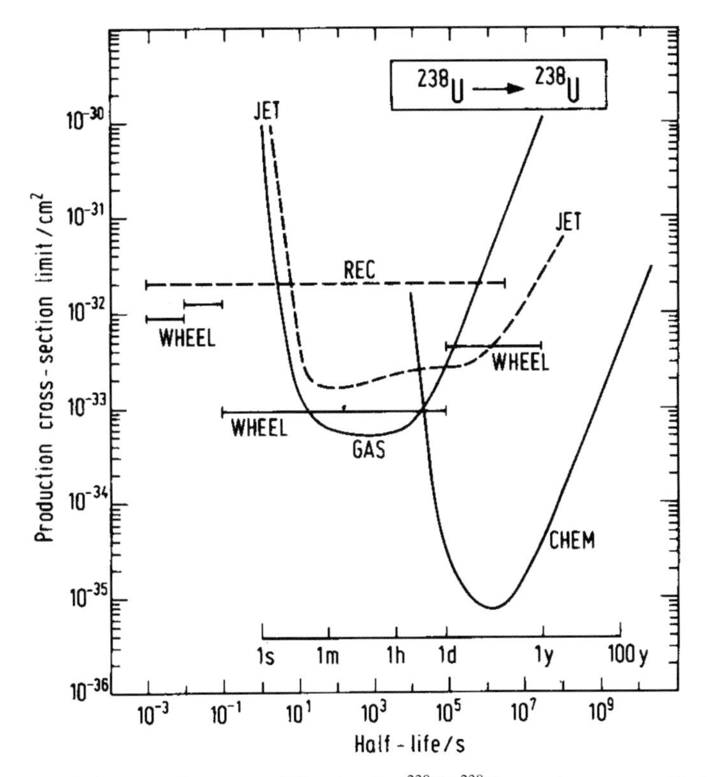

Fig. 13. Search for superheavy nuclides in the $^{238}U+^{238}U$ reaction: upper limits for the production cross section obtained with various techniques (see text), plotted versus half-life. Reproduced from N. Hildebrand et al. [108], Copyright (2002), with permission from Elsevier Science.

Direct searches for superheavy elements in the $^{238}U+^{238}U$ reaction were undertaken at the UNILAC by several groups. All these efforts remained without positive evidence. The data are summarized in Figure 13. The curve labeled CHEM [106] was obtained with off-line chemical separations [107] and an assay for α-and spontaneous fission activities; here, the 10 picobarn level was reached for half-lives between several days and years. Attempts to detect short-lived nuclides were less sensitive. The curve labeled GAS holds for an on-line search [108] for components volatile at room temperature. WHEEL [106] refers to fission track detection in the unseparated product mixture deposited on a rotating catcher, REC [109] to implantation of recoil atoms in a surface barrier detector, and JET to on-line transport from target to detector with a gas jet [91,110].

Attempts to find superheavy elements in the $^{238}U+^{248}Cm$ reaction failed, too [111], although the production cross sections for transcurium isotopes increased by three orders of magnitude [112] compared with the $^{238}U+^{238}U$ reaction.

4. Epilogue

In a condensed retrospective as presented here, the exploratory period of superheavy element research may appear as a sequence of frustrating failures. But for the participants it was an exciting time, with a broad range of approaches, from systematic scanning of large numbers of samples to quite unconventional experiments. Claims for success were raised from time to time and challenged for better experiments to prove them. Although in all cases the claims could not be substantiated, they forced the researchers to push physical and chemical detection techniques to levels not achieved before.

In the early eighties, the exploratory period came to an end. After so many unsuccessful attempts to discover the island of superheavy nuclei, extensive experimental and theoretical studies of the relevant nuclear processes were undertaken to provide a better knowledge. From then on expansion of the Periodic Table happened in a systematic element-after-element approach, initiated in 1981 at the UNILAC by the synthesis of element 107 [113]. Since then, the landscape sketched in Figure 3 has changed completely (see Ch. 1). Nowadays, a ridge of short-lived α-particle emitters extends from the peninsula of known nuclei to the region previously considered as an island. With an increasing data base for very heavy nuclei, the extrapolated half-lives of nuclides closer to the magic nuclei went down, e.g. to about ten days for Z=110, N=184 [114]. Hence, further searches for superheavy elements in Nature appear to be rather hopeless. The chemists involved in searches for short-lived superheavy elements have since changed their emphasis. Instead of using chemical methods as a tool to discover unknown elements, they now apply their know-how to explore the unknown chemistry of already known elements at the end of the Periodic Table.

References

1. Wheeler, J.A.: “Fission Physics and Nuclear Theory”. In: Proceedings of the “International Conference on the Peaceful Uses of Atomic Energy”, Geneva, 8-20 August 1955, Vol. 2, pp. 155-163, 220-226.
2. Scharff-Goldhaber, G.: Nucleonics **15**, (1957) No. 9, 122.
3. Myers, W.D., Swiatecki, W.J.: Nucl. Phys. **81**, (1966) 1.
4. Meldner, H.: “Predictions of New Magic Regions and Masses for Super-Heavy Nuclei from Calculations with Realistic Shell Model Single Particle Hamiltonians”. In Ref. [5], pp. 593-598.
5. Proceedings of the “International Symposium Why and How Should We Investigate Nuclides far off the Stability Line”, Lysekil, Sweden, 21-27 August 1966, Arkiv Fysik **36**, (1967).
6. Sobiczewski, A., Gareev, F.A., Kalinkin, B.N.: Phys. Lett. **22**, (1966) 500.
7. Nilsson, S.G., Nix, J.R., Sobiczewski, A., Szymański, Z., Wycech, S., Gustafson, C., Möller, P.: Nucl. Phys. **A115**, (1968) 545.
8. Grumann, J., Mosel, U., Fink, B., Greiner, W.: Z. Physik **228**, (1969) 371.
9. Nilsson, S.G., Tsang, C.F., Sobiczewski, A., Szymański, Z., Wycech, S., Gustafson, C., Lamm, I.-L., Möller, P., Nilsson, B.: Nucl. Phys. **A131**, (1969) 1.
10. Nilsson, S.G., Thompson, S.G., Tsang, C.F.: Phys. Lett. **28B**, (1969) 458.
11. Meldner, H., Herrmann, G.: Z. Naturforschung **24a**, (1969) 1429.
12. Fiset, E.O., Nix, J.R.: Nucl. Phys. **A193**, (1972) 647.
13. Flerov, G.N.: Cartoon in: Proceedings of the “Nobel Symposium 27. Super-Heavy Elements – Theoretical Predictions and Experimental Generation”, Ronneby, Sweden, 11-14 June 1974, Physica Scripta **10A**, (1974) 1.
14. Fricke, B., Greiner, W., Waber, J.T.: Theoret. Chim. Acta **21**, (1971) 235.
15. Fricke, B.: Structure & Bonding **21**, (1975) 89.
16. Keller, O.L., Burnett, J.L., Carlson, T.A., Nestor, C.W.: J. Phys. Chem. **74**, (1970) 1127.
17. Eichler, B.: Kernenergie **19**, (1976) 307.
18. Kratz, J.V., Liljenzin, J.O., Seaborg, G.T.: Inorg. Nucl. Chem. Lett. **10**, (1974) 951.
19. Pitzer, K.S.: J. Chem. Phys. **63**, (1975) 1032.
20. Waber, J.T., Cromer, D.T., Liberman, D.: J. Chem. Phys. **51**, (1969) 664.
21. Herrmann, G.: “Superheavy Elements”. In: *International Review of Science, Inorganic Chemistry Series Two, Vol. 8, Radiochemistry,* Ed. Maddock, A.G., Butterworths, London, (1975) 221-272.
22. Thompson, S.G., Tsang, C.F.: Science **178**, (1972) 1047.
23. Seaborg, G.T., Loveland, W., Morrissey, D.J.: Science **203**, (1979) 711.
24. Herrmann, G.: Nature **280**, (1979) 543.
25. Kratz, J.V.: Radiochimica Acta **32**, (1983) 25.
26. Flerov, G.N., Ter-Akopyan, G.M.: Rep. Prog. Phys. **46**, (1983) 817.
27. Herrmann, G.: Angew. Chem. Int. Ed. Engl. **27**, (1988) 1417; transl. from Angew. Chem. **100**, (1988) 1629.
28. Schramm, D.N., Fowler, W.A.: Nature **231**, (1971) 103.
29. Viola, V.E.: Nucl. Phys. **A139**, (1969) 188.
30. Schramm, D.N., Fiset, E.O.: Astrophys. J. **180**, (1973) 551.
31. Howard, W.M., Nix, J.R.: Nature **247**, (1974) 17.
32. Meyer, B.S., Möller, P., Howard, W.M., Mathews, G.J.: “Fission Barriers for *r*-Process Nuclei and Implications for Astrophysics”. In: Proceedings of the Conference “50 Years with Nuclear Fission”, Gaithersburg, Maryland, 25-28 April 1989, pp. 587-591.
33. Price, B.P., Fowler, P.H., Kidd, J.M., Kobetich, E.J., Fleischer, R.L., Nichols, G.E.: Phys. Rev. **D3**, (1971) 815.
34. Vdovenko, V.M., Sobotovich, M.: Sov. Phys. Doklady **14**, (1969), 1179; transl. from Dokl. Akad. Nauk SSSR Fiz. Ser. **189**, (1969) 980.

35. Haack, U.: Naturwissenschaften **60**, (1973) 65.
36. Flerov, G.N., Perelygin, V.P.: Sov. J. Atom. Energy **26**, (1969), 603; transl. from Atomnaya Energiya **26**, (1969) 520.
37. Flerov, G.N., Perelygin, V.P., Otgonsurén, O.: Sov. J. Atom. Energy **33**, (1972) 1144; transl. from Atomnaya Energiya **33**, (1972) 979.
38. Geisler, F.H., Phillips, P.R., Walker, R.M.: Nature **244**, (1973) 428.
39. Price, P.B., Fleischer, R.L., Woods, R.T.: Phys. Rev. **C1**, (1970) 1819.
40. Grimm, W., Herrmann, G., Schüssler, H.-D.: Phys. Rev. Lett. **26**, (1971) 1040, err. 1408.
41. Nix, J.R.: Phys. Lett. **30B**, (1969) 1.
42. Schmitt, H.W., Mosel, U.: Nucl. Phys. **A186**, (1972) 1.
43. Macklin, R.L., Glass, F.M., Halperin, J., Roseberry, R.T., Schmitt, H.W., Stoughton, R.W., Tobias, M.: Nucl. Instr. Meth. **102**, (1972) 181.
44. Ter-Akopyan, G.M., Popeko, A.G., Sokol, E.A., Chelnokov, L.P., Smirnov, V.I., Gorshkov, V.A.: Nucl. Instr. Meth. Phys. Res. **190**, (1981) 119.
45. Cheifetz, E., Jared, R.C., Giusti, E.R., Thompson, S.G.: Phys. Rev. **C6**, (1972) 1348.
46. Behringer, K., Grütter, A., von Gunten, H.R., Schmid, A., Wyttenbach, A., Hahn, B., Moser, U., Reist, H.W.: Phys. Rev. **C9**, (1974) 48.
47. McMinn, J., Ihle, H.R., Wagner, R.: Nucl. Instr. Meth. **139**, (1976) 175.
48. Flerov, G.N., Korotkin, Yu.S., Ter-Akopyan, G.M., Zvara, I., Oganessian, Yu.Ts., Popeko, A.G., Chuburkov, Yu.T., Chelnokov, I.P., Maslov, O.D., Smirnov, V.I., Gerstenberger, R.: Z. Physik **A292**, (1979) 43.
49. Chuburkov, Yu.T., Popeko, A.G., Skobelev, N.K.: Sov. Radiochem. **30**, (1988) 108; transl. from Radiokhimiya **30**, (1988) 112.
50. Ter-Akopyan, G.M., Sokol, E.A., Fam Ngoc Chuong, Ivanov, M.P., Popeko, G.S., Molzahn, D., Lund, T., Feige, G., Brandt, R.: Z. Physik **A316**, (1984) 213.
51. Gentry, R.V., Cahill, T.A., Fletcher, N.R., Kaufmann, H.C., Medsker, L.R., Nelson, J.W., Flocchini, R.G.: Phys. Rev. Lett. **37**, (1976) 11.
52. Gentry, R.V.: Science **169**, (1970) 670.
53. Gentry, R.V.: Ann. Rev. Nucl. Sci. **23**, (1973) 347.
54. Wolke, R.L.: Phys. Rev. Lett. **37**, (1976) 1098.
55. Bosch, F., El Goresy, A., Krätschmer, W., Martin, B., Povh, B., Nobiling, R., Traxel, K., Schwalm, D.: Phys. Rev. Lett. **37**, (1976) 1515.
56. Wölfli, W., Lang, J., Bonani, G., Suter, M., Stoller, Ch., Nissen, H.-U.: J. Phys. **G3**, (1977) L33.
57. Sparks, C.J., Raman, S., Yakel, H.L., Gentry, R.V., Krause, M.O.: Phys. Rev. Lett. **38**, (1977), 205.
58. Sparks, C.J., Raman, S., Ricci, E., Gentry, R.V., Krause, M.O.: Phys. Rev. Lett. **40**, (1978) 507, err. 1112.
59. Stéphan, C., Epherre, M., Cieślak, E., Sowiński, M., Tys, J.: Phys. Rev. Lett. **37**, (1976) 1534.
60. Stakemann, R., Heimann, R., Herrmann, G., Tittel, G., Trautmann, N.: Nature **297**, (1982) 136.
61. Reynolds, J.H., Turner, G.: J. Geophys. Res. **69**, (1964) 3263.
62. Anders, E., Heymann, D.: Science **164**, (1969) 821.
63. Dakowski, M.: Earth Planet. Sci. Lett. **6**, (1969) 152.
64. Anders, E., Larimer, J.W.: Science **175**, (1972) 981.
65. Anders, E., Higuchi, H., Gros, J., Takahashi, H., Morgan, J.W.: Science **190**, (1975) 1262.
66. Lewis, R.S., Srinivasan, B., Anders, E.: Science **190**, (1975) 1251.
67. Manuel, O.K., Hennecke, E.W., Sabu, D.D.: Nature Phys. Sci. **240**, (1972) 99.
68. Manuel, O.K., Sabu, D.D.: Science **195**, (1977) 208.
69. Begemann, F.: Rep. Prog. Phys. **43**, (1980) 1309.
70. Lewis, R.S., Anders, E., Shimamura, T., Lugmair, G.W.: Science **222** (1983) 1013.

71. Popeko, A.G., Skobelev, N.K., Ter-Akopyan, G.M., Goncharov, G.N.: Phys. Lett. **52B**, (1974) 417.
72. Flerov, G.N., Ter-Akopyan, G.M., Popeko, A.G., Fefilov, B.V., Subbotin, V.G.: Sov. J. Nucl. Phys. **26**, (1977) 237; transl. from Yadern. Fiz. **26**, (1977) 449.
73. Amirbekyan, A.V., Davtyan, L.S., Markaryan, D.V., Khudaverdyan, A.G.: Sov. J. Nucl. Phys. **36**, (1982) 786; transl. from Yadern. Fiz. 3**6**, (1982) 1356.
74. Zvara, I., Flerov, G.N., Zhuĭkov, B.L., Reetz, T., Shalaevskiĭ, M.R., Skobelev, N.K.: Sov. J. Nucl. Phys. **26**, (1977) 240; transl. from Yadern. Fiz. **26**, (1977) 455.
75. Lund, T., Becker, H.-J., Jungclas, H., Molzahn, D., Vater, P., Brandt, R.: Inorg. Nucl. Chem. Lett. **15**, (1979) 413.
76. Lund, T., Tress, G., Khan, E.U., Molzahn, D., Vater, P., Brandt, R.: J. Radioanal. Nucl. Chem. Lett. **93**, (1985) 363.
77. Shirk, E.K., Price, P.B.: Astrophys. J. **220**, (1978) 719.
78. Fowler, P.H., Walter, R.N.F., Masheder, M.R.W., Moses, R.T., Worley, A., Gay, A.M.: Astrophys. J. **314**, (1987) 739.
79. Otogonsurén, O., Perelygin, V.P., Stetsenko, S.G., Gavrilova, N.N., Fiéni, C., Pellas P.: Astrophys. J. **210**, (1976) 258.
80. Perelygin, V.P., Stetsenko, S.G.: JETP Lett. **32**, (1980) 622; transl. from Pisma Zh. Eksp. Teor. Fiz. **32**, (1980) 622.
81. Perron, C., Bourot-Denise, M., Perelygin, V.P., Birkholz, W., Stetsenko, S.G., Dersch, R., Zhu, T.C., Vater, P., Brandt, R.: Nucl. Tracks Radiat. Meas. **15** (1988) 231.
82. Perelygin, V.P., Stetsenko, S.G.: JETP Lett. **49**, (1989) 292; transl. from Pisma Zh. Eksp. Teor. Fiz. **49**, (1989) 257.
83. Otgonsurén, O., Perelygin, V.P., Flerov, G.N.: Sov. Phys. Doklady **14**, (1970) 1194; transl. from Dokl. Akad. Nauk SSSR Fiz. Ser. **189**, (1969) 1200.
84. Sikkeland, T.: “Synthesis of Nuclei in the Region of Z=126 and N=184”. In Ref. [5], pp. 539-552.
85. Wong, C.Y.: Nucl. Phys. **A103**, (1967) 625.
86. Lefort, M.: Ann. Physique **5**, (1970) 355.
87. Randrup, J., Larsson, S.E., Möller, P., Sobiczewski, A., Lukasiak, A.: Physica Scripta **10A**, (1974) 60.
88. Flerov, G.N., Oganessian, Yu.Ts., Lobanov, Yu.V., Pleve, A.A., Ter-Akopyan, G.M., Demin, A.G., Tretyakova, S.P., Chepigin, V.I., Tretyakov, Yu.P.: Sov. J. Nucl. Phys. **19**, (1974) 247; transl. from Yad. Fiz. **19**, (1974) 492.
89. Bimbot, R., Deprun, C., Gardès, D., Gauvin, H., Le Beyec, Y., Lefort, M., Pèter, J.: Nature **234**, (1971) 215.
90. Colombani, P., Gatty, B., Jacmart, J.C., Lefort, M., Pèter, J., Riou, M., Stéphan, C., Tarrago, X.: Phys. Lett. **42B**, (1972) 208.
91. Aumann, D.C., Faleschini, W., Friedmann, L., Weismann, D.: Phys. Lett. **82B**, (1979) 361.
92. Marinov, A., Batty, C.J., Kilvingston, A.I., Newton, G.W.A., Robinson, V.J., Hemingway, J.D.: Nature **229**, (1971) 464.
93. Unik, J.P., Horwitz, E.P., Wolf, K.L., Ahmad, I., Fried, S., Cohen, D., Fields, P.R., Bloomquist, C.A.A., Henderson, D.J., Nucl. Phys. **A191**, (1972) 233.
94. Westgaard, L., Erdal, B.R., Hansen, P.G., Kugler, E., Sletten, G., Sundell, S., Fritsch, T., Henrich, E., Theis, W., Wolf, G.K., Camplan, J., Klapisch, R., Meunier R., Poszkanzer, A.M., Stéphan, C., Tys, J.: Nucl. Phys. **A192**, (1972) 517.
95. Batty, C.J., Kilvingston, A.I., Weil, J.L., Newton, G.A.W., Skarestad, M., Hemingway, J.D.: Nature **244**, (1973) 429.
96. Hulet, E.K., Lougheed, R.W., Wild, J.F., Landrum, J.H., Stevenson, P.C., Ghiorso, A., Nitschke, J.M., Otto, R.J., Morrissey, D.J., Baisden, P.A., Gavin, B.F., Lee, D., Silva, R.J., Fowler, M.M., Seaborg, G.T.: Phys. Rev. Lett. **39**, (1977) 385.

97. Illige, J.D., Hulet, E.K., Nitschke, J.M., Dougan, R.J., Lougheed, R.W., Ghiorso, A., Landrum, J.H.: Phys. Lett. **78B**, (1978) 209.
98. Oganessian, Yu.Ts., Bruchertseifer, H., Buklanov, G.V., Chepigin, V.I., Choy Val Sek, Eichler, B., Gavrilov, K.A., Gäggeler, H., Korotkin, Yu.S., Orlova, O.A., Reetz, T., Seidel, W., Ter-Akopyan, G.M., Tretyakova, S.P., Zvara, I.: Nucl. Phys. **A294**, (1978) 213.
99. Ter-Akopyan, G.M., Bruchertseifer, H., Buklanov, G.V., Orlova, O.A., Pleve, A.A., Chepigin, V.I., Choy Val Sek: Sov. J. Nucl. Phys. **29**, (1979) 312; transl. from Yad. Fiz. **29**, (1979) 608.
100. Armbruster, P., Agarwal, Y.K., Brüchle, W., Brügger, M., Dufour, J.P., Gäggeler, H., Hessberger, F.P., Hofmann, S., Lemmertz, P., Münzenberg, G., Poppensieker, K., Reisdorf, W., Schädel, M., Schmidt, K.-H., Schneider, J.H.R., Schneider, W.F.W., Sümmerer, K., Vermeulen, D., Wirth, G., Ghiorso, A., Gregorich, K.E., Lee, D., Leino, M., Moody, K.J., Seaborg, G.T., Welch, R.B., Wilmarth, P., Yashita, S., Frink, C., Greulich, N., Herrmann, G., Hickmann, U., Hildebrand, N., Kratz, J.V., Trautmann, N., Fowler, M.M., Hoffman, D.C., Daniels, W.R., von Gunten, H.R., Dornhöfer, H.: Phys. Rev. Lett. **54**, (1985) 406.
101. Peuser, P., Tharun, U., Keim, H.-J., Trautmann, N., Herrmann, G., Wirth, G.: Nucl. Instr. Meth. Phys. Res. **A239**, (1985) 529.
102. Lougheed, R.W., Landrum, J.H., Hulet, E.K., Wild, J.F., Dougan, R.J., Dougan, A.D., Gäggeler, H., Schädel, M., Moody, K.J., Gregorich, K.E., Seaborg, G.T.: Phys. Rev. **C32**, (1985) 1760.
103. Oganessian, Yu.Ts., Utyonkov, V.K., Moody, K.J.: Phys. Atomic Nuclei **64,** (2001) 1349.
104. Schädel, M., Kratz, J.V., Ahrens, H., Brüchle, W., Franz, G., Gäggeler, H., Warnecke, I., Wirth, G., Herrmann, G., Trautmann, N., Weis, M.: Phys. Rev. Lett. **41**, (1978) 469.
105. Riedel, C., Nörenberg, W.: Z. Physik **A290**, (1979) 385.
106. Gäggeler, H., Trautmann, N., Brüchle, W., Herrmann, G., Kratz, J.V., Peuser, P., Schädel, M., Tittel, G., Wirth, G., Ahrens, H., Folger, H., Franz, G., Sümmerer K., Zendel, M.: Phys. Rev. Lett. **45**, (1980) 1824.
107. Herrmann, G.: Pure Appl. Chem. **53**, (1981) 949.
108. Hildebrand, N., Frink, C., Greulich, N., Hickmann, U., Kratz, J.V., Trautmann, N., Herrmann, G., Brügger, M., Gäggeler, H., Sümmerer, K., Wirth, G.: Nucl. Instr. Meth. Phys. Res. **A260**, (1987) 407.
109. Hildenbrand, K.D., Freiesleben, H., Pühlhofer, F., Schneider, W.F.W., Bock, R., v. Harrach, D., Specht, H.J.: Phys. Rev. Lett. **39**, (1977) 1065.
110. Jungclas, H., Hirdes, D., Brandt, R., Lemmertz, P., Georg, E., Wollnik, H.: Phys. Lett. **79B**, (1978) 58.
111. Kratz, J.V., Brüchle, W., Folger, H., Gäggeler, H., Schädel, M., Sümmerer, K., Wirth, G., Greulich, N., Herrmann, G., Hickmann, U., Peuser, P., Trautmann, N., Hulet, E.K., Lougheed, R.W., Nitschke, J.M., Ferguson, R.L., Hahn, R.L.: Phys. Rev. **C33**, (1986) 504.
112. Schädel, M., Brüchle, W., Gäggeler, H., Kratz, J.V., Sümmerer, K., Wirth, G., Herrmann, G., Stakemann, R., Tittel, G., Trautmann, N., Nitschke, J.M., Hulet, E.K., Lougheed, R.W., Hahn, R.L., Ferguson, R.L.: Phys. Rev. Lett. **48**, (1982) 852.
113. Münzenberg, G., Hofmann, S., Heßberger, F.P., Reisdorf, W., Schmidt, K.H., Schneider, J.H.R., Armbruster, P., Sahm, C.C., Thuma, B.: Z. Phys. **A300**, (1981) 107.
114. Patyk, Z., Skalski, J., Sobicsewski, A., Ćwiok, S.: Nucl. Phys. **A502**, (1989) 591c.

Index